Statistical Theory of the Analysis of Experimental Designs

STATISTICS

Textbooks and Monographs

A SERIES EDITED BY

D. B. OWEN, Coordinating Editor
Department of Statistics
Southern Methodist University
Dallas, Texas

PETER LEWIS
Naval Postgraduate School
Monterey, California

PAUL D. MINTON
Virginia Commonwealth University
Richmond, Virginia

JOHN W. PRATT
Harvard University
Boston, Massachusetts

VOLUME 1: The Generalized Jackknife Statistic
H. L. Gray and W. R. Schucany

VOLUME 2: Multivariate Analysis
Anant M. Kshirsagar

VOLUME 3: Statistics and Society
Walter T. Federer

VOLUME 4: Multivariate Analysis: A Selected and Abstracted Bibliography, 1957 - 1972
Kocherlakota Subrahmaniam and Kathleen Subrahmaniam

VOLUME 5: Design of Experiments: A Realistic Approach
Virgil L. Anderson and Robert A. McLean

VOLUME 6: Statistical and Mathematical Aspects of Pollution Problems
edited by John W. Pratt

VOLUME 7: Introduction to Probability and Statistics (in two parts). Part I: Probability. Part II: Statistics
Narayan C. Giri

VOLUME 8: Statistical Theory of the Analysis of Experimental Designs
J. Ogawa

OTHER VOLUMES IN PREPARATION

Statistical Theory of the Analysis of Experimental Designs

J. OGAWA
Department of Mathematics, Statistics and Computing Science
The University of Calgary
Calgary, Alberta, Canada

MARCEL DEKKER, INC. New York 1974

MARCEL DEKKER, INC.
270 Madison Avenue, New York, New York 10016

LIBRARY OF CONGRESS CATALOG CARD NUMBER: 73-90769

ISBN: 0-8247-6116-2

Current printing (last digit):
10 9 8 7 6 5 4 3 2 1

PRINTED IN THE UNITED STATES OF AMERICA

CONTENTS

PREFACE

This book is an outgrowth of the graduate course, "Design of Experiment and Analysis of Variance", given at the University of Calgary, 1970-1973. The book covers the fundamental portions of the statistical theory of the analysis of experimental designs and is strictly restricted to linear models and normal distribution. The author relies heavily on linear algebraic methods as his mathematical tools. The construction of designs is not covered because there are already a couple of excellent books on this subject.

The author was first introduced to this field of modern statistics when he had a fortunate chance to attend Professor R. C. Bose's lectures at the Department of Statistics, University of North Carolina, Chapel Hill, in 1956-1958. Thus the author is heavily indebted to Professor Bose, especially in Chapters IV and V, and Chapter V is a mere reproduction of Professor Bose's lecture. Needless to say, the author is solely responsible for any errors or shortcomings, if any, in the presentation.

The author should acknowledge his indebtness to Professor D. J. Finney's book: *An Introduction to The Theory of Experimental Design*, Chicago University Press, 1963.

One novel point of this book is a fairly satisfactory treatment of the so-called randomization procedure of block designs; the last chapter, VI, is devoted completely to the mathematically rigorous treatment of the randomization of a PBIB design. The author is indebted to his former colleagues Professor S. Ikeda and

Mr. M. Ogasawara for their collaboration in this line of research. Especially the author is grateful to Professor Ikeda, who spent three months during the summer of 1974 with the Department of Mathematics, Statistics and Computing Science, The University of Calgary as a visiting scientist under the sponsorship of National Research Council of Canada through GRANT NO. A 7683, for his painstaking proofreading of the whole manuscript and the addition of the appendix to Chapter 6.

Preparation of the manuscript was financially supported by the National Research Council of Canada through GRANT NO. A 7683.

The author's thanks are due to Miss Barbara Bemben, the editor of the Marcel Dekker, Inc., who took pains to produce this book and also to Mrs. Tae Hayashitani who has done the retyping job of the manuscript.

Calgary, Alberta, Canada.
June, 1974

Junjiro Ogawa

Statistical Theory of the Analysis of Experimental Designs

Chapter 1

ANALYSIS OF VARIANCE

Since the analysis of variance is the standard technique for handling the data or observations obtained from an experiment, a brief summary of its fundamental features should be presented in the beginning. Readers should make reference to *The Analysis of Variance* by Scheffé [1].

1.1. The Partition of a Sum of Squares Corrected by the Mean

Given n (numerical) objects $a_1, a_2, \ldots, a_n$, the deviations from the mean

$$\bar{a} = \frac{1}{n}(a_1 + a_2 + \cdots + a_n)$$

are given by

$$a_1 - \bar{a},\ a_2 - \bar{a}, \cdots,\ a_n - \bar{a}.$$

The sum of the squares of the deviations is given by

$$Q = \sum_{\alpha=1}^{n} (a_\alpha - \bar{a})^2 = \sum_{\alpha=1}^{n} a_\alpha^2 - n\bar{a}^2.$$

This is also called *the sum of squares corrected by the mean.*

Let us consider the sum of squares more closely for small values of n:

For $n = 2$,

$$Q = (a_1 - \bar{a})^2 + (a_2 - \bar{a})^2 = a_1^2 + a_2^2 - \frac{1}{2}(a_1 + a_2)^2$$

$$= \frac{1}{2}(a_1 - a_2)^2 = \left(\frac{1}{\sqrt{2}}\, a_1 - \frac{1}{\sqrt{2}}\, a_2\right)^2.$$

For $n = 3$,

$$Q = (a_1 - \bar{a})^2 + (a_2 - \bar{a})^2 + (a_3 - \bar{a})^2$$

$$= a_1^2 + a_2^2 + a_3^2 - \frac{1}{3}(a_1 + a_2 + a_3)^2$$

$$= \frac{1}{2}(a_1 - a_2)^2 + \frac{1}{2}a_1^2 + a_1a_2 + \frac{1}{2}a_2^2 + a_3^2 - \frac{1}{3}(a_1 + a_2 + a_3)^2$$

$$= \frac{1}{2}(a_1 - a_2)^2 + \frac{1}{6}(a_1^2 + a_2^2 + 4a_3^2 + 2a_1a_2 - 4a_2a_3 - 4a_3a_1)$$

$$= \frac{1}{2}(a_1 - a_2)^2 + \frac{1}{6}(a_1 + a_2 - 2a_3)^2$$

$$= \left(\frac{1}{\sqrt{2}}a_1 - \frac{1}{\sqrt{2}}a_2\right)^2 + \left(\frac{1}{\sqrt{6}}a_1 + \frac{1}{\sqrt{6}}a_2 - \frac{2}{\sqrt{6}}a_3\right)^2.$$

By a cyclical permutation of the a terms, one obtains the other two partitions:

$$= \left(\frac{1}{\sqrt{2}}a_2 - \frac{1}{\sqrt{2}}a_3\right)^2 + \left(\frac{-2}{\sqrt{6}}a_1 + \frac{1}{\sqrt{6}}a_2 + \frac{1}{\sqrt{6}}a_3\right)^2$$

$$= \left(\frac{1}{\sqrt{2}}a_3 - \frac{1}{\sqrt{2}}a_1\right)^2 + \left(\frac{1}{\sqrt{6}}a_1 - \frac{2}{\sqrt{6}}a_2 + \frac{1}{\sqrt{6}}a_3\right)^2.$$

A linear combination of the n objects $a_1, a_2, \ldots, a_n$,

$$L = c_1a_1 + c_2a_2 + \cdots + c_na_n,$$

is said to be a *contrast* of $a_1, a_2, \ldots, a_n$ if $\sum_{\alpha=1}^{n} c_\alpha = 0$. The contrast L is said to be *normalized* if $\sum_{\alpha=1}^{n} c_\alpha^2 = 1$. Two contrasts

$$L_1 = \sum_{\alpha=1}^{n} c_{1\alpha}a_\alpha \quad \text{and} \quad L_2 = \sum_{\alpha=1}^{n} c_{2\alpha}a_\alpha$$

are said to be *mutually orthogonal* if $\sum_{\alpha=1}^{n} c_{1\alpha}c_{2\alpha} = 0$.

For $n = 2$, the sum of squares corrected by the mean is a square of a normalized contrast. For $n = 3$, Q is partitioned into the sum of the squares of two mutually orthogonal, normalized contrasts.

For $n = 4$,

$$Q = (a_1 - \bar{a})^2 + (a_2 - \bar{a})^2 + (a_3 - \bar{a})^2 + (a_4 - \bar{a})^2$$

$$= a_1^2 + a_2^2 + a_3^2 + a_4^2 - \frac{1}{4}(a_1 + a_2 + a_3 + a_4)^2$$

$$= \left[a_1^2 + a_2^2 + a_3^2 - \frac{1}{3}(a_1 + a_2 + a_3)^2\right]$$

$$+ \left[\frac{1}{3}(a_1 + a_2 + a_3)^2 - \frac{1}{4}(a_1 + a_2 + a_3 + a_4)^2 + a_4^2\right].$$

Now since

$$\frac{1}{3}(a_1 + a_2 + a_3)^2 - \frac{1}{4}(a_1 + a_2 + a_3 + a_4)^2 + a_4^2$$

$$= \frac{1}{12}(a_1 + a_2 + a_3)^2 - \frac{6}{12}(a_1 + a_2 + a_3)a_4 + \frac{9}{12} a_4^2$$

$$= \frac{1}{12}(a_1 + a_2 + a_3 - 3a_4)^2,$$

one obtains a partition of Q as follows:

$$Q = \left(\frac{1}{\sqrt{2}} a_1 - \frac{1}{\sqrt{2}} a_2\right)^2 + \left(\frac{1}{\sqrt{6}} a_1 + \frac{1}{\sqrt{6}} a_2 - \frac{2}{\sqrt{6}} a_3\right)^2$$

$$+ \left(\frac{1}{\sqrt{12}} a_1 + \frac{1}{\sqrt{12}} a_2 + \frac{1}{\sqrt{12}} a_3 - \frac{3}{\sqrt{12}} a_4\right)^2,$$

and by a cyclical change of the a terms one gets

$$\left(\frac{1}{\sqrt{2}} a_1 - \frac{1}{\sqrt{2}} a_2\right)^2 + \left(\frac{1}{\sqrt{6}} a_1 + \frac{1}{\sqrt{6}} a_2 - \frac{2}{\sqrt{6}} a_4\right)^2$$

$$+ \left(\frac{1}{\sqrt{12}} a_1 + \frac{1}{\sqrt{12}} a_2 - \frac{3}{\sqrt{12}} a_3 + \frac{1}{\sqrt{12}} a_4\right)^2,$$

$$\left(\frac{1}{\sqrt{2}} a_1 - \frac{1}{\sqrt{2}} a_3\right)^2 + \left(\frac{1}{\sqrt{6}} a_1 - \frac{2}{\sqrt{6}} a_2 + \frac{1}{\sqrt{6}} a_3\right)^2$$

$$+ \left(\frac{1}{\sqrt{12}} a_1 + \frac{1}{\sqrt{12}} a_2 + \frac{1}{\sqrt{12}} a_3 - \frac{3}{\sqrt{12}} a_4\right)^2,$$

$$\left(\frac{1}{\sqrt{2}}\,a_1 - \frac{1}{\sqrt{2}}\,a_3\right)^2 + \left(\frac{1}{\sqrt{6}}\,a_1 + \frac{1}{\sqrt{6}}\,a_3 - \frac{2}{\sqrt{6}}\,a_4\right)^2$$

$$+ \left(\frac{1}{\sqrt{12}}\,a_1 - \frac{3}{\sqrt{12}}\,a_2 + \frac{1}{\sqrt{12}}\,a_3 + \frac{1}{\sqrt{12}}\,a_4\right)^2,$$

$$\left(\frac{1}{\sqrt{2}}\,a_1 - \frac{1}{\sqrt{2}}\,a_4\right)^2 + \left(\frac{1}{\sqrt{6}}\,a_1 - \frac{2}{\sqrt{6}}\,a_2 + \frac{1}{\sqrt{6}}\,a_4\right)^2$$

$$+ \left(\frac{1}{\sqrt{12}}\,a_1 + \frac{1}{\sqrt{12}}\,a_2 - \frac{3}{\sqrt{12}}\,a_3 + \frac{1}{\sqrt{12}}\,a_4\right)^2,$$

$$\left(\frac{1}{\sqrt{2}}\,a_1 - \frac{1}{\sqrt{2}}\,a_4\right)^2 + \left(\frac{1}{\sqrt{6}}\,a_1 - \frac{2}{\sqrt{6}}\,a_3 + \frac{1}{\sqrt{6}}\,a_4\right)^2$$

$$+ \left(\frac{1}{\sqrt{12}}\,a_1 - \frac{3}{\sqrt{12}}\,a_2 + \frac{1}{\sqrt{12}}\,a_3 + \frac{1}{\sqrt{12}}\,a_4\right)^2,$$

$$\left(\frac{1}{\sqrt{2}}\,a_2 - \frac{1}{\sqrt{2}}\,a_3\right)^2 + \left(\frac{-2}{\sqrt{6}}\,a_1 + \frac{1}{\sqrt{6}}\,a_2 + \frac{1}{\sqrt{6}}\,a_3\right)^2$$

$$+ \left(\frac{1}{\sqrt{12}}\,a_1 + \frac{1}{\sqrt{12}}\,a_2 + \frac{1}{\sqrt{12}}\,a_3 - \frac{3}{\sqrt{12}}\,a_4\right)^2,$$

$$\left(\frac{1}{\sqrt{2}}\,a_2 - \frac{1}{\sqrt{2}}\,a_3\right)^2 + \left(\frac{1}{\sqrt{6}}\,a_2 + \frac{1}{\sqrt{6}}\,a_3 - \frac{2}{\sqrt{6}}\,a_4\right)^2$$

$$+ \left(\frac{-3}{\sqrt{12}}\,a_1 + \frac{1}{\sqrt{12}}\,a_2 + \frac{1}{\sqrt{12}}\,a_3 + \frac{1}{\sqrt{12}}\,a_4\right)^2,$$

$$\left(\frac{1}{\sqrt{2}}\,a_2 - \frac{1}{\sqrt{2}}\,a_4\right)^2 + \left(\frac{-2}{\sqrt{6}}\,a_1 + \frac{1}{\sqrt{6}}\,a_2 + \frac{1}{\sqrt{6}}\,a_4\right)^2$$

$$+ \left(\frac{1}{\sqrt{12}}\,a_1 + \frac{1}{\sqrt{12}}\,a_2 - \frac{3}{\sqrt{12}}\,a_3 + \frac{1}{\sqrt{12}}\,a_4\right)^2,$$

$$\left(\frac{1}{\sqrt{2}}\, a_2 - \frac{1}{\sqrt{2}}\, a_4\right)^2 + \left(\frac{1}{\sqrt{6}}\, a_2 - \frac{2}{\sqrt{6}}\, a_3 + \frac{1}{\sqrt{6}}\, a_4\right)^2$$

$$+ \left(\frac{-3}{\sqrt{12}}\, a_1 + \frac{1}{\sqrt{12}}\, a_2 + \frac{1}{\sqrt{12}}\, a_3 + \frac{1}{\sqrt{12}}\, a_4\right)^2,$$

$$\left(\frac{1}{\sqrt{2}}\, a_3 - \frac{1}{\sqrt{2}}\, a_4\right)^2 + \left(\frac{-2}{\sqrt{6}}\, a_1 + \frac{1}{\sqrt{6}}\, a_3 + \frac{1}{\sqrt{6}}\, a_4\right)^2$$

$$\left(\frac{1}{\sqrt{12}}\, a_1 - \frac{3}{\sqrt{12}}\, a_2 + \frac{1}{\sqrt{12}}\, a_3 + \frac{1}{\sqrt{12}}\, a_4\right)^2,$$

$$\left(\frac{1}{\sqrt{2}}\, a_3 - \frac{1}{\sqrt{2}}\, a_4\right)^2 + \left(\frac{-2}{\sqrt{6}}\, a_2 + \frac{1}{\sqrt{6}}\, a_3 + \frac{1}{\sqrt{6}}\, a_4\right)^2$$

$$+ \left(\frac{-3}{\sqrt{12}}\, a_1 + \frac{1}{\sqrt{12}}\, a_2 + \frac{1}{\sqrt{12}}\, a_3 + \frac{1}{\sqrt{12}}\, a_4\right)^2.$$

These are all sums of the squares of three orthonormal contrasts. Indeed the three contrasts in the last partitions are

$$0a_1 + 0a_2 + \frac{1}{\sqrt{2}}\, a_3 - \frac{1}{\sqrt{2}}\, a_4, \qquad 0a_1 - \frac{2}{\sqrt{6}}\, a_2 + \frac{1}{\sqrt{6}}\, a_3 + \frac{1}{\sqrt{6}}\, a_4,$$

$$-\frac{3}{\sqrt{12}}\, a_1 + \frac{1}{\sqrt{12}}\, a_2 + \frac{1}{\sqrt{12}}\, a_3 + \frac{1}{\sqrt{12}}\, a_4.$$

One can see that there are other types of partition, as follows:

$$\left(\frac{1}{\sqrt{2}}\, a_1 - \frac{1}{\sqrt{2}}\, a_2\right)^2 + \left(\frac{1}{\sqrt{2}}\, a_3 - \frac{1}{\sqrt{2}}\, a_4\right)^2$$

$$+ \left(\frac{1}{\sqrt{4}}\, a_1 + \frac{1}{\sqrt{4}}\, a_2 - \frac{1}{\sqrt{4}}\, a_3 - \frac{1}{\sqrt{4}}\, a_4\right)^2,$$

$$\left(\frac{1}{\sqrt{2}}\, a_1 - \frac{1}{\sqrt{2}}\, a_3\right)^2 + \left(\frac{1}{\sqrt{2}}\, a_2 - \frac{1}{\sqrt{2}}\, a_4\right)^2$$

$$+ \left(\frac{1}{\sqrt{4}}\, a_1 - \frac{1}{\sqrt{4}}\, a_2 + \frac{1}{\sqrt{4}}\, a_3 - \frac{1}{\sqrt{4}}\, a_4\right)^2,$$

$$\left(-\frac{3}{\sqrt{20}}a_1 - \frac{1}{\sqrt{20}}a_2 + \frac{1}{\sqrt{20}}a_3 + \frac{3}{\sqrt{20}}a_4\right)^2 + \left(\frac{1}{\sqrt{4}}a_1 - \frac{1}{\sqrt{4}}a_2 - \frac{1}{\sqrt{4}}a_3 + \frac{1}{\sqrt{4}}a_4\right)^2$$
$$+ \left(-\frac{1}{\sqrt{20}}a_1 + \frac{3}{\sqrt{20}}a_2 - \frac{3}{\sqrt{20}}a_3 + \frac{1}{\sqrt{20}}a_4\right)^2,$$

$$\left(\frac{1}{\sqrt{4}}a_1 + \frac{1}{\sqrt{4}}a_2 - \frac{1}{\sqrt{4}}a_3 - \frac{1}{\sqrt{4}}a_4\right)^2 + \left(\frac{1}{\sqrt{4}}a_1 - \frac{1}{\sqrt{4}}a_2 + \frac{1}{\sqrt{4}}a_3 - \frac{1}{\sqrt{4}}a_4\right)^2$$
$$+ \left(\frac{1}{\sqrt{4}}a_1 - \frac{1}{\sqrt{4}}a_2 - \frac{1}{\sqrt{4}}a_3 + \frac{1}{\sqrt{4}}a_4\right)^2.$$

In this case we notice that

$$\left\| \begin{matrix} \frac{1}{\sqrt{4}} & \frac{1}{\sqrt{4}} & \frac{1}{\sqrt{4}} & \frac{1}{\sqrt{4}} \\ -\sqrt{\frac{3}{4}} & \frac{1}{\sqrt{4.3}} & \frac{1}{\sqrt{4.3}} & \frac{1}{\sqrt{4.3}} \\ 0 & -\sqrt{\frac{2}{3}} & \frac{1}{\sqrt{3.2}} & \frac{1}{\sqrt{3.2}} \\ 0 & 0 & -\sqrt{\frac{1}{2}} & \frac{1}{\sqrt{2.1}} \end{matrix} \right\| \quad \text{and} \quad \left\| \begin{matrix} \frac{1}{\sqrt{4}} & \frac{1}{\sqrt{4}} & \frac{1}{\sqrt{4}} & \frac{1}{\sqrt{4}} \\ \frac{1}{\sqrt{4}} & \frac{1}{\sqrt{4}} & -\frac{1}{\sqrt{4}} & -\frac{1}{\sqrt{4}} \\ \frac{1}{\sqrt{4}} & -\frac{1}{\sqrt{4}} & \frac{1}{\sqrt{4}} & -\frac{1}{\sqrt{4}} \\ \frac{1}{\sqrt{4}} & -\frac{1}{\sqrt{4}} & -\frac{1}{\sqrt{4}} & \frac{1}{\sqrt{4}} \end{matrix} \right\|$$

are orthogonal matrices. The former can be generalized for any n, but the latter can be generalized only for $n \equiv 0 \bmod 4$.

So far we have seen that *the sum of the squares of* n *deviations from the mean is partitioned into the sum of* $(n - 1)$ *squares of mutually orthonormal contrasts* for $n = 2$, 3, and 4. One can show that this is generally true. In fact, if we put

$$\sqrt{n}\,\bar{a} = \frac{1}{\sqrt{n}}(a_1 + a_2 + a_3 + \cdots + a_{n-1} + a_n),$$

$$L_1 = \sqrt{\frac{n-1}{n}}\,a_1 - \frac{1}{\sqrt{n(n-1)}}(a_2 + \cdots + a_n),$$

$$L_2 = \sqrt{\frac{n-2}{n-1}}\,a_2 - \frac{1}{\sqrt{(n-1)(n-2)}}(a_3 + \cdots + a_n),$$

. . . .

$$L_\alpha = \sqrt{\frac{n-\alpha}{n-\alpha+1}}\,a_\alpha - \frac{1}{\sqrt{(n-\alpha+1)(n-\alpha)}}(a_{\alpha+1} + \cdots + a_n),$$

. . . .

$$L_{n-1} = \frac{1}{\sqrt{2}}(a_{n-1} - a_n),$$

it is easy to check that this is an orthogonal transformation, called *the Helmert transformation*, and hence

$$a_1^2 + a_2^2 + \cdots + a_n^2 = n\bar{a}^2 + L_1^2 + \cdots + L_{n-1}^2.$$

and consequently

$$Q = a_1^2 + \cdots + a_n^2 - n\bar{a}^2 = L_1^2 + L_2^2 + \cdots + L_{n-1}^2.$$

If we take any orthogonal matrix of order $n - 1$

$$H = \left\| h_{\alpha\beta} \right\|,$$

and let

$$L'_\alpha = \sum_{\beta=1}^{n-1} h_{\alpha\beta} L_\beta \qquad (\alpha = 1, \ldots, n-1),$$

then

$$\sum_{\alpha=1}^{n-1} L_\alpha^2 = \sum_{\alpha=1}^{n-1} L_\alpha'^2.$$

Thus, although there are so many choices of the mutually orthogonal normalized contrasts, the number of the orthogonal contrasts is the dimension of the linear space

$$\sum c_\alpha = 0,$$

which is called *the contrast space.* It is invariant and is known as *the degree of freedom* of the sum of squares Q.

Each contrast is said *to carry a degree of freedom.* Degrees of freedom carried by orthogonal contrasts are said *to be orthogonal.*

Suppose the given n objects are divided into p groups with elements $r_1,\ldots,r_p$, respectively,

$$r_1 + r_2 + \cdots + r_p = n.$$

Let the subtotals of the groups be $U_1, U_2, \ldots, U_p$, and hence

$$U_1 + U_2 + \cdots + U_p = n\bar{a}.$$

Then *the sum of squares*

$$\sum_{j=1}^{p} \frac{U_j^2}{r_j} - \frac{(\Sigma_j U_j)^2}{n} = \sum_{j=1}^{p} \frac{U_j^2}{r_j} - n\bar{a}^2$$

carries $(p - 1)$ *degrees of freedom, and the difference*

$$\sum_{\alpha=1}^{n} a_\alpha^2 - \sum_{j=1}^{p} \frac{U_j^2}{r_j}$$

carries $(n - p)$ *degrees of freedom, which are orthogonal to the degrees of freedom carried by the sum of squares* $\Sigma_j U_j^2/r_j - n\bar{a}^2$.

Proof. Let $\underline{\zeta}_j' = (\zeta_{j1},\ldots,\zeta_{jn})$, where $\zeta_{jf} = 1$ if a_f is in the jth group and 0 otherwise; then

$$\underline{\zeta}_j'\underline{\zeta}_j = r_j \text{ and } \underline{\zeta}_i'\underline{\zeta}_j = 0 \qquad (i \neq j),$$

$$\sum_{j=1}^{p} \underline{\zeta}_j = \underline{1}$$

$$U_j = \underline{\zeta}_j'\underline{a} \qquad (j = 1,\ldots,p).$$

One can construct $r_j - 1$ mutually orthogonal and normalized contrasts that are linear combinations only of those a terms for which $\zeta_{jf} = 1$.

There are $\sum_{j=1}^{p} (r_j - 1) = n - p$ such contrasts that are orthogonal to those contrasts that are linear combinations of $U_1, \ldots, U_p$.

Now let

$$W_j = \sqrt{r_j}\left(\frac{U_j}{r_j} - \frac{U_1 + \cdots + U_p}{n}\right),$$

then

$$\sum_{j=1}^{p} \sqrt{r_j}\, W_j = 0$$

and

$$\sum_{j=1}^{p} W_j^2 = \sum_{j=1}^{p} r_j \left(\frac{U_j}{r_j} - \frac{U_1 + \cdots + U_p}{n}\right)^2 = \sum_{j=1}^{p} \frac{U_j^2}{r_j} - \frac{(U_1 + \cdots + U_p)^2}{n}.$$

Therefore the sum of the squares of $U_1, \ldots, U_p$ can be written as

$$\sum_{j=1}^{p} \frac{U_j^2}{r_j} - \frac{(U_1 + \cdots + U_p)^2}{n}$$

$$= W_1^2 + \cdots + W_{p-1}^2 + \frac{1}{r_p}\left(\sum_{j=1}^{p-1} \sqrt{r_j}\, W_j\right)^2$$

$$= (W_1, \ldots, W_{p-1}) \left\| \begin{matrix} 1 + \frac{r_1}{r_p} & \frac{\sqrt{r_1 r_2}}{r_p} & \cdots & \frac{\sqrt{r_1 r_{p-1}}}{r_p} \\ \frac{\sqrt{r_1 r_2}}{r_p} & 1 + \frac{r_2}{r_p} & \cdots & \frac{r_2 r_{p-1}}{r_p} \\ \cdots & \cdots & \cdots & \cdots \\ \frac{\sqrt{r_1 r_{p-1}}}{r_p} & \frac{\sqrt{r_2 r_{p-1}}}{r_p} & \cdots & 1 + \frac{r_{p-1}}{r_p} \end{matrix} \right\| \begin{pmatrix} W_1 \\ \vdots \\ W_{p-1} \end{pmatrix}.$$

If one makes the transformation

$$L_1 = \sqrt{\frac{r_2 + \cdots + r_p}{r_1 + r_2 + \cdots + r_p}}\, W_1$$

$$- \frac{\sqrt{r_1}}{\sqrt{(r_1 + \cdots + r_p)(r_2 + \cdots + r_p)}} \left(\sqrt{r_2}\, W_2 + \cdots + \sqrt{r_p}\, W_p\right)$$

$$= \sqrt{\frac{r_1 + r_2 + \cdots + r_p}{r_2 + \cdots + r_p}}\, W_1,$$

$$L_2 = \sqrt{\frac{r_3 + \cdots + r_p}{r_2 + r_3 + \cdots + r_p}}\, W_2$$

$$- \frac{\sqrt{r_2}}{\sqrt{(r_2 + \cdots + r_p)(r_3 + \cdots + r_p)}} \left(\sqrt{r_3}\, W_3 + \cdots + \sqrt{r_p}\, W_p\right)$$

$$= \frac{\sqrt{r_1 r_2}}{\sqrt{(r_2 + \cdots + r_p)(r_3 + \cdots + r_p)}}\, W_1 + \sqrt{\frac{r_2 + r_3 + \cdots + r_p}{r_3 + \cdots + r_p}}\, W_2,$$

$$L_3 = \frac{\sqrt{r_1 r_3}}{\sqrt{(r_3 + \cdots + r_p)(r_4 + \cdots + r_p)}}\, W_1$$

$$+ \frac{\sqrt{r_2 r_3}}{\sqrt{(r_3 + \cdots + r_p)(r_4 + \cdots + r_p)}}\, W_2$$

$$+ \sqrt{\frac{r_3 + r_4 + \cdots + r_p}{r_4 + \cdots + r_p}}\, W_3,$$

. . . .

$$L_{p-1} = \frac{\sqrt{r_1 r_{p-1}}}{\sqrt{(r_{p-1} + r_p)r_p}} W_1 + \frac{\sqrt{r_2 r_{p-1}}}{\sqrt{(r_{p-1} + r_p)r_p}} W_2$$

$$+ \cdots + \frac{\sqrt{r_{p-2} r_{p-1}}}{\sqrt{(r_{p-1} + r_p)r_p}} W_{p-2} + \sqrt{\frac{r_{p-1} + r_p}{r_p}} W_{p-1},$$

then it can be seen that

$$\sum_{j=1}^{p} \frac{U_j^2}{r_j} - \frac{(U_1 + \cdots + U_p)^2}{n} = L_1^2 + \cdots + L_{p-1}^2 .$$

The L_j terms are all orthonormal contrasts of the a terms and they are linear combinations of $U_1, \ldots, U_p$.

Indeed, let

$$W_j = \sqrt{r_j} \left(\frac{U_j}{r_j} - \bar{a} \right) = -\frac{\sqrt{r_j}}{n} U_1 - \cdots - \frac{\sqrt{r_j}}{n} U_{j-1}$$

$$+ \left(\frac{\sqrt{r_j}}{r_j} - \frac{\sqrt{r_j}}{n} \right) U_j - \frac{\sqrt{r_j}}{n} U_{j+1} - \cdots - \frac{\sqrt{r_j}}{n} U_p$$

$$= c_{j1}a_1 + c_{j2}a_2 + \cdots + c_{jn}a_n,$$

for example. Then

$$\sum_{\alpha=1}^{n} c_{j\alpha} = -\frac{\sqrt{r_j}}{n} r_1 - \cdots - \frac{\sqrt{r_j}}{n} r_{j-1} + \left(\frac{\sqrt{r_j}}{r_j} - \frac{\sqrt{r_j}}{n} \right) r_j$$

$$-\frac{\sqrt{r_j}}{n} r_{j+1} - \cdots - \frac{\sqrt{r_j}}{n} r_p = \sqrt{r_j}\,\frac{n - r_j}{n} - \sqrt{r_j}\,\frac{n - r_j}{n} = 0,$$

$$\sum_{\alpha=1}^{n} c_{j\alpha}^2 = \frac{r_j}{n^2} r_1 + \cdots + \frac{r_j}{n^2} r_{j-1} + \left(\frac{\sqrt{r_j}}{r_j} - \frac{\sqrt{r_j}}{n} \right)^2 r_j + \frac{r_j}{n^2} r_{j-1}$$

$$+ \cdots + \frac{r_j}{n^2} r_p = \frac{r_j}{n^2}(n - r_j) + \frac{(n - r_j)^2}{n^2} = \frac{n - r_j}{n},$$

$$\sum_{\alpha=1}^{n} c_{j\alpha} c_{j'\alpha} = \frac{\sqrt{r_j r_{j'}}}{n^2} r_1 + \cdots - \frac{\sqrt{r_j}}{n}\left(\frac{\sqrt{r_{j'}}}{r_{j'}} - \frac{\sqrt{r_{j'}}}{n}\right) r_{j'} + \frac{\sqrt{r_j r_j}}{n^2} r_{j'+1}$$

$$- \cdots - \left(\frac{\sqrt{r_j}}{r_j} - \frac{\sqrt{r_j}}{n}\right)\frac{\sqrt{r_{j'}}}{n} r_j + \cdots + \frac{\sqrt{r_j r_{j'}}}{n^2} r_p$$

$$= \frac{\sqrt{r_j r_{j'}}}{n^2}(n - r_j - r_j') - \frac{\sqrt{r_j r_{j'}}}{n}\frac{n - r_{j'}}{n} = \frac{\sqrt{r_j r_{j'}}}{n}\frac{n - r_j}{n}$$

$$= \frac{\sqrt{r_j r_{j'}}}{n^2}\left[(n - r_j - r_j') - (n - r_j') - (n - r_j)\right]$$

$$= -\frac{\sqrt{r_j r_{j'}}}{n}, \qquad (j \neq j').$$

Now let

$$L_\lambda = \sqrt{\frac{n - r_1 - \cdots - r_\lambda}{n - r_1 - \cdots - r_{\lambda-1}}}\, W_\lambda$$

$$- \frac{\sqrt{r_\lambda}(\sqrt{r_{\lambda+1}} W_{\lambda+1} + \cdots + \sqrt{r_p} W_p)}{\sqrt{(n - r_1 - \cdots - r_{\lambda-1})(n - r_1 - \cdots - r_\lambda)}}$$

$$= l_{\lambda 1} a_1 + l_{\lambda 2} a_2 + \cdots + l_{\lambda n} a_n,$$

for example. Then

$$l_{\lambda\alpha} = \sqrt{\frac{n - r_1 - \cdots - r_\lambda}{n - r_1 - \cdots - r_{\lambda-1}}}\, c_{\lambda\alpha}$$

$$- \frac{\sqrt{r_\lambda}(\sqrt{r_{\lambda+1}}\, c_{\lambda+1\alpha} + \cdots + \sqrt{r_p}\, c_{p\alpha})}{\sqrt{(n - r_1 - \cdots - r_{\lambda-1})(n - r_1 - \cdots - r_\lambda)}},$$

and hence

$$\sum_{\alpha=1}^{n} l_{\lambda\alpha} = \sqrt{\frac{n - r_1 - \cdots - r_\lambda}{n - r_1 - \cdots - r_{\lambda-1}}} \sum_{\alpha=1}^{n} c_{\lambda\alpha}$$

$$- \frac{\sqrt{r_\lambda}\left(\sqrt{r_{\lambda+1}} \sum_{\alpha=1}^{n} c_{\lambda+1\alpha} + \cdots + \sqrt{r_p} \sum_{\alpha=1}^{n} c_{p\alpha}\right)}{\sqrt{(n - r_1 - \cdots - r_{\lambda-1})(n - r_1 - \cdots - r_\lambda)}}$$

$$= 0.$$

$$\sum_{\alpha=1}^{n} l_{\lambda\alpha}^2 = \frac{n - r_1 - \cdots - r_\lambda}{n - r_1 - \cdots - r_{\lambda-1}} \sum_{\alpha=1}^{n} c_{\lambda\alpha}^2$$

$$+ \frac{r_\lambda\left(r_{\lambda+1} \sum_{\alpha=1}^{n} c_{\lambda+1\alpha}^2 + \cdots + r_p \sum_{\alpha=1}^{n} c_{p\alpha}^2\right)}{(n - r_1 - \cdots - r_{\lambda-1})(n - r_1 - \cdots - r_\lambda)}$$

$$- \frac{2}{n - r_1 - \cdots - r_{\lambda-1}} \sum_{j=\lambda+1}^{p} \sqrt{r_\lambda r_j} \sum_{\alpha=1}^{n} c_{\lambda\alpha} c_{j\alpha}$$

$$+ 2\, \frac{r_\lambda \sum_{j<j'} \sqrt{r_j r_{j'}} \sum_{\alpha=1}^{n} c_{j\alpha} c_{j'\alpha}}{(n - r_1 - \cdots - r_{\lambda-1})(n - r_1 - \cdots - r_\lambda)}$$

$$= \frac{n - r_1 - \cdots - r_\lambda}{n - r_1 - \cdots - r_{\lambda-1}} \, \frac{n - r_\lambda}{n}$$

$$+ \frac{r_\lambda}{(n - r_1 - \cdots - r_{\lambda-1})(n - r_1 - \cdots - r_\lambda)} \sum_{j=\lambda+1}^{p} r_j \, \frac{n - r_j}{n}$$

$$+ \frac{2}{n - r_1 - \cdots - r_{\lambda-1}} \sum_{j=\lambda+1}^{p} \frac{r_\lambda r_j}{n}$$

$$- \frac{2r_\lambda}{(n - r_1 - \cdots - r_{\lambda-1})(n - r_1 - \cdots - r_\lambda)} \sum_{j<j'} \frac{r_j r_{j'}}{n}$$

$$= \frac{n - r_1 - \cdots - r_\lambda}{n - r_1 - \cdots - r_{\lambda-1}} \frac{n - r_\lambda}{n}$$

$$+ \frac{r_\lambda}{(n - r_1 - \cdots - r_{\lambda-1})(n - r_1 - \cdots - r_\lambda)} \Big[(n - r_1 - \cdots - r_\lambda)$$

$$- \frac{1}{n} (n - r_1 - \cdots - r_\lambda)^2\Big]$$

$$+ \frac{2}{n - r_1 - \cdots - r_{\lambda-1}} \frac{r_\lambda (n - r_1 - \cdots - r_\lambda)}{n}$$

$$= \frac{n - r_1 - \cdots - r_\lambda}{n - r_1 - \cdots - r_{\lambda-1}} \left[\frac{n - r_\lambda}{n} + \frac{r_\lambda}{(n - r_1 - \cdots - r_\lambda)} - \frac{r_\lambda}{n} + \frac{2r_\lambda}{n}\right]$$

$$= \frac{n - r_1 - \cdots - r_\lambda}{n - r_1 - \cdots - r_{\lambda-1}} \left[1 + \frac{r_\lambda}{n - r_1 - \quad - r_\lambda}\right] = 1,$$

and, for the sake of simplicity, one calculates

$$\sum_{\alpha=1}^{n} l_{\lambda\alpha} l_{\lambda+1\alpha} = \sum_{\alpha=1}^{n} \Bigg[\sqrt{\frac{n - r_1 - \cdots - r_\lambda}{n - r_1 - \cdots - r_{\lambda-1}}}\, c_{\lambda\alpha}$$

$$- \frac{\sqrt{r_1}(\sqrt{r_{\lambda+1}} c_{\lambda+1\alpha} + \cdots + \sqrt{r_p} c_{p\alpha})}{(n - r_1 - \cdots - r_{\lambda-1})(n - r_1 - \cdots - r_\lambda)}\Bigg]$$

$$\left[\sqrt{\frac{n - r_1 - \cdots - r_{\lambda+1}}{n - r_1 - \cdots - r_\lambda}}\, c_{\lambda+1\alpha}\right.$$

$$\left. - \frac{\sqrt{r_{\lambda+1}}(\sqrt{r_{\lambda+2}}\, c_{\lambda+2\alpha} + \cdots + r_p c_{p\alpha})}{\sqrt{(n - r_1 - \cdots - r_\lambda)(n - r_1 - \cdots - r_{\lambda-1})}}\right]$$

$$= - \sqrt{\frac{n - r_1 - \cdots - r_{\lambda+1}}{n - r_1 - \cdots - r_{\lambda-1}}}\, \frac{\sqrt{r_\lambda r_{\lambda+1}}}{n} - \sqrt{\frac{n - r_1 - \cdots - r_{\lambda+1}}{n - r_1 - \cdots - r_{\lambda-1}}}$$

$$\left(\frac{\sqrt{r_\lambda r_{\lambda+1}}}{n - r_1 - \cdots - r_\lambda}\, \frac{n - r_{\lambda+1}}{n} - \frac{\sqrt{r_\lambda r_{\lambda+2}}}{n - r_1 - \cdots - r_\lambda}\, \frac{\sqrt{r_{\lambda+1} r_{\lambda+2}}}{n}\right.$$

$$\left. - \cdots - \frac{\sqrt{r_\lambda r_p}}{n - r_1 - \cdots - r_\lambda}\, \frac{\sqrt{r_{\lambda+1} r_p}}{n}\right) + \sqrt{\frac{n - r_1 - \cdots - r_{\lambda+1}}{n - r_1 - \cdots - r_{\lambda-1}}}$$

$$\left(\frac{\sqrt{r_{\lambda+1} r_{\lambda+2}}}{n - r_1 - \cdots - r_{\lambda+1}}\, \frac{\sqrt{r_\lambda r_{\lambda+2}}}{n} + \cdots + \frac{\sqrt{r_{\lambda+1} r_p}}{n - r_1 - \cdots - r_{\lambda+1}}\, \frac{\sqrt{r_\lambda r_p}}{n}\right)$$

$$+ \frac{\sqrt{r_\lambda r_{\lambda+1}}}{\sqrt{(n - r_1 - \cdots - r_{\lambda-1})(n - r_1 - \cdots - r_{\lambda+1})}}\, \frac{1}{n - r_1 - \cdots - r_\lambda}$$

$$\left(- \sqrt{r_{\lambda+1} r_{\lambda+2}}\, \frac{\sqrt{r_{\lambda+1} r_{\lambda+2}}}{n} + r_{\lambda+2}\, \frac{n - r_{\lambda+2}}{n} - \frac{r_{\lambda+2} r_{\lambda+3}}{n} - \cdots - \frac{r_{\lambda+2} r_p}{n}\right.$$

$$- \frac{r_{\lambda+1} r_{\lambda+3}}{n} - \frac{r_{\lambda+2} r_{\lambda+3}}{n} + r_{\lambda+3}\, \frac{n - r_{\lambda+3}}{n} - \frac{r_{\lambda+3} r_{\lambda+4}}{n} - \cdots - \frac{r_{\lambda+1} r_p}{n}$$

$$+ r_{p-1}\, \frac{n - r_{p-1}}{n} - \frac{r_{p-1} r_p}{n} + r_p\, \frac{n - r_p}{n}$$

$$= \sqrt{\frac{n - r_1 - \cdots - r_{\lambda+1}}{n - r_1 - \cdots - r_{\lambda-1}}} \sqrt{r_\lambda r_{\lambda+1}} \left[-\frac{1}{n} - \frac{n - r_{\lambda+1}}{(n - r_1 - \cdots - r_\lambda)n} \right.$$

$$+ \frac{r_{\lambda+2}}{(n - r_1 - \cdots - r_\lambda)n} + \cdots + \frac{r_p}{(n - r_1 - \cdots - r_\lambda)n}$$

$$\left. + \frac{1}{n - r_1 - \cdots - r_{\lambda+1}} \left(\frac{r_{\lambda+2}}{n} + \cdots + \frac{r_p}{n} \right) \right]$$

$$+ \frac{\sqrt{r_\lambda r_{\lambda-1}}}{\sqrt{(n - r_1 - \cdots - r_{\lambda-1})(n - r_1 - \cdots - r_{\lambda+1})}} \frac{1}{n - r_1 - \cdots - r_\lambda}$$

$$\left(r_{\lambda+2} - \frac{r_{\lambda+2}}{n} (r_{\lambda+1} + r_{\lambda+2} + \cdots + r_p) + r_{\lambda+3} \right.$$

$$\left. - \frac{r_{\lambda+3}}{n} (r_{\lambda+1} + r_{\lambda+2} + \cdots + r_p) + \cdots + r_p - \frac{r_p}{n} (r_{\lambda+1} + \cdots + r_p) \right)$$

$$= \sqrt{r_\lambda r_{\lambda+1}} \sqrt{\frac{n - r_1 - \cdots - r_{\lambda+1}}{n - r_1 - \cdots - r_{\lambda-1}}} \left[-\frac{1}{n} - \frac{r_1 + \cdots + r_\lambda}{n(n - r_1 - \cdots - r_\lambda)} + \frac{1}{n} \right]$$

$$+ \sqrt{r_\lambda r_{\lambda+1}} \sqrt{\frac{n - r_1 - \cdots - r_{\lambda+1}}{n - r_1 - \cdots - r_{\lambda-1}}} \times$$

$$\times \frac{1}{(n - r_1 - \cdots - r_\lambda)(n - r_1 - \cdots - r_{\lambda+1})}$$

$$\times \left(r_{\lambda+2} - \frac{r_{\lambda+2}}{n} (n - r_1 - \cdots - r_\lambda) + r_{\lambda+3} - \frac{r_{\lambda+3}}{n} (n - r_1 - \cdots - r_\lambda) \right.$$

$$+ \cdots + r_p - \frac{r_p}{n}(n - r_1 - \cdots - r_\lambda) \Big\} = 0,$$

because the expression inside the last bracket becomes

$$n - r_1 - \cdots - r_{\lambda+1} - \frac{1}{n}(n - r_1 - \cdots - r_{\lambda+1})(n - r_1 - \cdots - r_\lambda)$$

$$= \frac{1}{n}(r_1 + r_2 + \cdots + r_\lambda)(n - r_1 - \cdots - r_{\lambda+1})$$

and the left hand side of the last equality contains the factor

$$-\frac{1}{n}\frac{r_1 + \cdots + r_\lambda}{n - r_1 - \cdots - r_\lambda} + \frac{1}{n}\frac{r_1 + \cdots + r_\lambda}{n - r_1 - \cdots - r_\lambda} = 0.$$

Thus we have seen that L_λ, $\lambda = 1, 2, \ldots, p - 1$ are normalized orthogonal contrasts of $a_1, \ldots, a_n$ and those are linear combinations of $U_1, \ldots, U_p$.

Suppose that $n = pq$. The n objects are divided into p groups with q objects each, and they are also divided into q groups with p objects each in such a manner that each of the q objects in each group of the former classification appears in one of the q groups of the latter classification, and vice versa. Let the group totals of the former classification be $U_1, \ldots, U_p$ and let the group totals of the latter classification be $V_1, \ldots, V_q$. Then the sums of squares

$$Q_1 = \sum_{j=1}^{p} \frac{1}{q} U_j^2 - \frac{(U_1 + \cdots + U_p)^2}{n}$$

and

$$Q_2 = \sum_{j=1}^{q} \frac{1}{p} V_j^2 - \frac{(V_1 + \cdots + V_q)^2}{n}$$

carry $(p - 1)$ degrees of freedom and $(q - 1)$ degrees of freedom, respectively, and they are *orthogonal to each other*. The residual sum of squares

$$Q - Q_1 - Q_2 = \sum_{\alpha=1}^{n} a_\alpha^2 - \frac{1}{q}\sum_{j=1}^{p} U_j^2 - \frac{1}{p}\sum_{j=1}^{q} V_j^2 + n\bar{a}^2$$

carries $n - p - q + 1$ *degrees of freedom, which are orthogonal to those mentioned above.*

Proof. Let $\underline{\zeta}_j' = (\zeta_{j1}, \ldots \zeta_{jn})$, where $\zeta_{jf} = 1$ if a_f belongs to the jth group of the former classification and 0 otherwise, $j = 1, \ldots, p$. Let

$$\underline{\zeta}_j'\underline{\zeta}_j = q \text{ and } \underline{\zeta}_i'\underline{\zeta}_j = 0 \qquad (i \neq j),$$

$$\sum_{j=1}^{p} \underline{\zeta}_j = \underline{1} \text{ and } U_j = \underline{\zeta}_j'\underline{a}.$$

Similarly let $\underline{\eta}_j' = (\eta_{j1}, \ldots, \eta_{jn})$, where $\eta_{jf} = 1$ if a_f belongs to the jth group of the latter classification and 0 otherwise, $j = 1, \ldots, q$;

$$\underline{\eta}_j'\underline{\eta}_j = p \text{ and } \underline{\eta}_i'\underline{\eta}_j = 0 \qquad (i \neq j),$$

$$\sum_{j=1}^{q} \underline{\eta}_j = \underline{1};\ V_j = \underline{\eta}_j'\underline{a};\ \underline{\zeta}_i'\underline{\eta}_j = 1.$$

Contrasts making up Q_1 and Q_2 must be linear combinations of the U and V terms respectively, and in this case they turn out to be contrasts in terms of U and V respectively.

The contrast of $U_1, \ldots, U_p$,

$$L_1 = (c_1, \ldots, c_p) \begin{pmatrix} \underline{\zeta}_1' \\ \vdots \\ \underline{\zeta}_p' \end{pmatrix} \underline{a},\quad c_1 + \cdots + c_p = 0.$$

The contrast of $V_1, \ldots, V_q$,

$$L_2 = (d_1, \ldots, d_q) \begin{pmatrix} \underline{\eta}_1' \\ \vdots \\ \underline{\eta}_q' \end{pmatrix} \underline{a},\quad d_1 + \cdots + d_q = 0.$$

Hence the inner product of the coefficient vectors is given by

$$(c_1,\dots,c_p)\begin{pmatrix}\underline{\zeta}_1'\\ \vdots \\ \underline{\zeta}_p'\end{pmatrix}[\underline{\eta}_1,\dots,\underline{\eta}_q]\begin{pmatrix}d_1\\ \vdots\\ d_q\end{pmatrix}$$

$$= (c_1,\dots,c_p)\left\|\begin{matrix}1 & \dots & 1\\ \vdots & & \vdots\\ 1 & \dots & 1\end{matrix}\right\|\begin{pmatrix}d_1\\ \vdots\\ d_q\end{pmatrix} = 0$$

One can easily generalize this for the case of m-way classification.

Problem 1.1.1. (E. Schmidt). Given a set of vectors $\underline{c}_1,\underline{c}_2,\dots,\underline{c}_n$, show that the set of vectors

$$\underline{\zeta}_i = \frac{1}{\sqrt{A_{i-1}\cdot A_i}}\begin{vmatrix} a_{11} & a_{12} & \dots & a_{1i}\\ a_{21} & a_{22} & \dots & a_{2i}\\ \vdots & \vdots & & \vdots\\ a_{i-1,1} & a_{i-1,2} & \dots & a_{i-1,i}\\ \underline{c}_1 & \underline{c}_2 & \dots & \underline{c}_i\end{vmatrix},$$

where

$$A_i = \begin{vmatrix} a_{11} & a_{12} & \dots & a_{1i}\\ a_{21} & a_{22} & \dots & a_{2i}\\ \vdots & \vdots & & \vdots\\ a_{i1} & a_{i2} & \dots & a_{ii}\end{vmatrix} \quad \text{and}$$

$$a_{ij} = \underline{c}_i'\underline{c}_j,$$

is an orthonormal system.

Problem 1.1.2. Write down the orthogonal matrix whose first row is

$$\frac{\eta_1}{\sqrt{\eta_1^2 + \cdots + \eta_n^2}} \quad \frac{\eta_2}{\sqrt{\eta_1^2 + \cdots + \eta_n^2}} \quad \cdots \quad \frac{\eta_n}{\sqrt{\eta_1^2 + \cdots + \eta_n^2}}$$

by the use of the orthogonalization method due to Schmidt and show that in the particular case where $\eta_1 = \eta_2 = \cdots = \eta_n$ this becomes the Helmert orthogonal matrix.

Problem 1.1.3. State a generalization of the mutually orthogonal classifications.

1.2 Normal Theory of Distribution Problems

Suppose that $x_1, x_2, \ldots, x_n$ are mutually independent in the stochastic sense and identically distributed as the normal distribution $N(m, \sigma^2)$. Then the probability element is given by

$$\left(\frac{1}{\sqrt{2\pi}\,\sigma}\right)^n exp\left(-\frac{1}{2\sigma^2}\sum_{\alpha=1}^{n}(x_\alpha - m)^2\right) dx_1 dx_2 \cdots dx_n$$

$$= C\ exp\left(-\frac{1}{2\sigma^2}\sum_{\alpha=1}^{n}(x_\alpha - \bar{x})^2 - \frac{n}{2\sigma^2}(\bar{x} - m)^2\right) dx_1 \cdots dx_n$$

We consider the joint distribution of

$$s^2 = \frac{1}{n}\sum_{\alpha=1}^{n}(x_\alpha - \bar{x})^2 \quad \text{and} \quad \bar{x} = \frac{1}{n}\sum_{\alpha=1}^{n} x_\alpha$$

and let its density function be $f(s^2, \bar{x})$. Then

$$f(s^2,\bar{x})\ ds^2\ d\bar{x}$$

$$\doteqdot P\left[s^2 \le \frac{1}{n}\sum_{\alpha=1}^{n}(x_\alpha - \bar{x})^2 < (s+ds)^2 \doteqdot s^2 + ds^2,\ \bar{x} \le \frac{1}{n}\Sigma_\alpha x_\alpha < \bar{x} + d\bar{x}\right]$$

$$= C\ exp\left[-\frac{ns}{2\sigma^2} - \frac{n}{2\sigma^2}(\bar{x} - m)^2\right]\int\cdots\int dx_1 dx_2 \cdots dx_n$$

where the domain of the integral is given by

$$s^2 \le \frac{1}{n}\Sigma_\alpha(x_\alpha - \bar{x})^2 < (s+ds)^2 \doteqdot s^2 + ds^2$$

$$\bar{x} \le \frac{1}{n}\Sigma_\alpha x_\alpha < \bar{x} + d\bar{x}$$

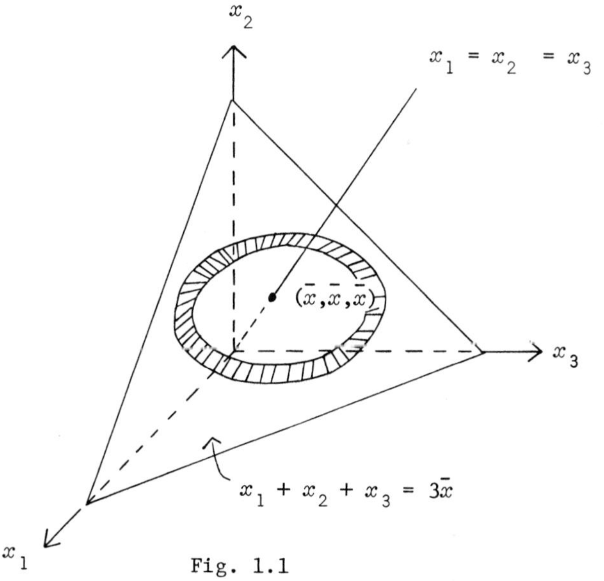

Fig. 1.1

Geometrically speaking (see Fig. 1.1), the preceding integral represents the volume bounded by a cylinder shell of height $d\bar{x}$ and a spherical shell of $(n - 1)$ dimensions whose center is $(\bar{x},\ldots,\bar{x})$ having the radius $\sqrt{n}\ s$ as its base.

The whole volume of the n-1 dimensional sphere is proportional to

$$s^{n-1},$$

and hence the volume of the spherical shell is proportional to

$$s^{n-2}\, ds\, d\bar{x}$$

or proportional to

$$(s^2)^{(n-3)/2}\, ds^2\, d\bar{x}$$

Hence

$$f(s^2,\bar{x})\, ds^2\, d\bar{x}$$

$$= \frac{\sqrt{n}}{\sqrt{2\pi}\,\sigma}\, exp\left(-\frac{n}{2\sigma^2}(\bar{x} - m)^2\right) d\bar{x}\, \frac{1}{\sigma^{n-1}\Gamma\left(\frac{n-1}{2}\right)}\, exp\left(-\frac{ns^2}{2\sigma^2}\right)$$

$$\times \left(\frac{ns^2}{2\sigma^2}\right)^{(n-1)/2-1} d\left(\frac{ns^2}{2\sigma^2}\right),$$

This can be written

$$\bar{x} \sim N(m, \frac{\sigma^2}{n}), \quad \Sigma_\alpha (x_\alpha - \bar{x})^2 = \sigma^2 \chi^2_{n-1} .$$

Lemma 1.2.1. (R. A. Fisher). Let $x_1,\ldots,x_n$ be a random sample of size n from $N(0, \sigma^2)$ and let

$$L_\alpha = \sum_{\beta=1}^{n} c_{\alpha\beta} x_\beta \qquad (\alpha = 1,\ldots,p < n),$$

$$\Sigma_\gamma c_{\alpha\gamma} c_{\beta\gamma} = \delta_{\alpha\beta} = \begin{cases} 1 & \text{if } \alpha = \beta, \\ 0 & \text{if } \alpha \neq \beta. \end{cases}$$

Then

$$\text{(i)} \quad Q(x_1,\ldots,x_n) = \sum_{\alpha=1}^{n} x_\alpha^2 - L_1^2 - \cdots - L_p^2$$

is independent of $L_1,\ldots, L_p$, and

$$\text{(ii)} \quad Q(x_1,\ldots,x_n) = \sigma^2\chi_{n-p}^2 .$$

Noncentral χ-Square. Suppose x_α is distributed as $N(a_\alpha,\sigma^2)$ and $x_1,\ldots,x_n$ are mutually independent. The probability element of the joint distribution of $x_1,\cdots,x_n$ is given by

$$\left(\frac{1}{\sqrt{2\pi}\,\sigma}\right)^n exp\left(-\frac{1}{2\sigma^2}\sum_\alpha (x_\alpha - a_\alpha)^2\right) dx_1\ldots dx_n$$

$$= \left(\frac{1}{\sqrt{2\pi}\,\sigma}\right)^n exp\left(-\frac{1}{2\sigma^2}\sum_\alpha a_\alpha^2\right) exp\left(-\frac{1}{2\sigma^2}\left(\sum_\alpha x_\alpha^2 - 2\sum_\alpha a_\alpha x_\alpha\right)\right) dx_1\ldots dx_n .$$

Let

$$\xi = \frac{a_1}{\sqrt{a_1^2 + \cdots + a_n^2}}\, x_1 + \frac{a_2}{\sqrt{a_1^2 + \cdots + a_n^2}}\, x_2 + \cdots + \frac{a_n}{\sqrt{a_1^2 + \cdots + a_n^2}}\, x_n,$$

$$\xi_1 = \sqrt{\frac{a_2^2 + \cdots + a_n^2}{a_1^2 + a_2^2 + \cdots + a_n^2}}\, x_1$$

$$- \frac{a_1}{\sqrt{(a_1^2 + \cdots + a_n^2)\,(a_2^2 + \cdots + a_n^2)}}\,(a_2x_2 + \cdots + a_nx_n),$$

. . . .

$$\xi_{n-1} = \sqrt{\frac{a_n^2}{a_{n-1}^2 + a_n^2}}\, x_{n-1} - \frac{a_{n-1}}{\sqrt{(a_{n-1}^2 + a_n^2)a_n^2}}\, a_nx_n .$$

Then, since this is an orthogonal transformation,

$$\Sigma_\alpha x_\alpha^2 = \xi^2 + \xi_1^2 + \cdots + \xi_{n-1}^2 .$$

Since $\xi \sim N\left(\sqrt{\Sigma a_\alpha^2}, \sigma^2\right)$, $\xi_\alpha \sim N(0,\sigma^2)$, and they are mutually independent,

$$\sum_{\alpha=1}^{n-1} \xi_\alpha^2 = \sigma^2 \chi_{n-1}^2$$

and

$$\Sigma x_\alpha^2 - 2\Sigma a_\alpha x_\alpha = \xi^2 - 2\sqrt{\Sigma a_\alpha^2}\, \xi + \sigma^2 \chi_{n-1}^2 .$$

We want to find the probability-density function of

$$\sigma^2 \chi_n'^2 = \sum_{\alpha=1}^{n} x_\alpha^2 = \xi^2 + \xi_1^2 + \cdots + \xi_{n-1}^2$$

The joint distribution of ξ and $Z^2 = \xi_1^2 + \cdots + \xi_{n-1}^2$ is given by

$$\frac{1}{\sqrt{2\pi}\,\sigma} exp\left[-\frac{1}{2\sigma^2}(\xi - \sigma\lambda)^2\right] d\xi \frac{1}{\Gamma\left(\frac{n-1}{2}\right)} exp\left(-\frac{Z^2}{2\sigma^2}\right)^{(n-3)/2} d\left(\frac{Z^2}{2\sigma^2}\right),$$

where we have put $\sigma^2\lambda^2 = \Sigma_\alpha a_\alpha^2$. It becomes

$$\frac{exp\ (-\lambda^2/2)}{\sqrt{2\pi}\,\Gamma\left(\frac{n-1}{2}\right)} exp\left(-\frac{\chi'^2}{2}\right) \sum_{\nu=0}^{\infty} \frac{\lambda^\nu}{\nu!} \left(\frac{1}{2}\right)^{(n-3)/2} \left(\frac{z}{\sigma}\right)^{n-2} \left(\frac{\xi}{\sigma}\right)^{\nu} d\left(\frac{z}{\sigma}\right) d\left(\frac{\xi}{\sigma}\right);$$

where

$$\frac{Z}{\sigma} = \chi' cos\ \Theta, \quad \frac{\xi}{\sigma} = \chi' sin\ \Theta,, \quad -\frac{\pi}{2} \le \Theta \le \frac{\pi}{2} ;$$

$$\frac{\partial(z/\sigma, \xi/\sigma)}{(\chi', \theta)} = \begin{vmatrix} cos\ \theta & -\chi'\ sin\ \theta \\ sin\ \theta & \chi'\ cos\ \theta \end{vmatrix} = \chi'$$

$$= \frac{exp(-\lambda^2/2)}{\sqrt{2\pi}\ \Gamma\left(\frac{n-1}{2}\right)}\ exp\left(-\frac{\chi'^2}{2}\right) \sum_{\nu=0}^{\infty} \frac{\lambda^\nu}{\nu!} \left(\frac{1}{2}\right)^{(n-3)/2} (\chi')^{n+\nu-1}\ cos^{n-2}\ \theta$$

$$\times\ sin^\nu\ \theta\ d\theta\ d\chi'\ ,$$

since for $\nu = 2\mu + 1$ (odd),

$$\int_{-\pi/2}^{\pi/2} cos^{n-2}\theta\ sin^{2\mu}\theta d(cos\ \theta) = \int_{-\pi/2}^{\pi/2} cos^{n-2}\theta(1 - cos^2\theta)^\mu d(cos\ \theta)$$

$$= \int_{-\pi/2}^{0} + \int_{0}^{\pi/2} cos^{n-2}\theta(1 - cos^2\theta)^\mu d(cos\ \theta)$$

$$= \int_0^1 x^{n-2}(1 - x^2)^\mu dx + \int_1^0 x^{n-2}(1 - x^2)^\mu dx = 0,$$

For $\nu = 2\mu$ (even),

$$\int_{-\pi/2}^{\pi/2} cos^{n-2}\ \theta\ sin^{2\mu}\ \theta\ d\theta = \int_{-1}^{1} (1 - u^2)^{(n-3)/2}\ u^{2\mu}\ du \quad (u = sin\ \theta)$$

$$= \int_0^1 (1 - v)^{(n-1)/2-1} v^{\mu+\frac{1}{2}-1}\ dv$$

$$(u = v,\ 2du = v^{-\frac{1}{2}}\ dv)$$

$$= B\left(\frac{n-1}{2},\ \mu + \frac{1}{2}\right)$$

$$= \frac{\Gamma\left(\frac{n-1}{2}\right)\ \Gamma\left(\mu + \frac{1}{2}\right)}{\Gamma\left(\frac{n}{2} + \mu\right)}$$

and

$$(2\mu)! = 2\mu\cdot(2\mu - 1)\cdot(2\mu - 2)\cdot(2\mu - 3)\cdot \;\ldots\; \cdot 2\cdot 1$$

$$= 2^{2\mu}\cdot\mu!\cdot\left(\mu - \frac{1}{2}\right)\left(\mu - \frac{3}{2}\right)\cdot \;\ldots\; \cdot \frac{1}{2} ,$$

$$\Gamma\left(\mu + \frac{1}{2}\right) = \left(\mu - \frac{1}{2}\right)\left(\mu - \frac{3}{2}\right) \cdots \frac{1}{2}\,\Gamma\left(\frac{1}{2}\right) = \left(\mu - \frac{1}{2}\right)\left(\mu - \frac{3}{2}\right) \cdots \frac{1}{2}\sqrt{\pi} .$$

Therefore

$$(2\mu)! = \frac{2^{2\mu}\cdot\mu!}{\sqrt{\pi}}\,\Gamma\left(\mu + \frac{1}{2}\right) ,$$

$$\frac{1}{(2\mu)!}\int_{-\pi/2}^{\pi/2} \cos^{n-2}\theta \; \sin^{2\mu}\theta \; d\theta = \frac{\sqrt{\pi}}{2^{2\mu}\mu!}\,\frac{\Gamma[(n - 1)/2]}{\Gamma[(n/2) + \mu]} .$$

Hence our probability element of χ'^2 is

$$\frac{\exp(-\lambda^2/2)}{\sqrt{2\pi}\;\Gamma\left(\frac{n-1}{2}\right)}\,\exp\left(-\frac{\chi'^2}{2}\right)\sum_{\mu=0}^{\infty}\frac{(\lambda^2)^{\mu}}{(2\mu)!}\left(\frac{1}{2}\right)^{(n-3)/2}(\chi')^{n+2\mu-1}\,d\chi'$$

$$\times\int_{-\pi/2}^{\pi/2}\cos^{n-2}\theta\;\sin^{2\mu}\;d\theta$$

$$= \frac{\exp(-\lambda^2/2)}{\sqrt{2}}\,\exp\left(-\frac{\chi'^2}{2}\right)\sum_{\mu=0}^{\infty}\frac{(\lambda^2)^{\mu}}{2^{2\mu}\mu!\,\Gamma\left(\frac{n}{2} + \mu\right)}(\chi'^2)^{(n/2)+\mu-1}\left(\frac{1}{2}\right)^{(n-1)/2}d\chi'^2$$

$$= \exp(-\lambda^2/2)\sum_{\mu=0}^{\infty}\frac{\left(\frac{\lambda^2}{2}\right)^{\mu}}{\mu!}\,\frac{\left(\frac{\chi'^2}{2}\right)^{(n/2)+\mu-1}}{\Gamma\left(\frac{n}{2} + \mu\right)}\,\exp\left(-\frac{\chi'^2}{2}\right)\,d\left(\frac{\chi'^2}{2}\right) .$$

This is called the *noncentral* χ-*square distribution* with n degrees of freedom and with the *noncentrality parameter* $\lambda^2 = \Sigma a_\alpha^2/\sigma^2$. If

$\lambda = 0$, this reduces to

$$\frac{1}{\Gamma(n/2)} \left(\frac{\chi^2}{2}\right)^{(n/2)-1} exp -\left(\frac{\chi^2}{2}\right) d\left(\frac{\chi^2}{2}\right) ,$$

that is, the central χ-square distribution with n degrees of freedom.

We calculate the characteristic function or Fourier transform of the noncentral χ-square distribution :

$$\phi(t) = exp\left(-\frac{\lambda^2}{2}\right) \sum_{\mu=0}^{\infty} \frac{(\lambda^2/2)^{\mu}}{\mu!} \frac{1}{\Gamma[(f/2)+\mu]} \int_0^{\infty} \left(\frac{\chi'^2}{2}\right)^{(f/2)+\mu-1}$$

$$\times \; exp\left[-\frac{\chi'^2}{2}(1 - 2it)\right] d\left(\frac{\chi'^2}{2}\right)$$

$$= exp\left(-\frac{\lambda^2}{2}\right) \sum_{\mu=0}^{\infty} \frac{(\lambda^2/2)^{\mu}}{\mu!} (1 - 2it)^{(-f/2)-\mu}$$

$$= (1 - 2it)^{(-f/2)} exp -\left(\frac{\lambda^2}{2}\right) \sum_{\mu=0}^{\infty} \frac{[\lambda^2/2(1-2it)]^{\mu}}{\mu!}$$

$$= (1 - 2it)^{(-f/2)} exp\left(\frac{it\lambda^2}{1 - 2it}\right).$$

Note : For $t < 0$, we apply the Cauchy integral theorem to

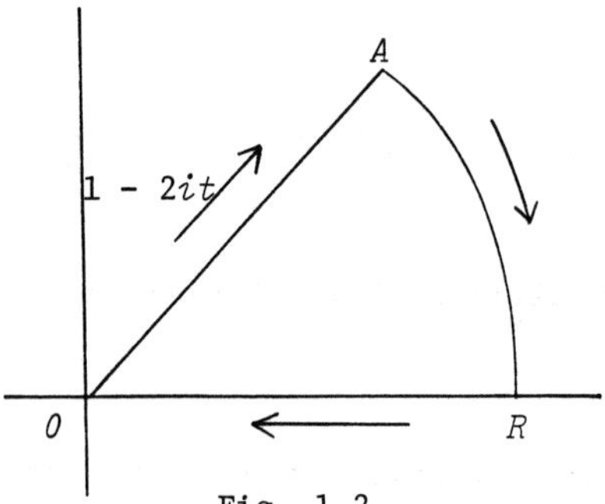

Fig. 1.2

$$\int z^{(f/2)+\mu-1} e^{-z}\, dz,$$

where

$$z = \frac{\chi'^2}{2}(1 - 2it) \text{ (see Fig.1.2).}$$

$$\int_{\overrightarrow{OA}} + \int_{\overset{\frown}{AR}} + \int_{\overrightarrow{RO}} = 0$$

On the arc AR the modulus of the integrand is

$$\left| z^{(f/2)+\mu-1} e^{-z} \right| = R^{(f/2)+\mu-1} e^{-R} \to 0 \text{ as } R \to \infty.$$

Therefore

$$\int_{\overrightarrow{OA}} z^{(f/2)+\mu-1} e^{-z}\, dz = \int_0^{\infty} x^{(f/2)+\mu-1} e^{-x}\, dx = \Gamma\left(\frac{f}{2} + \mu\right).$$

Hence, if $\chi_\alpha'^2$ is the noncentral χ-square with f_α degrees of freedom and with the noncentrality parameter λ_α^2, and if they are mutually independent, then the characteristic function of $\sum_\alpha \chi_\alpha'^2$ is

$$E\left[exp\left(it \sum_\alpha \chi_\alpha'^2\right)\right] = \prod_\alpha E\left[exp\ (it\chi_\alpha'^2)\right] = (1 - 2it)^{(-1/2)\Sigma f_\alpha}\ exp\ \frac{2t\Sigma\lambda_\alpha^2}{1 - 2it},$$

that is, $\Sigma\chi_\alpha'^2$ is also the noncentral χ-square with Σf_α degrees of and with the noncentrality parameter $\Sigma\lambda_\alpha^2$. This is called *the additive property*.

Distribution of Quadratic Forms. Suppose that $x_1,\dots,x_n$ is a random sample of size n from a normal population $N(0,1)$. We consider a quadratic form

$$Q(x_1,\dots,x_n) = \underline{x}'A\underline{x} = \sum_{\alpha,\beta=1}^{n} a_{\alpha\beta}x_\alpha x_\beta \qquad (a_{\alpha\beta} = a_{\beta\alpha}).$$

We can choose an orthogonal matrix P such that

$$P'AP = D_\lambda,$$

where

$$D_\lambda = \begin{Vmatrix} \lambda_1 & & 0 \\ & \ddots & \\ 0 & & \lambda_n \end{Vmatrix}$$

Furthermore, $\underline{y} = P'\underline{x}$ is also a random sample of size n from $N(0,1)$, and

$$Q = \sum_{\alpha=1}^{n} \lambda_\alpha y_\alpha^2.$$

The characteristic function of $Q(\underline{x})$ can be calculated as follows

$$\phi(t) = \left(\frac{1}{\sqrt{2\pi}}\right)^n \int_{-\infty}^{\infty} \cdots \int exp\left\{-\frac{1}{2}\left[\Sigma_\alpha x_\alpha^2 - 2itQ(x)\right]\right\} dx_1 \cdots dx_n$$

$$\underline{x}'(I - 2itA)\underline{x} = \underline{x}'PP'(I - 2itA)PP'\underline{x} = \underline{y}'(I - 2itP'AP)\underline{y}$$

$$= \sum_{\alpha=1}^{n} (I - 2it\lambda_\alpha)y_\alpha^2,$$

$$\frac{\partial(\underline{x})}{\partial(\underline{y})} = |P| = \pm 1.$$

Since one can show that

$$\frac{1}{\sqrt{2\pi}} \int_{-\infty}^{\infty} exp\left\{-\frac{1}{2}(1 - 2it)y^2\right\} dy = (1 - 2it)^{-\frac{1}{2}},$$

consequently one obtains

$$\phi(t) = \prod_{\alpha=1}^{n} (1 - 2it\lambda_\alpha)^{-\frac{1}{2}} = |I - 2itA|^{-\frac{1}{2}}.$$

Thus the necessary and sufficient condition for the quadratic form $Q(\underline{x})$ to be a χ-square is that the nonzero characteristic roots of A be all unity or $A^2 = A$. If $A^2 = A$, then the rank f of A is equal to its trace, and consequently $Q(\underline{x}) = \chi_f^2$.

<u>Problem</u>. Consider the same problem for the population $N(\mu,\Sigma)$: $A\Sigma A = A$ (See Carpenter, O. [2]).

The distribution of the statistic $Q(\underline{x}) = \underline{x}'A\underline{x}$ is reduced to that of $\sum_{\alpha=1}^{f} \lambda_\alpha y_\alpha^2$. To find the distribution of $\Sigma\lambda_\alpha y_\alpha^2$, one should refer to the papers by Kotz, Johnson and Boyd, [3,4] on the series representations of the distribution of quadratic forms in normal variables.

Let us consider two quadratic forms,

$$Q_1 = \underline{x}'A\underline{x} \quad \text{and} \quad Q_2 = \underline{x}'B\underline{x}.$$

The characteristic function of their joint distribution is given by

$$\phi(t_1,t_2) = E\left\{exp[i(t_1Q_1 + t_2Q_2)]\right\} = |I - 2it_1A - 2it_2B|^{-\frac{1}{2}}.$$

The necessary and sufficient condition for the independence of Q_1 and Q_2 is, in terms of the characteristic functions

$$\phi(t_1,t_2) = \phi(t_1,0)\phi(0,t_2),$$

or

$$\left|I - 2it_1A - 2it_2B\right| = \left|I - 2it_1A\right| \cdot \left|I - 2it_2B\right|.$$

One can show that this is *equivalent to the condition* $AB = 0$. If $AB = 0$, then

$$(I - t_1A)(I - t_2B) = I - t_1A - t_2B,$$

and hence

$$|I - t_1A||I - t_2B| = |I - t_1A - t_2B|.$$

We have to show the converse. Suppose that A is of rank r and the nonzero characteristic roots of A are $\lambda_1,\ldots,\lambda_r$. Transforming A by a suitable orthogonal matrix P into the diagonal form such that

$$P'AP = \begin{Vmatrix} \lambda_1 & & & & 0 \\ & \ddots & & & \\ & & \lambda_r & & \\ & & & 0 & \\ & & & & \ddots \\ 0 & & & & 0 \end{Vmatrix}$$

and letting

$$C = P'BP = \begin{Vmatrix} C_{11} & C_{12} \\ C_{12}' & C_{22} \end{Vmatrix},$$

we have

$$\left|I - t_1 A - t_2 B\right| = \left|P'P - t_1 P'AP - t_2 P'BP\right|$$

$$= \left|\begin{array}{ccc|ccc}
1 - t_1\lambda_1 - t_2 c_{11} & \cdots & -t_2 c_{1r} & -t_2 c_{1\,r+1} & \cdots & -t_2 c_{1n} \\
\vdots & & \vdots & \vdots & & \vdots \\
-t_2 c_{r1} & \cdots & 1 - t_1\lambda_r - t_2 c_{rr} & -t_2 c_{r\,r+1} & & -t_2 c_{rn} \\
\hline
-t_2 c_{r+1\,1} & \cdots & -t_2 c_{r+1\,r} & 1 - t_2 c_{r+1\,r+1} & \cdots & -t_2 c_{r+1\,n} \\
\vdots & & \vdots & \vdots & & \vdots \\
-t_2 c_{n\,1} & \cdots & -t_2 c_{n\,r} & -t_2 c_{n\,r+1} & \cdots & 1 - t_2 c_{n\,n}
\end{array}\right|$$

Considering this as a polynomial in t_1, the coefficient of the term of highest degree is

$$(-1)^r \lambda_1 \cdots \lambda_r \left|\begin{array}{ccc}
1 - t_2 c_{r+1\,r+1} & \cdots & -t_2 c_{r+1\,n} \\
\vdots & & \vdots \\
-t_2 c_{n\,r+1} & \cdots & 1 - t_2 c_{n\,n}
\end{array}\right| .$$

This must be equal to $(-1)^r\lambda_1 \cdots {}_r|I - t_2C|$. This implies that the nonvanishing eigenvalues of C are those of C_{22} and let them be $\mu_1,\ldots,\mu_s$ $(s \le n - r)$. Since for any symmetric matrix $K = \|k_{\alpha\beta}\|$, this can be brought into a diagonal form by a suitable orthogonal matrix Q :

$$Q'KQ = D_\lambda,$$

$$Q'KKQ = D_{\lambda^2},$$

$$tr(Q'KKQ) = trK^2 = \sum_{\alpha\ \beta} k^2_{\alpha\beta} = \begin{array}{c}\text{sums of squares of}\\ \text{nonzero eigenvalues.}\end{array}$$

Hence one gets

$$trC^2 = tr\left(C^2_{11} + 2C_{12}C'_{12} + C^2_{22}\right) = \mu_1^2 + \cdots + \mu_s^2$$

and

$$trC^2_{22} = \mu_1^2 + \cdots + \mu_s^2,$$

whence one is led to

$$trC^2_{11} = 0 \text{ and } trC_{12}C'_{12} = 0.$$

Hence

$$c_{\alpha\beta} = 0; \quad \alpha = 1,\ldots,r, \quad \beta = 1,\ldots,n.$$

Consequently

$$P'AP \cdot P'BP = \left\|\begin{array}{ccc|c} \lambda_1 & & & \\ & \ddots & & 0 \\ & & \lambda_r & \\ \hline & & & 0 \\ & 0 & & \ddots \\ & & & & 0 \end{array}\right\| \left\|\begin{array}{c|c} 0 & 0 \\ \hline 0 & C_{22} \end{array}\right\| = 0$$

or

$$AB = 0.$$

Lemma 1.2.2. Given that $C = A + B$, where all matrices are symmetric and $r_C = r_A + r_B$, if C is idempotent, then

$$A^2 = A, \quad B^2 = B, \quad AB = 0.$$

Proof. The image space of the matrix C is apparently the union of those of A and B, and hence

$$\dim C(L) = \dim A(L) + \dim B(L) - \dim A(L) \cap B(L).$$

Due to the condition $r_C = r_A + r_B$, we obtain

$$\dim A(L) \cap B(L) = 0.$$

Suppose A has a nonvanishing eigenvalue λ that is different from 1; then there exists a nonzero eigenvector $\underline{x}$,

$$A\underline{x} = \lambda\underline{x},$$

and since $\underline{x}$ must belong to $C(L)$, we get

$$\underline{x} = \lambda\underline{x} + B\underline{x}$$

or

$$B\underline{x} = (1 - \lambda)\underline{x};$$

that is $\underline{x} \neq 0$ belongs to the image space $B(L)$. Therefore this is a contradiction, and any nonzero eigenvalue of A must be unity. In other words, A must be idempotent. Similarly one can show that B is also idempotent.

Since C is idempotent,

$$A + B = (A + B)^2 = A + B + AB + BA,$$

$$AB = - BA.$$

If $\underline{x} \in A(L)$, then

$$AB\underline{x} = - BA\underline{x} = - B\underline{x} .$$

Hence

$$B\underline{x} = 0,$$

and therefore

$$AB\underline{x} = 0.$$

It is evident that, for $\underline{x} \in A(L)^{\perp}$ = kernel of A,

$$AB\underline{x} = - BA\underline{x} = 0.$$

Consequently

$$AB = BA = 0.$$

The reader is referred to Matsushita [10] and Ogawa [8] for further details.

Cochran's Theorem

$$\underline{x} \sim N(0, \sigma^2 I),$$
$$Q_\alpha = \underline{x}' A_\alpha \underline{x} \qquad (\alpha = 1,\ldots,s),$$
$$\sum_{\alpha=1}^{s} Q_\alpha = \sum_{\alpha=1}^{n} x_\alpha^2.$$

If the rank of Q_α, $r_\alpha \le n_\alpha$, $\alpha = 1,\ldots,s$, then Q_α are independent Q_α is the χ-square with degrees of freedom n_α if and only if $\Sigma_\alpha n_\alpha = n$.

In terms of matrices, one can rephrase Cochran's theorem as follows :

Theorem 1.2.1. Let $A_1, A_2,\ldots, A_s$ be real, symmetric matrices with respective ranks $r_1, r_2,\ldots, r_s$ such that

$$A_1 + A_2 + \cdots + A_s = I.$$

Then the necessary and sufficient condition for

$$A_\alpha^2 = A_\alpha \text{ and } A_\alpha A_\beta = 0, \ \alpha \neq \beta$$

is that

$$r_1 + r_2 + \cdots + r_s = n.$$

Proof. First we will show that, if $\Sigma r_\alpha = n$, then $A_\alpha^2 = A_\alpha$ and $A_\alpha A_\beta = 0$, $\alpha \neq \beta$.

Let $B_1 = A_2 + \cdots + A_s$. Then

$$r(B_1) \leq r_2 + \cdots + r_s,$$

$$n \leq r(A_1) + r(B_1) \leq r_1 + r_2 + \cdots + r_s = n.$$

Hence

$$r(B_1) = r_2 + \cdots + r_s.$$

Applying Lemma 1.2.2, we obtain

$$A_1^2 = A_1; \quad B_1^2 = B_1; \quad 0 = A_1 B_1 = A_1 A_2 + \cdots + A_1 A_s.$$

Next, let $B_2 = A_3 + \cdots + A_s$, then $B_1 = A_2 + B_2$, we can see that

$$r(B_2) = r_3 + \cdots + r_s.$$

Applying Lemma 1.2.2, we obtain

$$A_2^2 = A_2; \quad B_2^2 = B_2; \quad 0 = A_2 B_2 = A_2 A_3 + \cdots + A_2 A_s.$$

Proceeding this way we finally obtain

$$A_{s-1}^2 = A_{s-1}; \quad A_s^2 = A_s; \quad A_{s-1} A_s = 0.$$

Rearranging the order of $A_1, \ldots, A_s$ in such a way that any specified pair A_α, B_β comes to the last position and using the same reasoning as before, we can show that

$$A_\alpha^2 = A_\alpha; \quad A_\beta^2 = A_\beta; \quad A_\alpha A_\beta = 0 \qquad (\alpha \neq \beta).$$

The proof of the converse is very easy indeed.

The reader is referred to the papers by Banerjee [6] and Loynes [7] for further details.

1.3. Testing The Hypothesis on Regression Coefficients

We are considering the following situation : $x_1, \ldots, x_n$ are the independent observations from normal populations with means

$$E(x_\alpha) = g_{\alpha 1}\beta_1 + \cdots + g_{\alpha s}\beta_s \qquad (\alpha = 1, \ldots, n)$$

and with a common unknown variance σ^2, where

$$G = \begin{Vmatrix} g_{11} & \cdots & g_{s} \\ \vdots & & \\ g_{n1} & & g_{ns} \end{Vmatrix}$$

are known constants and the rank of G is assumed to be *s-full rank case*. The gramian matrix $G'G$ is a positive definite matrix. In vector notation we define the error vector

$$\underline{e} = \begin{pmatrix} e_1 \\ \vdots \\ e_n \end{pmatrix} = \underline{x} - G\underline{\beta}.$$

The least-squares estimate (LSE)$\underline{b}$ of $\underline{\beta}$ is the value of $\underline{\beta}$ that minimizes the sum of squares $\underline{e}'\underline{e} = \Sigma e_\alpha^2 = (\underline{x} - G\underline{\beta})'(\underline{x} - G\underline{\beta})$ considered as a function $\underline{\beta}$. It is determined by the *normal equation*

$$G'(\underline{x} - G\underline{b}) = 0,$$

where

$$\underline{b} = (G'G)^{-1} G'\underline{x}.$$

Hence $E(\underline{b}) = (G'G)^{-1} G'E(\underline{x}) = (G'G)^{-1} G'G\,\underline{\beta} = \underline{\beta}$, and thus

$$\underline{b} - \underline{\beta} = (G'G)^{-1}G'\underline{e}.$$

The residual variate is

$$\underline{y} = \underline{x} - G\underline{b} = [\mathrm{I} - G(G'G)^{-1}G']\underline{e}.$$

Hence the (absolute) minimum value Q_0 of the sum of squares Σe_α^2 is given by

$$Q_0 = \underline{y}'\underline{y} = \underline{e}'[\mathrm{I} - G(G'G)^{-1}G']\underline{e} = \underline{x}'\underline{x} - \underline{b}'G'G\underline{b}.$$

Since $\mathrm{I} - G(G'G)^{-1}G'$ is an idempotent matrix and

$$tr\{\mathrm{I} - G(G'G)^{-1}G'\} = n - tr\, G(G'G)^{-1}G' = n - s,$$

$$Q_\alpha = \sigma^2\chi^2_{n-s},$$

$$(\underline{b} - \underline{\beta})'(G'G)(\underline{b} - \underline{\beta}) = \underline{e}\; G(G'G)^{-1}G'\underline{e} = \sigma^2\chi^2_s\ ,$$

and these two χ-squares are stochastically independent, (see pp. 30-42). Hence

$$F = \frac{n - s}{s}\;\frac{(\mathrm{b} - \beta)'G'G(\mathrm{b} - \beta)}{\underline{x}'\underline{x} - \underline{b}'G'G\underline{b}}$$

is distributed as an F^s_{n-s}-distribution, that is *the central F-distribution.*

We consider a somewhat more complicated situation:

$$G = \left\|G_1\ G_2\right\|; \quad G_1 = \left\|\begin{matrix} g_{11} & \cdots & g_{1r} \\ \vdots & & \vdots \\ g_{n1} & \cdots & g_{nr}\end{matrix}\right\|; \quad G_2 = \left\|\begin{matrix} g_{1r+1} & \cdots & g_{1s} \\ \vdots & & \vdots \\ g_{nr+1} & \cdots & g_{ns}\end{matrix}\right\|;$$

$$r(G_1) = r; \qquad r(G_2) = s - r;$$

$$\underline{\beta}' = (\underline{\beta}_1' \ \underline{\beta}_2'); \qquad \underline{\beta}_1' = (\beta_1 \ldots \beta_r); \qquad \underline{\beta}' = (\beta_{r+1} \ldots \beta_s).$$

Suppose we are now required to test the statistical hypothesis

$$H_0 : \underline{\beta}_2 = \underline{\beta}_2^0.$$

One should note that any linear hypothesis can be reduced to this form. The least-squares estimate $\underline{b}_1^*$ of $\underline{\beta}_1$ under the null hypothesis H_0 is given by minimizing

$$Q^* = (\underline{x} - G_1\underline{\beta}_1 - G_2\underline{\beta}_2^0)'(\underline{x} - G_1\underline{\beta}_1 - G_2\underline{\beta}_2^0).$$

The normal equation for $\underline{b}_1^*$ is

$$G_1'G_1\underline{b}_1^* = G'(\underline{x} - G_2\underline{\beta}_2^0),$$

and hence

$$\underline{b}_1^* = (G_1'G_1)^{-1} G'(\underline{x} - G_2\underline{\beta}_2^0),$$

$$E(\underline{b}_1^*) = \underline{\beta} + (G_1'G)^{-1} G_1'G_2(\underline{\beta}_2 - \underline{\beta}_2^0),$$

and

$$\underline{b}_1^* - \underline{\beta}_1 = (G_1'G_1)^{-1} G_1'\underline{e} + (G_1'G_1)^{-1} G_1'G_2(\underline{\beta}_2 - \underline{\beta}_2^0)$$

if the null hypothesis H_0 is not true. The residual variate is

$$\underline{y}^* = \underline{x} - G_1\underline{b}_1^* - G_2\underline{\beta}_2^0 = \left[I - G_1(G_1'G_1)^{-1} G_1'\right]\underline{e} + \left[I - G_1(G_1'G_1)^{-1} G_1'\right] \times G_2(\underline{\beta}_2 - \underline{\beta}_2^0).$$

Hence $\underline{y}^*$ and $\underline{b}_1^*$ are stochastically independent.

Case I : in this case we have $G_1'G_2 = 0$.

$$G(G'G)^{-1}G' = \begin{Vmatrix} G_1(G_1'G_1)^{-1}G_1' & 0 \\ 0 & G_2(G_2'G_2)^{-1}G_2' \end{Vmatrix},$$

and hence

$$\underline{y} = \left[I - G(G'G)^{-1}G'\right]\underline{e} = \left[I - G_1(G_1'G_1)^{-1}G_1'\right]\underline{e} - G_2(G_2'G_2)^{-1}G_2'\underline{e}$$

$$= \underline{y}^* - G_2(G_2'G_2)^{-1}G_2'\underline{e} - G_2(\underline{\beta}_2 - \underline{\beta}_2^0)$$

or

$$\underline{y}^* = \underline{y} + G_2(G_2'G_2)^{-1}G_2'\underline{e} + G_2(\underline{\beta}_2 - \underline{\beta}_2^0).$$

Now we introduce new variates by

$$\underline{\zeta}_1' = (\zeta_1 \ldots \zeta_r) = \underline{e}'G_1(G_1'G_1)^{-1/2},$$

$$\underline{\zeta}_2' = (\zeta_{r+1} \cdots \zeta_s) = \underline{e}'G_2(G_2'G_2)^{-1/2},$$

$$\underline{\zeta}_3' = (\zeta_{s+1} \cdots \zeta_n) = \underline{e}'U',$$

where the $(n - s) \times n$ matrix U is chosen in such a way that

$$C' = \left\| G_1(G_1'G_1)^{-1/2} \vdots G_2(G_2'G_2)^{-1/2} \vdots U' \right\|$$

is orthogonal. Then

$$E(\underline{\zeta\zeta}') = CE(\underline{ee}')C' = \sigma^2 I_n,$$

and

$$\underline{e} = C'\underline{\zeta} = G_1(G_1'G_1)^{-1/2}\underline{\zeta}_1 + G_2(G_2'G_2)^{-1/2}\underline{\zeta}_2 + U'\underline{\zeta}_3.$$

Thus we obtain the $\underline{b}$, $\underline{y}$ and $\underline{b}^*$, $\underline{y}^*$ in terms of $\underline{\zeta}$ as follows :

$$\underline{b} - \underline{\beta} = \begin{pmatrix} (G_1'G_1)^{-1/2} & 0 \\ 0 & (G_1'G_1)^{-1/2} \end{pmatrix} \begin{pmatrix} \underline{\zeta}_1 \\ \underline{\zeta}_2 \end{pmatrix},$$

$$\underline{y} = U'\underline{\zeta}_3,$$

$$\underline{b}_1^* - \underline{\beta}_1 = (G_1'G_1)^{-1/2}\underline{\zeta}_1,$$

$$\underline{y}^* = G_2(G_2'G_2)^{-1/2}\underline{\zeta}_2 + U'\underline{\zeta}_3 + G_2(\underline{\beta}_2 - \underline{\beta}_2^0).$$

Hence

$$\underline{y}^* - \underline{y} = G_2(G_2'G_2)^{-1/2}\underline{\zeta}_2 + G_2(\underline{\beta}_2 - \underline{\beta}_2^0),$$

and therefore

$$(\underline{y}^* - \underline{y})'\underline{y} = 0 \rightarrow \underline{y}^{*\prime}\underline{y} = \underline{y}'\underline{y} = Q_a$$

$$Q_r - Q_a = \underline{y}^{*\prime}\,\underline{y}^* - \underline{y}'\underline{y} = (\underline{y}^* - \underline{y})'(\underline{y}^* - \underline{y})$$

$$= \underline{\zeta}_2'\underline{\zeta}_2 + 2(\underline{\beta}_2 - \underline{\beta}_2^0)'(G_2'G_2)^{1/2}\underline{\zeta}_2 + (\underline{\beta}_2 - \underline{\beta}_2^0)'G_2'G_2(\underline{\beta}_2 - \underline{\beta}_2^0).$$

Hence one can see that $Q_r - Q_a = \sigma^2\chi'^2_{s-r}$, where χ'^2_{s-r} is the noncentral χ-square with $s - r$ degrees of freedom and with the noncentrality parameter

$$\lambda^2 = \frac{1}{\sigma^2}(\underline{\beta}_2 - \underline{\beta}_2^0)'G_2'G_2(\underline{\beta}_2 - \underline{\beta}_2^0).$$

Since $Q_a = \underline{\zeta}_3'\underline{\zeta}_3 = \sigma^2\chi^2_{n-s}$, and this is independent of $Q_r - Q_a$, the probability element of

$$G = \frac{Q_r - Q_a}{Q_a} = \frac{s - r}{n - s} F$$

is given by

$$exp\left(-\frac{\lambda^2}{2}\right) \sum_{\nu=0}^{\infty} \frac{\left(\frac{\lambda^2}{2}\right)^{\nu}}{\nu\,!} \frac{\Gamma\left(\frac{n-r}{2} + \nu\right)}{\Gamma\left(\frac{s-r}{2} + \nu\right) \Gamma\left(\frac{n-s}{2}\right)} G^{[(s-r)/2]+\nu-1}$$

$$\times \left(\frac{1}{1 + G}\right)^{[(n-r)/2]+\nu} dG.$$

If $\lambda^2 = 0$ (i.e., H is true), then this reduces to the central $(s - r)/(n - s)F$ distribution.

<u>Case II</u>: $G_1'G_2 \neq 0.$

This case can be reduced to Case I by the Schmidt procedure of orthogonalization. One gets the same conclusion with

$$\lambda^2 = \frac{1}{\sigma^2}(\underline{\beta}_2 - \underline{\beta}_2^0)'G_2'\left[I - G_1(G_1'G_1)^{-1} G_1'\right]G_2(\underline{\beta}_2 - \underline{\beta}_2^0).$$

Readers should carry out the whole process by themselves. For details the reader is referred to J. Ogawa [8].

<u>Problem 1.3.1.</u> Show that in any Euclidean vector space V (i.e., a finite-dimensional vector space over the real field endowed with a positive symmetric inner product) a linear transformation $\sigma : V \to V$ is a perpendicular projection if and only if $\sigma = \sigma^2 = \sigma^*$, where σ^* denotes the transformation *adjoint* to σ.

$$(\sigma\underline{x},\ \underline{y}) = (\underline{x},\ \sigma^*\underline{y}) \quad \text{for all } \underline{x},\underline{y} \in V.$$

Vectors in V are represented relative to an orthonormal basis by n-tuples, and linear transformations of V are represented by square matrices acting on column vectors. If W is a subspace of V spanned by m linearly independent vectors, represented by column vectors $\underline{x}_1,\ldots,\underline{x}_m$, and X is the $n \times m$ matrix

$$X = \|\underline{x}_1 \ldots \underline{x}_m\|,$$

show that the matrix of the perpendicular projection of V on W is $X(X'X)^{-1}X'$, where X' is the transpose of X.

Problem 1.3.2. Find the matrix of the perpendicular projection of a four-dimensional space on the subspace spanned by the vectors

$$\begin{bmatrix} 1 \\ 0 \\ 2 \\ -1 \end{bmatrix} \quad \text{and} \quad \begin{bmatrix} 0 \\ 1 \\ 1 \\ -1 \end{bmatrix}.$$

1.4. General Linear Hypotheses

We make the following assumptions :

(1) $x_1,\ldots,x_n$ are normally and independently distributed with unknown common variance σ^2,

(2) $\mu_\alpha = E(x_\alpha)$, $\alpha = 1,\ldots,n$ are linear functions of s unknown parameters $\beta_1,\ldots,\beta_s$, *i.e.*,

$$(1.4.1)\quad \underline{\mu} = \begin{bmatrix} \mu_1 \\ \vdots \\ \mu_n \end{bmatrix} = \begin{Vmatrix} g_{11} & \cdots & g_{1s} \\ \vdots & & \vdots \\ g_{n1} & \cdots & g_{ns} \end{Vmatrix} \cdot \begin{bmatrix} \beta_1 \\ \vdots \\ \beta_s \end{bmatrix} \equiv G\underline{\beta},$$

and the rank of G is equal to s.

Eliminating $\underline{\beta}$ from (1.4.1), the assumption 2 is seen to be equivalent to the following conditions

$$\text{(i)} \quad \sum_{\alpha=1}^{n} \lambda_{i\alpha}\mu_{\alpha} = 0, \qquad i = 1,\ldots,n - s.$$

$$\text{(ii) The rank of } \|\lambda_{i\alpha}\| = n - s.$$

Suppose we are required to test that $\underline{\beta}$ satisfies $s - r$ linearly independent linear restrictions

$$\sum_{j=1}^{k} k_{ij}\beta_j = 0 \qquad (i = 1,\ldots,s - r. \quad s < r). \tag{1.4.2}$$

These can be transformed into equivalent restrictions under the assumption (1.4.1.):

$$\sum_{\alpha=1}^{n} \rho_{j\alpha}\mu_{\alpha} = 0 \qquad (j = 1,\ldots,s - r), \tag{1.4.3}$$

and the rank of the following matrix is equal to $n - r$:

$$\left\| \begin{matrix} \lambda_{11} & \cdots & \lambda_{1n} \\ \vdots & & \vdots \\ \lambda_{n-s\ 1} & \cdots & \lambda_{n-s\ n} \\ \rho_{11} & \cdots & \rho_{1n} \\ \vdots & & \vdots \\ \rho_{s-r\ 1} & \cdots & \rho_{s-r\ n} \end{matrix} \right\|$$

By applying the Schmidt orthogonalization procedure one can orthonormalize the $n - r$ row vectors of the above matrix. We assume that this procedure has already been applied and the end result has been written down.

Let

$$P = \begin{Vmatrix} \lambda_{11} & \cdots & \lambda_{1n} \\ \vdots & & \vdots \\ \lambda_{n-s\,1} & \cdots & \lambda_{n-s\,n} \\ \rho_{11} & \cdots & \rho_{1n} \\ \vdots & & \vdots \\ \rho_{s-r\,1} & \cdots & \rho_{s-r\,n} \\ \zeta_{11} & \cdots & \zeta_{1n} \\ \vdots & & \vdots \\ \zeta_{r\,1} & \cdots & \zeta_{r\,n} \end{Vmatrix} \tag{1.4.4}$$

be an orthogonal matrix of order n and let

$$\underline{x}^* = P\underline{x}, \qquad \underline{\mu}^* = P\underline{\mu}. \tag{1.4.5}$$

Then

$$\begin{aligned} Q = \sum_{\alpha=1}^{n} (x_\alpha - \mu_\alpha)^2 &= (\underline{x} - \underline{\mu})'(\underline{x} - \underline{\mu}) = (\underline{x}^* - \underline{\mu}^*)'(\underline{x}^* - \underline{\mu}^*) \\ &= \sum_{\alpha=1}^{n} (x_\alpha^* - \mu_\alpha^*)^2 \\ &= \sum_{\alpha=1}^{n-s} x_\alpha^{*2} + \sum_{\beta=n-s+1}^{n} (x_\beta^* - \mu_\beta^*)^2 \quad \text{under assumption 2} \\ &= \sum_{\alpha=1}^{n-r} x_\alpha^{*2} + \sum_{\beta=n-r+1}^{n} (x_\beta^* - \mu_\beta^*) \quad \text{under assumption 2 plus the null hypothesis.} \end{aligned} \tag{1.4.6}$$

Hence the absolute minimum of Q is

$$Q_a = \sum_{\alpha=1}^{n-s} x_\alpha^{*2} = \sigma^2\chi^2_{n-s}, \tag{1.4.7}$$

and the relative minimum of Q under the null hypothesis is

$$Q_r = \sum_{\alpha=1}^{n-r} x_\alpha^{*2} = \sigma^2\chi^2_{n-r}. \tag{1.4.8}$$

Thus

$$Q_r - Q_a = \sum_{\alpha=n-s+1}^{n-r} x_\alpha^{*2} = \sigma^2\chi^2_{s-r} \tag{1.4.9}$$

is independent of Q_a. Therefore

$$F = \frac{n-s}{s-r}\,\frac{Q_r - Q_a}{Q_a} \tag{1.4.10}$$

is distributed as the F-distribution with $(s - r,\ n - s)$ degrees of freedom.

The regression value $\underline{Y} = G\underline{b} = \underline{x}-\underline{y}$ satisfies the relation

$$\underline{y} = \underline{x} - \underline{Y}.$$

Since $\underline{y}'\underline{Y} = \underline{y}'G\underline{b} = 0$, it follows that

$$\underline{y}'(\underline{x} - \underline{y}) = 0 \Rightarrow \underline{x}'\underline{y} = \underline{y}'\underline{y},$$

and hence

$$(\underline{x} - \underline{Y})'\underline{Y} = 0 \Rightarrow \underline{x}'\underline{Y} = \underline{Y}'\underline{Y}.$$

The absolute minimum value of Q is

$$\begin{aligned} Q_a &= (\underline{x}-G\underline{b})'(\underline{x}-G\underline{b}) = (\underline{x}-\underline{Y})'(\underline{x}-\underline{Y}) = (\underline{x}-\underline{Y})'\underline{x} = \underline{x}'\underline{x} - \underline{x}'\underline{Y} \\ &= \underline{x}'\underline{x} - \underline{Y}'\underline{Y} \\ &= \sum_{\alpha=1}^{n} x_\alpha^2 - \sum_{\alpha=1}^{n} Y_\alpha^2 \end{aligned} \tag{1.4.11}$$

Let the regression value $G\underline{b}^*$ under the null hypothesis be $\underline{Y}^*$. Then by a similar argument we obtain

$$Q_r = \sum_{\alpha=1}^{n} x_\alpha^2 - \sum_{\alpha=1}^{n} Y_\alpha^{*2}, \tag{1.4.12}$$

whence we may have

$$Q_r - Q_a = \sum_\alpha Y_\alpha^2 - \sum_\alpha Y_\alpha^{*2} = \sum_\alpha (Y_\alpha - Y_\alpha^*)^2, \tag{1.4.13}$$

We have to show that

$$\sum_\alpha (Y_\alpha - Y_\alpha^*)Y_\alpha^* = 0 \quad \text{or} \quad (\underline{Y} - \underline{Y}^*)'\underline{Y}^* = 0$$

Indeed since (1.4.2) is equivalent to the condition that the $\beta_1,\ldots,\beta_s$ are expressed linearly by r parameters $\gamma_1,\ldots,\gamma_r$,

$$\beta_i = \sum_{j=1}^{r} h_{ij}\gamma_j \qquad (i = 1,\ldots,s). \tag{1.4.14}$$

If we denote the least-squares estimates of $\gamma_1,\ldots,\gamma_r$ by $c_1,\ldots,c_r$, respectively, then

$$Y_i^* = \sum_{l=1}^{r} c_l \sum_{t=1}^{s} g_{it}h_{tl}\ . \tag{1.4.15}$$

From $(\underline{x} - \underline{Y})'\underline{Y} = 0 \Leftrightarrow \underline{x}'[I - G(G'G)^{-1}G']G\underline{b} = 0 \Leftrightarrow \underline{x}'[I - G(G'G)^{-1}G']G = 0$,

$$\sum_{\alpha=1}^{n} (x_\alpha - Y_\alpha)g_{\alpha t} = 0$$

for all $t = 1, 2,\ldots,s$. Multiplying this by $h_{tl}c_l$ and summing up with respect to t and l, we get

$$(\underline{x} - \underline{Y})'\underline{Y}^* = 0. \tag{1.4.16}$$

By the same reason leading to $(\underline{x} - \underline{Y})'\underline{Y} = 0$, we obtain

$$(1.4.17) \qquad (\underline{x} - \underline{Y}^*)'\underline{Y}^* = 0.$$

Hence by subtraction, we obtain $(\underline{Y}^* - \underline{Y})'\underline{Y}^* = 0$, as was to be proved.

We have an important partition of the sum of squares :

$$(1.4.18) \qquad \sum_{\alpha=1}^{n} x_\alpha^2 = \sum_{\alpha=1}^{n} (x_\alpha - Y_\alpha)^2 + \sum_{\alpha=1}^{n} (Y_\alpha - Y_\alpha^*)^2 + \sum_{\alpha=1}^{n} Y_\alpha^{*2} .$$

Now we can generalize this as follows : Let

$$H_1 : \mu_\alpha = \sum_{j=1}^{s} g_{\alpha j}\beta_j \qquad (\alpha = 1,\ldots,n),$$

$$H_2 : H_1 \ \& \ \sum_{j=1}^{b} a_{ij\ j} = 0;\ s_1 = 0 \qquad (i = 1,\ldots,s_2),$$

. . . .

$$H_t : H_{t-1} \ \& \ \sum_{j=1}^{b} a_{ij}\beta_j = 0;\ i = s_{t-1} + 1,\ldots,s_t \qquad (s_t \leq s).$$

Let the regression value of $\underline{x}$ on G under H_s be $Y^{(s)}$. Then

$$(1.4.19) \qquad \sum_\alpha x_\alpha^2 = \sum_\alpha (x_\alpha - Y_\alpha^{(1)})^2 + \sum_2 (Y_\alpha^{(1)} - Y_\alpha^{(2)})^2 + \cdots$$

$$+ \sum_\alpha (Y_\alpha^{(t-1)} - Y_\alpha^{(t)})^2 + \sum_\alpha Y_\alpha^{(t)2} .$$

Example 1.4.1. Let us consider the regression equation

$$E(y) = \alpha + \beta x,$$

where the value of y has been observed for a certain value of x. For example, y may be the length of a steel rod and x the temperature at which this length is measured.

Given a set of observations $y_1,\ldots,y_n$ at $x_1,\ldots,x_n$, one might with to test whether β has some hypothetical value β^0 (frequently equal to zero). The sum of squares to be minimuzed is

$$Q = \sum_{\alpha=1}^{n} (y_\alpha - \alpha - \beta x_\alpha)^2.$$

The normal equations for the least-squares estimates are

$$a + b\bar{x} = \bar{y}$$

and

$$na\bar{x} + b \sum_{\alpha=1} x_\alpha^2 = \sum_{\alpha=1}^{n} x_\alpha y_\alpha.$$

Hence

$$b = \frac{\sum_\alpha x_\alpha y_\alpha - n\bar{x}\bar{y}}{\sum_\alpha x_\alpha^2 - n\bar{x}^2}$$

$$E(b) = \sum_\alpha \frac{(x_\alpha - \bar{x})}{\sum_\alpha x_\alpha^2 - nx^2} E(y_\alpha) = \sum_\alpha \frac{(x_\alpha - \bar{x})^2}{\sum_\alpha x_\alpha^2 - n\bar{x}^2} \beta = \beta,$$

and

$$V(b) = E(b - \beta)^2 = \sum_\alpha \frac{(x_\alpha - \bar{x})}{\left[\sum_\alpha (x_\alpha - \bar{x})^2\right]^2} E(y_\alpha - \alpha - \beta x_\alpha)^2 = \frac{\sigma^2}{\sum_\alpha (x_\alpha - \bar{x})^2}.$$

The minimum value of Q is

$$Q_a = \sum_{\alpha=1}^{n} (y_\alpha - a - bx_\alpha)^2 = \sum_\alpha \left[y_\alpha - \bar{y} - b(x - \bar{x})\right]^2$$

$$= \sum_\alpha (y_\alpha - \bar{y})^2 - b^2 \sum_\alpha (x_\alpha - \bar{x})^2$$

$$= \sum_\alpha (y_\alpha - \bar{y})^2 - \frac{\left(\sum_\alpha x_\alpha y_\alpha - n\bar{x}\bar{y}\right)^2}{\sum_\alpha (x_\alpha - \bar{x})^2}$$

or

$$\sum_\alpha y_\alpha^2 - n\bar{y}^2 = Q_a + \frac{\left(\sum_\alpha x_\alpha y_\alpha - n\bar{x}\bar{y}\right)^2}{\sum_\alpha (x_\alpha - \bar{x})^2}.$$

Hence one can make use of the t-distribution with $n - 2$ degrees of freedom :

$$t_{n-2} = \sqrt{n - 2}\, \frac{\sum_\alpha x_\alpha y_\alpha - n\bar{x}\bar{y}}{\sqrt{Q_a \sum_\alpha (x_\alpha - \bar{x})^2}}$$

Example 1.4.2. : One-way classification or completely randomized design

$E(x_{ij}) = \mu + \zeta_i,$

$j = 1,\dots,n_i,\quad i = 1,\dots,s,$ and $\sum_{j=1}^{s} n_j = n.$

The total sum of squares is

$$Q = \sum_{i=1}^{s} \sum_{j=1}^{n_i} x_{ij}^2 - n\bar{x}^2.$$

We introduce the following notations :

$$x_{i\cdot} = \sum_{j=1}^{n_i} x_{ij},$$

$$\bar{x}_{i\cdot} = \frac{1}{n_i} x_{i\cdot}$$

$$x_{\cdot\cdot} = \sum_{i=1}^{s} x_{i\cdot} = \sum_{i=1}^{s} n_i \bar{x}_{i\cdot} = \sum_{i=1}^{s} \sum_{j=1}^{n_i} x_{ij},$$

$$\bar{x} = \frac{1}{n} x_{\cdot\cdot}$$

Since

$$x_{ij} - \bar{x} = x_{ij} - \bar{x}_{i\cdot} + \bar{x}_{i\cdot} - \bar{x},$$

we get the identity

$$\sum_{j=1}^{n_i} (x_{ij} - \bar{x})^2 = \sum_{j=1}^{n_i} (x_{ij} - \bar{x}_{i\cdot})^2 + n_i(\bar{x}_{i\cdot} - \bar{x})^2,$$

and hence

$$Q = \sum_{i=1}^{s} \sum_{j=1}^{n_i} (x_{ij} - \bar{x})^2 = \sum_{i=1}^{s} \sum_{j=1}^{n_i} (x_{ij} - \bar{x}_{i\cdot})^2 + \sum_{i=1}^{s} n_i(\bar{x}_{i\cdot} - \bar{x})^2$$

or,

$$\sum_{i=1}^{s} \sum_{i=1}^{n_i} x_{ij}^2 = n\bar{x}^2 + \sum_{i=1}^{s} \sum_{j=1}^{n_i} (x_{ij} - \bar{x}_{i\cdot})^2 + \sum_{i=1}^{s} n_i(\bar{x}_{i\cdot} - \bar{x})^2$$

$$= Q_1 + Q_2 + Q_3, \quad \text{say,}$$

where

$$Q_1 = (\sqrt{n}\,\bar{x})^2, \quad r(Q_1) = 1;$$

$$Q_1 = \sum_{i=1}^{s} \sum_{j=1}^{n_i} (x_{ij} - \bar{x}_{i\cdot})^2, \quad \sum_{j=1}^{n_i} (x_{ij} - \bar{x}_{i\cdot}) = 0 \text{ for } i = 1, 2, \ldots, s,$$

$$r(Q_2) \le n - s$$

$$Q_3 = \sum_{i=1}^{s} n_i(\bar{x}_{i\cdot} - \bar{x})^2, \quad \sum_{i=1}^{s} n_i(\bar{x}_{i\cdot} - \bar{x}) = 0, \quad r(Q_3) \le s - 1,$$

whereas

$$n = r(Q_1 + Q_2 + Q_3) \le r(Q_1) + r(Q_2) + r(Q_3) \le n.$$

Hence

$$r(Q_1) = 1; \quad r(Q_2) = n - s; \quad r(Q_3) = s - 1.$$

Thus by Cochran's theorem $Q_3 = \sigma^2\chi'^2_{s-1}$, with the noncentrality parameter $\sum_{i=1}^{s} n_i(\tau_i - \bar{\tau})^2/\sigma^2$, $\bar{\tau} = \frac{1}{n}\sum_{i=1}^{s} n_i\tau_i$, and $Q_2 = \sigma^2\chi^2_{n-s}$, and they are independent. Hence

$$F = \frac{n-s}{s-1} \frac{\sum_i n_i\,\bar{x}_{i\cdot}^2 - n\bar{x}^2}{\sum_i \sum_j x_{ij}^2 - \sum_i n_i\bar{x}_{i\cdot}^2}$$

is distributed as the noncentral F-distribution.

The so-called Analysis of Variance Table for the one-way classified data is given in the following Table 1.1.:

Table 1.1.

Analysis of Variance Table for a Completely Randomized Design

Source of variation	Degrees of freedom	Sum of squares	Expection of mean sum of squares
Between treatments	$s - 1$	$\sum_{i=1}^{s} \frac{x_{i\cdot}^2}{n_{i}} - \frac{x_{\cdot\cdot}^2}{n}$	$\sigma^2 + \sum_{i=1}^{s} n_{i} \frac{(\tau_{i} - \bar{\tau})^2}{s - 1}$
Within treatments	$n - s$	$\sum_{i=1}^{s} \sum_{j=1}^{n_{i}} x_{ij}^2 - \sum_{i=1}^{s} \frac{x_{i\cdot}^2}{n_{i}}$	σ^2
Total sum of squares	$n - 1$	$\sum_{i=1}^{s} \sum_{j=1}^{n_{i}} x_{ij}^2 - \frac{x_{\cdot\cdot}^2}{n}$	

$$\bar{\tau} = \frac{1}{n} \sum_{i=1}^{s} n_{i}\tau_{i}.$$

We shall be dealing with the randomization starting with Chapter II and most seriously in Chapter VI. Here we have just followed the traditional usage of the word.

1.5. Analysis of Variance in an m-Way Classification Design

To give an example of a three-way classification, suppose we have 10 weather stations. The mean rainfall was recorded by these 10 stations every month in 5 successive years. Every observation is then characterized by three numbers or labels : the number of the weather station, the month, and the year in which the observation was made. Thus the observations can be denoted by x_{a_1, a_2, a_3}, $a_1 = 1, 2,\ldots, 10$; $a_2 = 1, 2,\ldots, 12$, $a_3 = 1, 2,\ldots, 5$, where a_1 is the number of the weather station, a_2 is the month, and a_3 is the year of observation. One may, for instance, want to know whether the rainfall was different in different locations or in different years. It is of interest to know also whether the combination of a certain location with a certain month has any bearing on the amount of rainfall.

We take the model such that the mean rainfall in one particular station during one particular month in one particular year is made up of the effects of station, month, and year as well as of the effects of the interactions of month and year, month and station, year and station, and finally one effect due to the interaction of month, station, and year.
Thus

$$(1.5.1)\quad E(x_{a_1a_2a_3}) = \mu(1,2,3;a_1,a_2,a_3)+\mu(1,2;a_1,a_2)+\mu(2,3;a_2,a_3)$$

$$= \mu(1,3;a_1,a_3)+\mu(1;a_1)+\mu(2;a_2)+\mu(3;a_3)+\mu,$$

where we impose the restrictions

$$\sum_{a_1} \mu(1,2,3;a_1,a_2,a_3) = \sum_{a_2} \mu(1,2,3;a_1,a_2,a_3)$$

$$= \sum_{a_3} \mu(1,2,3;a_1,a_2,a_3) = 0,$$

$$(1.5.2)\quad \sum_{a_{i_2}} \mu(i_1,i_2;a_{i_1},a_{i_2}) = \sum_{a_{i_2}} \mu(i_1,i_2;a_{i_1},a_{i_2}) = 0$$

$$\sum_{a_{i_1}} \mu(i_1;\, a_{i_1}) = 0,$$

for the parameter representing main effects. On the surface there are t_1 main effects of the first factor $\mu(1;\, a_1)$, subjecting to a linear restriction

$$\sum_{a_1=1}^{t_1} \mu(1;\, a_1) = \mu(1;\, \cdot) = 0,$$

and hence there are $t_1 - 1$ independent parameters. Similarly there are $t_2 - 1$ and $t_3 - 1$ independent main effects of the second and third factors, respectively. As for the parameters representing the two-factor interactions between the first and second factors, $\mu(1,2;a_1,a_2)$, we are imposing the restrictions

$$\mu(1,2;\, \cdot,\, a_2) = 0 \qquad (a_2 = 1,\ldots,t_2),$$

$$\mu(1,2;\, a_1,\, \cdot) = 0 \qquad (a_1 = 1,\ldots,t_1).$$

One can derive the relation $\mu(1,2;\cdot\,,\cdot\,) = 0$ from both systems, so there are only

$$t_1 - 1 + t_2 - 1 + 1$$

linearly independent restrictions. Therefore there are

$$t_1t_2 - (t_1 - 1\,) - (t_2 - 1\,) - 1 = t_1t_2 - t_1 - t_2 + 1 = (t_1 - 1)(t_2 - 1)$$

independent parameters $\mu(1,2;\, a_1,a_2)$ representing the two-factor interactions between the first and second factors. Similarly there are $(t_1 - 1)\,(t_3 - 1)$ parameters $\mu(1,3;\, a_1,a_3)$ and $(t_2 - 1)(t_3 - 1)$ parameters $\mu(2,3;a_2,a_3)$.

As for the three-factor interactions, we are imposing the following restrictions :

(1) $\mu(1,2,3;\ \cdot\ ,a_2,a_3) = 0\ (a_2 = 1,\ldots,t_2;\ a_3 = 1,\ldots,t_3)$,

(2) $\mu(1,2,3;\ a_1,\ \cdot,a_3) = 0\ (a_1 = 1,\ldots,t_1;\ a_3 = 1,\ldots,t_3)$,

(3) $\mu(1,2,3;\ a_1,a_2,\ \cdot) = 0\ (a_1 = 1,\ldots,t_1;\ a_2 = 1,\ldots,t_2)$.

from which we can derive the following:

(1) $\longrightarrow$
$$\mu(1,2,3;\cdot,a_2,\cdot) = 0,\ a_2 = 1,\ldots,t_2, \rightarrow \mu(1,2,3;\cdot,\cdot,\cdot) = 0,$$
$$\mu(1,2,3;\cdot,\cdot,a_3) = 0,\ a_3 = 1,\ldots,t_3, \rightarrow \mu(1,2,3;\cdot,\cdot,\cdot) = 0.$$

(2) $\longrightarrow$
$$\mu(1,2,3;\cdot,\cdot,a_3) = 0,\ a_3 = 1,\ldots,t_3, \rightarrow \mu(1,2,3;\cdot,\cdot,\cdot) = 0,$$
$$\mu(1,2,3;a_1,\cdot,\cdot) = 0,\ a_1 = 1,\ldots,t_1, \rightarrow \mu(1,2,3;\cdot,\cdot,\cdot) = 0.$$

(3) $\longrightarrow$
$$\mu(1,2,3;a_1,\cdot,\cdot) = 0,\ a_1 = 1,\ldots,t_1, \rightarrow \mu(1,2,3;\cdot,\cdot,\cdot) = 0,$$
$$\mu(1,2,3;\cdot,a_2,\cdot) = 0,\ a_2 = 1,\ldots,t_2, \rightarrow \mu(1,2,3;\cdot,\cdot,\cdot) = 0.$$

Consider the relations

$$\mu(1,2,3;\cdot,\cdot,\cdot) = 0,$$
$$\mu(1,2,3;a_1,\cdot,\cdot) = 0 \qquad (a_1 = 1,\ldots,t_1 - 1),$$
$$\mu(1,2,3;\cdot,a_2,\cdot) = 0 \qquad (a_2 = 1,\ldots,t_2 - 1),$$
$$\mu(1,2,3;\cdot,\cdot,a_3) = 0 \qquad (a_3 = 1,\ldots,t_3 - 1).$$

Since there are $t_1t_2 - (t_1 - 1) - (t_2 - 1) - 1 = (t_1 - 1)(t_2 - 1)$ linearly independent relations

$\mu(1,2,3;a_1,a_2,\cdot) = 0\quad (a_1 = 1,\ldots,t_1 - 1;\ a_2 = 1,\ldots,t_2 - 1)$, and similarly

$$\mu(1,2,3;a_1,\cdot,a_3) = 0 \quad (a_1 = 1,\ldots,t_1 - 1; \quad a_3 = 1,\ldots,t_3 - 1),$$

$$\mu(1,2,3;\cdot,a_2,a_3) = 0 \quad (a_2 = 1,\ldots,t_2 - 1, \quad a_3 = 1,\ldots,t_3 - 1),$$

We can see that the number of independent parameters representing the three-factor interactions is

$$\begin{aligned} t_1t_2t_3 - (t_1 - 1)(t_2 - 1) - (t_1 - 1)(t_3 - 1) - (t_2 - 1)(t_3 - 1) \\ - (t_1 - 1) - (t_2 - 1) - (t_3 - 1) - 1 \\ = t_1t_2t_3 - t_1t_2 - t_1t_3 - t_2t_3 + t_1 + t_2 + t_3 - 1 \\ = (t_1 - 1)(t_2 - 1)(t_3 - 1). \end{aligned}$$

The identity

$$\begin{aligned} t_1t_2t_3 - 1 = (t_1 - 1) + (t_2 - 1) + (t_3 - 1) + (t_1 - 1)(t_2 - 1) \\ + (t_2 - 1)(t_3 - 1) + (t_3 - 1)(t_1 - 1) \\ + (t_1 - 1)(t_2 - 1)(t_3 - 1) \end{aligned}$$

represents the partition of the degrees of freedom carried by the total sum of squares to be mentioned shortly.

Thus our model can be expressed as

$$\begin{aligned} x_{a_1a_2a_3} &= \mu(1,2,3;a_1,a_2,a_3) + \mu(1,2;a_1,a_2) \\ &+ \mu(2,3;a_2,a_3) + \mu(1,3;a_1,a_3) + \mu(1;a_1) \qquad (1.5.4) \\ &+ \mu(2;a_2) + \mu(3;a_3) + \mu + e_{a_1a_2a_3}, \end{aligned}$$

where the $e_{a_1a_2a_3}$ are assumed to be normally and independently distributed with means zero and a constant unknown variance .

In the example mentioned in the beginning of this section, one has

$$t_1 = 10, \quad t_2 = 12, \quad t_3 = 5.$$

$$x_{a_1a_2\cdot} = \sum_{a_3=1}^{5} x_{a_1a_2a_3},$$

$$x_{a_1\cdot a_3} = \sum_{a_2=1}^{12} x_{a_1a_2a_3},$$

$$x_{\cdot a_2a_3} = \sum_{a_1=1}^{10} x_{a_1a_2a_3},$$

$$x_{a_1\cdot\cdot} = \sum_{a_2=1}^{12} \sum_{a_3=1}^{5} x_{a_1a_2a_3},$$

$$x_{\cdot a_2\cdot} = \sum_{a_1=1}^{10} \sum_{a_3=1}^{5} x_{a_1a_2a_3},$$

$$x_{\cdot\cdot a_3} = \sum_{a_1=1}^{10} \sum_{a_2=1}^{12} x_{a_1a_2a_3},$$

$$x_{\cdot\cdot\cdot} = \sum_{a_1=1}^{10} \sum_{a_2=1}^{12} \sum_{a_3=1}^{5} x_{a_1a_2a_3} = 600\,\bar{x},$$

The total sum of squares is

$$Q = \sum_{a_1=1}^{t_1} \sum_{a_2=1}^{t_2} \sum_{a_3=1}^{t_3} x^2_{a_1 a_2 a_3} - t_1 t_2 t_3 \bar{x}^2.$$

The sum of squares due to the first category or factor is

$$Q_1 = \sum_{a_1=1}^{t_1} x^2_{a_1 \cdot\cdot} \Big/ t_2 t_3 - t_1 t_2 t_3 \bar{x}^2.$$

The sum of squares due to the second category or factor is

$$Q_2 = \sum_{a_2=1}^{t_2} x^2_{\cdot a_2 \cdot} \Big/ t_1 t_3 - t_1 t_2 t_3 \bar{x}^2.$$

The sum of squares due to the third category or factor is

$$Q_3 = \sum_{a_3=1}^{t_3} x^2_{\cdot\cdot a_3} \Big/ t_1 t_2 - t_1 t_2 t_3 \bar{x}^2.$$

The sum of squares due to the interaction between the first and second factors is

$$\sum_{a_1=1}^{t_1} \sum_{a_2=1}^{t_2} x^2_{a_1 a_2 \cdot} \Big/ t_3 - \sum_{a_1=1}^{t_1} x^2_{a_1 \cdot\cdot} \Big/ t_2 t_3 - \sum_{a_2=1}^{t_2} x^2_{\cdot a_2 \cdot} \Big/ t_3 t_1$$

$$+ x^2_{\cdot\cdot\cdot} \Big/ t_1 t_2 t_3.$$

We are now going into a more detailed analysis under this model following [9]:

For an m-way classification,

$$x_{a_1,a_2,\ldots,a_m} = \sum_{\alpha=0}^{m} \sum_{1\ldots m} \mu(i_1\cdots i_\alpha;a_{i_1}\cdots a_{i_\alpha}) + e_{a_1a_2\cdots a_m}, \tag{1.5.5}$$

where

$$a_i = 1,2,\ldots,t_i, \qquad i = 1,\ldots,m;$$

$$\sum_{a_{i_j}=1}^{t_{i_j}} \mu(i_1\cdots i_\alpha;a_{i_1}\cdots a_{i_\alpha}) = 0.$$

The summation

$$\sum_{\alpha=0}^{m} \sum_{1\cdots m} \mu(i_1\cdots i_\alpha;a_{i_1}\cdots a_{i_\alpha})$$

is defined as a constant μ for $\alpha = 0$ and for $\alpha > 0$, and

$$\sum_{1\cdots m} \mu(i_1\cdots i_\alpha;a_{i_1}\cdots a_{i_\alpha})$$

denotes the summation over all combinations $i_1,\ldots,i_\alpha$ taken from $1,\ldots,m$ with $i_1 < i_2 < \cdots < i_\alpha$. The quantity $\mu(i_1\cdots i_\alpha;a_{i_1}\cdots a_{i_\alpha})$ is called an $(\alpha - 1)$-th order or α factor interaction.

Let the least-squares estimate of $\mu(i_1\cdots i_\alpha;a_{i_1}\cdots a_{i_\alpha})$ be $A(i_1\cdots i_\alpha;a_{i_1}\cdots a_{i_\alpha})$. This should be obtained by minimizing the sum of squares

$$Q = \sum_{a_1} \cdots \sum_{a_m} \left[x_{a_1 \cdots a_m} - \sum_{\alpha=0}^{m} \sum_{1 \cdots m} \mu(i_1 \cdots i_\alpha; a_{i_1} \cdots a_{i_\alpha}) \right]^2$$

under the restrictions given in (1.5.5). We proceed somewhat loosely, neglecting the Lagrange multipliers. The normal equations are given by

$$(1.5.6) \quad x(i_1 \cdots i_\alpha; a_{i_1} \cdots a_{i_\alpha}) = \sum_{\beta=0}^{\alpha} \sum_{i_1 \ldots i_\alpha} A(k_1 \cdots k_\beta\, a_{k_1} \cdots a_{k_\beta})$$

where $x(i_1 \cdots i_\alpha; a_{i_1} \cdots a_{i_\alpha})$: the mean of $x_{a_1 \cdots a_m}$ taken over all suffices except for $a_{i_1} \cdots a_{i_\alpha}$.

For $\alpha = 0$, $\quad \bar{x} = A$

For $\alpha = 1$, $\quad x(i_1; a_{i_1}) = A(i_1; a_{i_1}) + A,$

$$A(i_1; a_{i_1}) = x(i_1; a_{i_1}) - \bar{x}.$$

For $\alpha = 2$,

$$x(i_1, i_2; a_{i_1}, a_{i_2}) = A + A(i_1; a_{i_1}) + A(i_2; a_{i_2}) + A(i_1, i_2; a_{i_1}, a_{i_2}),$$

$$A(i_1, i_2; a_{i_1}, a_{i_2}) = x(i_1, i_2; a_{i_1}, a_{i_2}) - x(i_1; a_{i_1}) - x(i_2; a_{i_2}) + \bar{x},$$

One can show in general that

$$(1.5.7) \quad A(i_1, \ldots, i_\alpha; a_{i_1}, \ldots, a_{i_\alpha}) = \sum_{\beta=0}^{\alpha} (-1)^{\alpha-\beta} \times \sum_{i_1 \ldots i_\alpha} x(k_1 \cdots k_\beta; a_{k_1} \cdots a_{k_\beta}).$$

We prove (1.5.7) by mathematical induction. Assuming that (1.5.7) is true for all $\alpha' \leq \alpha$, we can show that (1.5.7) is true for $\alpha' = \alpha + 1$. We have

$$x(i_1 \cdots i_{\alpha+1}; a_{i_1} \cdots a_{i_{\alpha+1}}) = \sum_{\beta=0}^{\alpha+1} \sum_{i_1 \cdots i_{\alpha+1}} A(k_1 \cdots k_\beta; a_{k_1} \cdots a_{k_\beta}).$$

Hence

$$\begin{aligned}
&A(i_1 \cdots i_{\alpha+1}; a_{i_1} \cdots a_{i_{\alpha+1}}) \\
&= x(i_1 \cdots i_{\alpha+1}; a_{i_1} \cdots a_{i_{\alpha+1}}) - \sum_{\beta=0}^{\alpha} \sum_{i_1 \cdots i_{\alpha+1}} A(k_1 \cdots k_\beta; a_{k_1} \cdots a_{k_\beta}) \\
&= x(i_1 \cdots i_{\alpha+1}; a_{i_1} \cdots a_{i_{\alpha+1}}) \\
&\quad - \sum_{\beta=0}^{\alpha} \sum_{i_1 \cdots i_{\alpha+1}} \sum_{\gamma=0}^{\alpha} (-1)^{\beta-\gamma} \sum_{k_1 \cdots k_\beta} x(b_1 \cdots b_\gamma; a_{b_1} \cdots a_{b_\gamma}) \\
&= x(i_1 \cdots i_{\alpha+1}; a_{i_1} \cdots a_{i_{\alpha+1}}) \\
&\quad - \sum_{\gamma=0}^{\alpha} \sum_{i_1 \cdots i_{\alpha+1}} \sum_{\beta=\gamma}^{\alpha} (-1)^{\beta-\gamma} \sum_{k_1 \cdots k_\beta} x(b_1 \cdots b_\gamma; a_{b_1} \cdots a_{b_\gamma}).
\end{aligned}$$

Interchange of the order of summations is indicated diagramatically in the following Fig. 1.3.

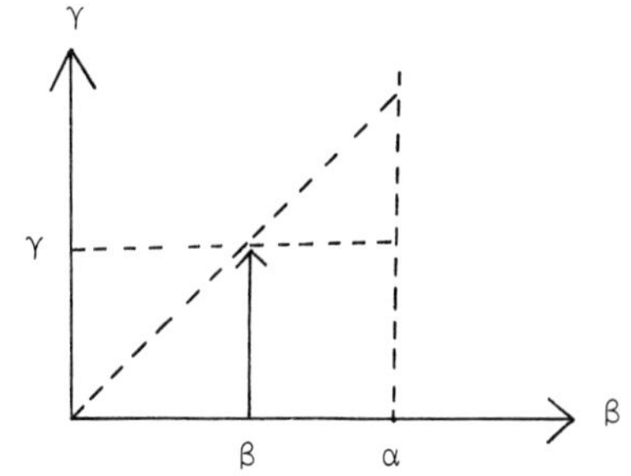

$$\sum_{\beta=0}^{\alpha} \sum_{\gamma=0}^{\beta} = \sum_{\gamma=0}^{\alpha} \sum_{\beta=\gamma}^{\alpha}$$

Fig. 1.3

The term $x(b_1 \cdots b_\gamma;\ a_{b_1} \cdots a_{b_\gamma})$ appears in the last summation above for every choice of $k_1 \cdots k_\beta$ that contains $b_1 \cdots b_\gamma$. Out of $\alpha + 1$ numbers $i_1 \cdots i_{\alpha+1}$ for fixed β, there are exactly

$$\binom{\alpha + 1 - \gamma}{\beta - \gamma}$$

choices; *i.e.*, for a fixed set $b_1 \cdots b_\gamma$ the number of combinations $k_1 \cdots k_\beta$ containing the set $b_1 \cdots b_\gamma$ out of $i_1 \cdots i_{\alpha+1}$ is $\binom{\alpha + 1 - \gamma}{\beta - \gamma}$. Hence the coefficient of $x(b_1 \cdots b_\gamma;\ a_{b_1} \cdots a_{b_\gamma})$ in the expression $A(i_1 \cdots i_{\alpha+1};\ a_{i_1} \cdots a_{i_{\alpha+1}})$ becomes

$$- \sum_{\beta=\gamma}^{\alpha} (-1)^{\beta-\gamma} \binom{\alpha + 1 - \gamma}{\beta - \gamma}$$

$$= - \sum_{\beta=0}^{\alpha-\gamma} (-1)^{\beta} \binom{\alpha + 1 - \gamma}{\beta}$$

$$= - \left[\sum_{\beta=0}^{\alpha+1-\gamma} (-1)^{\beta} \binom{\alpha + 1 - \gamma}{\beta} - (-1)^{\alpha+1-\gamma} \right]$$

$$= (-1)^{\alpha+1-\gamma}.$$

Consequently

$$A(i_1 \cdots i_{\alpha+1};\ a_{i_1} \cdots a_{i_{\alpha+1}})$$

$$= x(i_1 \cdots i_{\alpha+1};\ a_{i_1} \cdots a_{i_{\alpha+1}})$$

$$+ \sum_{\gamma=0}^{\alpha} (-1)^{\alpha+1-\gamma} \sum_{i_1 \cdots i_{\alpha+1}} x(b_1 \cdots b_\gamma;\ a_{b_1} \cdots a_{b_\gamma}).$$

One can also show by mathematical induction that $A(i_1 \cdots i_\alpha;\ a_{i_1} \cdots a_{i_\alpha})$ satisfies the restriction imposed by (1.5.5). Indeed, since

$$A(1 \cdots \alpha+1;\ a_1 \cdots a_{\alpha+1})$$

$$= x(1 \cdots \alpha+1;\ a_1 \cdots a_{\alpha+1}) - \sum_{\beta=0}^{\alpha} \sum_{1 \cdots \alpha+1} A(k_1 \cdots k_\beta;\ a_{k_1} \cdots a_{k_\beta}),$$

Hence

$$\sum_{a_1=1}^{t_1} A(1\cdots\alpha+1;\ a_1\cdots a_{\alpha+1})$$

$$= t_1 x(2\cdots\alpha+1;\ a_2\cdots a_{\alpha+1}) - t_1 \sum_{\beta=0}\ \sum_{2\cdots\alpha+1} A(k_1\cdots k_\beta;\ a_{k_1}\cdots a_{k_\beta}) = 0$$

There are $(t_{i_1} - 1)\cdots(t_{i_\alpha} - 1)$ independent parameters of the $\mu(i_1\cdots i_\alpha;\ a_{i_1}\cdots a_{i_\alpha})$, and hence there are

$$\sum_{\alpha=0}^{m}\ \sum_{1\cdots m} (t_{i_1} - 1)\cdots(t_{i_\alpha} - 1) = t_1\cdots t_m$$

independent parameters, the same as the number of observations. Hence one can see that $A(i_1\cdots i_\alpha;\ a_{i_1}\cdots a_{i_\alpha})$ is the unique solution, even taking account the restrictions imposed in (1.5.5).

Let us consider the statistical hypotheses

$$H_{(i_1\cdots i_\alpha)}:\ \mu(i_1\cdots i_\alpha;\ a_{i_1}\cdots a_{i_\alpha}) = 0 \quad \text{for all } a_{i_1}\cdots a_{i_\alpha}\,.$$

We arrange these hypotheses in a sequence in such a way that higher order interactions precede lower order ones. Interactions of the same order may be arranged in any arbitrary way but once for all.

$\gamma = 3$;

$$H_1 = H_{(1,2,3)},\qquad H_2^* = H_{(1,2)},\qquad H_3^* = H_{(1,3)},$$

$$H_4^* = H_{(2,3)},\qquad H_5^* = H_{(1)},\qquad H_6^* = H_{(2)},\qquad H_7^* = H_{(3)}.$$

Let

$$H_2 = H_1 \mathbin{\&} H_2^*, \quad H_3 = H_1 \mathbin{\&} H_2^* \mathbin{\&} H_3^* = H_1 \mathbin{\&} H_2 \mathbin{\&} H_3^*, \dots .$$

Now for the general case

$$H_h = H_1 \mathbin{\&} H_2 \mathbin{\&} \cdots \mathbin{\&} H_{h-1} \mathbin{\&} H_{(i_1 \cdots i_h)} .$$

Then

$$(1.5.8) \qquad Y^{(h-1)}_{a_1 \cdots a_m} - Y^{(h)}_{a_1 \cdots a_m} = A(i_1 \cdots i_h;\ a_{i_1} \cdots a_{i_h}).$$

In the sum

$$\sum_{a_1} \cdots \sum_{a_m} \left(Y^{(h-1)}_{a_1 \cdots a_m} - Y^{(h)}_{a_1 \cdots a_m} \right)^2 ,$$

the term $A(i_1 \cdots i_h;\ a_{i_1} \cdots a_{i_h})$ appears $(t_1 \cdots t_m)/(t_{i_1} \cdots t_{i_h})$ times for all $a_{i_1} \cdots a_{i_h}$. Thus one obtains the partition of the sum of squares :

$$(1.5.9) \qquad \sum_{a_1} \cdots \sum_{a_m} x^2_{a_1 \cdots a_m}$$

$$= \sum_{\alpha=0}^{m} \sum_{1 \cdots m} \frac{t_1 \cdots t_m}{t_{i_1} \cdots t_{i_\alpha}} \sum_{\alpha_{i_1}} \cdots \sum_{\alpha_{i_\alpha}} \left[A(i_1 \cdots i_\alpha;\ a_{i_1} \cdots a_{i_\alpha}) \right]^2 .$$

This can be shown directly from the relation - the first of the normal equations -

$$x_{a_1 \cdots a_m} = \sum_{\alpha=0}^{m} \sum_{1 \cdots m} A(i_1 \cdots i_\alpha;\ a_{i_1} \cdots a_{i_\alpha})$$

Moreover, one can show by the same reasoning that

$$(1.5.10)\quad \sum_{a_1}\cdots\sum_{a_m}\left\{x_{a_1\cdots a_m} - \sum_{\alpha=0}^{m}\sum_{1\cdots m}\mu(i_1\cdots i_\alpha;\ a_{i_1}\cdots a_{i_\alpha})\right\}^2$$

$$= \sum_{\alpha=0}^{m}\sum_{1\cdots m}\sum_{a_{i_1}}\cdots\sum_{a_{i_\alpha}}\frac{t_1\cdots t_m}{t_{i_1}\cdots t_{i_\alpha}}$$

$$\left\{A(i_1\cdots i_\alpha;\ a_{i_1}\cdots a_{i_\alpha}) - \mu(i_1\cdots i_\alpha;\ a_{i_1}\cdots a_{i_\alpha})\right\}^2.$$

In fact, the cross-product terms vanish, as one can verify.

One can classify the interactions into the following three types :

Interactions of type I : these can be assumed to be zero.

Interactions of type II : unknown.

Interactions of type III : we want to test hypotheses concerning individual values.

In order to calculate Q_a and Q_r, one must put in (1.5.10)

$$\mu(i_1\cdots i_\alpha;\ a_{i_1}\cdots a_{i_\alpha}) = 0$$

for interactions of type I and

$$\mu(i_1\cdots i_\alpha;\ a_{i_1}\cdots a_{i_\alpha}) = A(i_1\cdots i_\alpha;\ a_{i_1}\cdots a_{i_\alpha})$$

for interactions of type II. For interactions of type III we have to minimize

$$\sum_{a_{j_1}} \cdots \sum_{a_{j_k}} \left[A(j_1 \cdots j_k;\ a_{j_1} \cdots a_{j_k}) - \mu(j_1 \cdots j_k;\ a_{j_1} \cdots a_{j_k})\right]^2$$

for a particular $(j_1 \cdots j_k)$ under the restrictions

$$\sum_{a_{j_e}} \mu(j_1 \cdots j_e \cdots j_k;\ a_{j_1} \cdots a_{j_e} \cdots a_{j_k}) = 0.$$

<u>Example</u>. In the three-way classification, we assume that all three factor interactions are zero. We wish to test the hypothesis that all interactions between the first and second classifications are zero. The assumption then is

$$\mu(1,2,3;a_1,a_2,a_3) = 0,\ a_1 = 1,\ldots,t_1;\ a_2 = 1,\ldots,t_2;\ a_3 = 1,\ldots,t_3.$$

The hypothesis to be tested is

$$\mu(1,2;a_1,a_2) = 0,\quad a_1 = 1,\ldots,t_1;\quad a_2 = 1,\ldots,t_2.$$

Hence

$$Q_a = \sum_{a_1} \sum_{a_2} \sum_{a_3} \left[A(1,2,3;\ a_1,a_2,a_3)\right]^2$$

and

$$Q_r = Q_a + t_3 \sum_{a_1} \sum_{a_2} \left[A(1,2;a_1,a_2)\right]^2.$$

The F-statistic for testing this hypothesis is therefore

$$F = \frac{(t_1 - 1)(t_2 - 1)(t_3 - 1)}{(t_1 - 1)(t_2 - 1)} \; \frac{t_3 \sum_{a_1} \sum_{a_2} [A(1,2;a_1,a_2)]^2}{\sum_{a_1} \sum_{a_2} \sum_{a_3} [A(1,2,3;a_1,a_2,a_3)]^2} .$$

Suppose now that under the same assumptions as before we wish to test the hypothesis

$$\mu(1,2;\ 1,1) = \mu(1,2;\ 1,2).$$

To find Q_r in this case we would have to minimize

$$[A(1,2;\ 1,1) - \mu(1,2;\ 1,1)]^2 + [A(1,2;\ 1,2) - \mu(1,2;\ 1,1)]^2$$

$$+ \sum_{a_1=2}^{t_1} \sum_{a_2=1}^{t_2} [A(1,2;\ a_1,a_2) - \mu(1,2;\ a_1,a_2)]^2$$

$$+ \sum_{a_2=3}^{t_2} [A(1,2;\ 1,a_2) - \mu(1,2;\ 1,a_2)]^2$$

under the restrictions

$$\left.\begin{aligned} &\sum_{a_1} \mu(1,2;\ a_1,a_2) = 0, \quad a_2 = 1,\ldots,t_2 \\ &\sum_{a_2} \mu(1,2;\ a_1,a_2) = 0, \quad a_1 = 1,\ldots,t_1 \end{aligned}\right\} \mu(1,2;\ 1,1) = \mu(1,2;\ 1,2).$$

However, it will be easier to proceed as follows in this case. We consider the linear contrast $\mu(1,2;\ 1,1) - \mu(1,2;\ 1,2)$. Since

$$A(1,2;\ 1,1) = x(1,2;\ 1,1) - x(1;\ 1) - x(2;\ 1) + \bar{x}$$

and

$$A(1,2;\ 1,2) = x(1,2;\ 1,2) - x(1;\ 1) - x(2;\ 2) + \bar{x},$$

it follows that

$$A(1,2;\ 1,1) - A(1,2;\ 1,2) = x(1,2;\ 1,1) - x(1,2;\ 1,2) - x(2;\ 1) - x(2;\ 2),$$

whence

$$V[A(1,2;\ 1,1) - A(1,2;\ 1,2)]$$

$$= V[x(1,2;\ 1,1)] + V[x(1,2;\ 1,2)] + V[x(2;\ 1)] + V[x(2;\ 2)]$$

$$- 2C[x(1,2;\ 1,1)x(2;\ 1)] - 2C[x(1,2;\ 1,2)x(2;\ 2)]$$

$$= \sigma^2 \left[\frac{1}{t_3} + \frac{1}{t_3} + \frac{1}{t_1 t_3} + \frac{1}{t_1 t_3} - 2\,\frac{t_3}{t_1 t_3^2} - 2\,\frac{t_3}{t_1 t_3^2}\right]$$

$$= 2\sigma^2\,\frac{t_1 - 1}{t_1 t_3}.$$

The F-statistic to test the hypothesis $\mu(1,2;\ 1,1) = \mu(1,2;\ 1,2)$ is

$$F = \frac{(t_1 - 1)(t_2 - 1)(t_3 - 1)\ t_1 t_3}{2(t_1 - 1)}\ \frac{[A(1,2;\ 1,1) - A(1,2;\ 1,2)]^2}{\sum\limits_{a_1}\sum\limits_{a_2}\sum\limits_{a_3} [A(1,2,3;a_1,a_2,a_3)]^2}$$

Equation (1.5.9) can be generalized to yield

$$(1.5.11) \qquad \sum_{a_{i_1}} \cdots \sum_{a_{i_\alpha}} \left[x(i_1 \cdots i_\alpha; a_{i_1} \cdots a_{i_\alpha})\right]^2$$

$$= \sum_{\beta=0}^{\alpha} \sum_{i_1 \cdots i_\alpha} \frac{t_{i_1} \cdots t_{i_\alpha}}{t_{k_1} \cdots t_{k_\beta}} \sum_{a_{k_1}} \cdots \sum_{a_{k_\beta}} \left| A(k_1 \cdots k_\beta;\ a_{k_1} \cdots a_{k_\beta}) \right|^2$$

To facilitate the computation of the sums of squares of interaction the following identity will be proved :

$$\text{(1.5.12)} \qquad \sum_{a_{i_1}} \cdots \sum_{a_{i_k}} \left[A(i_1 \cdots i_k;\ a_{i_1} \cdots a_{i_k}) \right]^2$$

$$= \sum_{\alpha=0}^{k} (-1)^{k-\alpha} \sum_{i_1 \cdots i_k} \sum_{a_{j_1}} \cdots \sum_{a_{j_\alpha}} \frac{t_{i_1} \cdots t_{i_k}}{t_{j_1} \cdots t_{j_\alpha}} \left[\bar{x}(j_1 \cdots j_\alpha;\ a_{j_1} \cdots a_{j_\alpha}) \right]^2,$$

for example,

$$\sum_{a_1} \sum_{a_2} \sum_{a_3} [A(1,2,3;\ a_1,a_2,a_3,)]^2$$

$$= \sum_{a_1} \sum_{a_2} \sum_{a_3} [x(1,2,3;\ a_1,a_2,a_3,)]^2 - t_1 \sum_{a_2} \sum_{a_3} [x(2,3;\ a_2,a_3)]^2$$

$$- t_2 \sum_{a_1} \sum_{a_3} [x(1,3;\ a_1,a_3)]^2 - t_3 \sum_{a_1} \sum_{a_2} [x(1,2;\ a_1,a_2)]^2$$

$$+ t_1 t_2 \sum_{a_3} [x(3;\ a_3)]^2 + t_1 t_3 \sum_{a_2} [x(2;\ a_2)]^2$$

$$+ t_2 t_3 \sum_{a_1} [x(1;\ a_1)]^2 - t_1 t_2 t_3\, \bar{x}^2.$$

We now prove (1.5.12) by mathematical induction.

For $k = 0$,

$$A^2 = \bar{x}^2,$$

$$k = 1,$$

$$\sum_{a_{i_1}} \left[A(i_1;a_{i_1})\right]^2 = \sum_{a_{i_1}} \left[x(i_1;a_{i_1}) - \bar{x}\right]^2 = \sum_{a_{i_1}} \left[x(i_1;a_{i_1})\right]^2 - t_1\bar{x}^2.$$

Suppose that (1.5.12) is true for all $k' < k$. From (1.5.11) we have

$$\sum_{a_{i_1}} \cdots \sum_{a_{i_k}} \left[A(i_1\cdots i_k;\ a_{i_1}\cdots a_{i_k})\right]^2 \tag{1.5.13}$$

$$= \sum_{a_{i_1}} \cdots \sum_{a_{i_k}} \left[x(i_1\cdots i_k;\ a_{i_1}\cdots a_{i_k})\right]^2$$

$$- \sum_{\alpha=0}^{k-1} \sum_{i_1\cdots i_k} \frac{t_{i_1}\cdots t_{i_k}}{t_{k_1}\cdots t_{k_\alpha}} \sum_{a_{k_1}} \cdots \sum_{a_{k_\alpha}} \left[A(k_1\cdots k_\alpha;\ a_{k_1}\cdots a_{k_\alpha})\right]^2.$$

Applying (1.5.12) to (1.5.13) for $\alpha < k$, we find

$$\sum_{a_{i_1}} \cdots \sum_{a_{i_k}} \left[A(i_1\cdots i_k;\ a_{i_1}\cdots a_{i_k})\right]^2$$

$$= \sum_{a_{i_1}} \cdots \sum_{a_{i_k}} \left[x(i_1\cdots i_k;\ a_{i_1}\cdots a_{i_k})\right]^2$$

$$- \sum_{\alpha=0}^{k-1} \sum_{i_1 \cdots i_k} \frac{t_{i_1} \cdots t_{i_k}}{t_{k_1} \cdots t_{k_\alpha}} \sum_{\beta=0}^{\alpha} (-1)^{\alpha-\beta} \sum_{k_1 \cdots k_\alpha} \sum_{a_{j_1}} \cdots \sum_{a_{j_\beta}} \frac{t_{k_1} \cdots t_{k_\alpha}}{t_{j_1} \cdots t_{j_\beta}} \times \left[x(j_1 \cdots j_\beta ; a_{j_1} \cdots a_{j_\beta}) \right]^2 .$$

The term

$$\left[x(j_1 \cdots j_\beta ;\, a_{j_1} \cdots a_{j_\beta}) \right]^2 \frac{t_{i_1} \cdots t_{i_k}}{t_{j_1} \cdots t_{j_\beta}}$$

occurs in this expansion as often as we can make a choice of the indices $j_1, \ldots, j_\beta$ out of indices $k_1 \cdots k_\alpha$ with $\beta \le \alpha < k$. Since the indices $j_1 \cdots j_\beta$ are fixed, there remain $k - \beta$ indices to choose from. With α fixed there are then $\alpha - \beta$ indices to choose out of $k - \beta$ indices. Hence for fixed α the terms

$$\left[x(j_1 \cdots j_\beta ;\, a_{j_1} \cdots a_{j_\beta}) \right] \frac{t_{i_1} \cdots t_{i_k}}{t_{j_1} \cdots t_{j_\beta}}$$

occur ${}_{k-\beta}C_{\alpha-\beta}$ times. Hence the coefficient of the above term becomes

$$- \sum_{\alpha=\beta}^{k-1} (-1)^{\alpha-\beta} \binom{k-\beta}{\alpha-\beta} = - \sum_{l=0}^{k-\beta-1} (-1)^l \binom{k-\beta}{l}$$

$$= - \sum_{l=0}^{k-\beta} (-1)^l \binom{k-\beta}{l} + (-1)^{k-\beta}$$

and so,

$$\sum_{a_{i_1}} \cdots \sum_{a_{i_k}} \left[A(i_1 \cdots i_k;\, a_{i_1} \cdots a_{i_k})\right]^2$$

$$= \sum_{a_{i_1}} \cdots \sum_{a_{i_k}} \left[x(i_1 \cdots i_k;\, a_{i_1} \cdots a_{i_k})\right]^2$$

$$+ \sum_{\beta=0}^{k-1} (-1)^{k-\beta} \sum_{i_1 \cdots i_k} \sum_{a_{j_1}} \cdots \sum_{a_{j_\beta}} \frac{t_{i_1} \cdots t_{i_k}}{t_{j_1} \cdots t_{j_\beta}} \left[x(j_1 \cdots j_\beta;\, a_{j_1} \cdots a_{j_\beta})\right]^2.$$

An important special case arises if in an m*-way classification design we take several observations in every one of the multiple classifications.* Such a design may be treated as an $(m + 1)$-way classification design by simply numbering the variables in every subclass in an arbitrary manner. We shall then be justified in assuming that the $(m + 1)$-classification has no effect on the mean value, or

$$\mu(i_1 \cdots i_k m + 1;\, a_{i_1} \cdots a_{i_k} a_{m+1}) = 0$$

for all choices $i_1 < \cdots < i_k < m + 1$ out of $1, 2, \ldots, m, m + 1$. Hence the absolute minimum of the sum of squares becomes

$$Q_a = \sum_{\alpha=0}^{m} \sum_{1 \cdots m} \frac{t_1 \cdots t_m t_{m+1}}{t_{j_1} \cdots t_{j_\alpha}} \sum_{a_{j_1}} \cdots \sum_{a_{j_\alpha}} \sum_{a_{m+1}} \left[A(i_1 \cdots j_\alpha m + 1;\, a_{j_1} \cdots a_{j_\alpha})\right]^2$$

$$= \sum_{a_1} \cdots \sum_{a_{m+1}} \left[x(1\cdots m+1;\ a_1\cdots a_{m+1})\right]^2$$

$$- t_{m+1} \sum_{\alpha=0}^{m} \sum_{1\cdots m} \frac{t_1\cdots t_m}{t_{j_1}\cdots t_{j_\alpha}} \sum_{a_{j_1}} \cdots \sum_{a_{j_\alpha}} \left[A(j_1\cdots j_\alpha;\ a_{j_1}\cdots a_{j_\alpha})\right]^2$$

$$= \sum_{a_1} \cdots \sum_{a_{m+1}} \left[x(1\cdots m+1;\ a_1\cdots a_{m+1})\right]^2$$

$$- t_{m+1} \sum_{a_1} \cdots \sum_{a_m} \left[x(1\cdots m;\ a_1\cdots a_m)\right]^2$$

$$= \sum_{a_1} \cdots \sum_{a_{m+1}} \left[x(1\cdots m\ m+1;\ a_1\cdots a_m a_{m+1}) - x(1\cdots m;\ a_1\cdots a_m)\right]^2.$$

1.6. Analysis of Variance in a Two-Way Classification in Matrix Language

Suppose we are given a set of data

$$x_{a_1,a_2}, \quad a_1 = 1,\ldots,t_1; \quad a_2 = 1,\ldots,t_2; \quad n = t_1 t_2.$$

First we label the whole data from 1 to n in any order, and denote the observation by an n-dimensional vector $\underline{x}$. The grand total of the observations is given by

$$x_{\cdot\cdot} = \sum_{a_1=1}^{t_1} \sum_{a_2=1}^{t_2} x_{a_1 a_2} = \underline{1}'\underline{x},$$

where $\underline{1}' = (1 \cdots 1)$. Hence the total sum of squares corrected by the mean is

$$Q = \sum_{a_1=1}^{t_1} \sum_{a_2=1}^{t_2} x^2_{a_1 a_2} - \frac{1}{n} x^2_{\cdot\cdot} = \underline{x}'\underline{x} - \frac{1}{n} \underline{x}'\underline{1}\,\underline{1}'\underline{x} = \underline{x}'(I_n - \frac{1}{n}G)\underline{x},$$

where I_n is the unit matrix of order n and $G = \underline{1}\,\underline{1}'$ is the matrix whose elements are all unity.

We define the incidence vectors for the ith classification as follows:

$$\underline{\zeta}_\alpha^{(i)} = \begin{bmatrix} \zeta_{\alpha 1}^{(i)} \\ \vdots \\ \zeta_{\alpha n}^{(i)} \end{bmatrix},$$

where

$$\zeta_{\alpha f}^{(i)} = \begin{cases} 1, & \text{if the number } f \text{ corresponds to the } (a_1, a_2) \text{ in which } a_i = \alpha \\ & \alpha = 1, \ldots, t_i, \quad i = 1,2, \\ 0, & \text{otherwise.} \end{cases}$$

Then

$$\sum_{\alpha=1}^{t_i} \underline{\zeta}_\alpha^{(i)} = \underline{1} \qquad (i = 1,2).$$

Let

$$\Phi_i = \left\| \underline{\zeta}_1^{(i)} \cdots \underline{\zeta}_{t_i}^{(i)} \right\| \qquad (i = 1,2).$$

Then

$$\Phi_i'\Phi_i = \frac{n}{t_i} I_{t_i} \qquad (i = 1,2)$$

and

$$\Phi_1'\Phi_2 = J_{t_1,t_2},$$

where J_{t_1,t_2} stands for the rectangular matrix of the type $t_1 \times t_2$ whose elements are all unity.

Let us define matrices of order n

$$C_i = \Phi_i'\Phi_i \qquad (i = 1,2).$$

Then we can show that

$$E_0 = \frac{1}{n} G; \qquad E_1 = \frac{t_1}{n} C_1 - \frac{1}{n} G; \qquad E_2 = \frac{t_2}{n} C_2 - \frac{1}{n} G$$

are mutually orthogonal idempotent matrices with respective rank

$$\alpha_0 = trE_0 = 1; \qquad \alpha_1 = trE_1 = t_1 - 1; \qquad \alpha_2 = trE_2 = t_2 - 1.$$

Indeed, since

$$C_1G = \Phi_1\Phi_1'\underline{1}_n\underline{1}_n' = t_2\Phi_1\underline{1}_{t_1}\underline{1}_n' = t_2\underline{1}_n\underline{1}_n' = t_2G = GC_1,$$

$$C_2G = \Phi_2\Phi_2'\underline{1}_n\underline{1}_n' = t_1\Phi_2\underline{1}_{t_2}\underline{1}_n' = t_1\underline{1}_n\underline{1}_n' = t_1G = GC_2,$$

$$C_1C_2 = \Phi_1\Phi_1'\Phi_2\Phi_2' = \Phi_1J_{t_1t_2}\Phi_2' = \Phi_1\underline{1}_{t_1}\underline{1}_{t_2}'\Phi_2' = \underline{1}_n\underline{1}_n' = G = C_2C_1,$$

it follows that

$$E_0^2 = (\frac{1}{n}G)^2 = \frac{1}{n}G = E_0,$$

$$E^2 = (\frac{t_1}{n}C_1)^2 - \frac{t_1}{n^2}C_1G - \frac{t_1}{n^2}GC_1 + (\frac{1}{n}G)^2 = \frac{t_1}{n}C_1 - \frac{1}{n}G = E_1,$$

$$E_2^2 = (\frac{t_2}{n}C_2)^2 - \frac{t_2}{n^2}C_2G - \frac{t_2}{n^2}GC_2 + (\frac{1}{n}G)^2 = \frac{t_2}{n}C_2 - \frac{1}{n}G = E_2,$$

and

$$E_0E_1 = \frac{t_1}{n^2}GC_1 - (\frac{1}{n}G)^2 = \frac{1}{n}G - \frac{1}{n}G = 0 = E_1E_0,$$

$$E_0E_1 = \frac{t_1}{n^2}GC_1 - (\frac{1}{n}G)^2 = \frac{1}{n}G - \frac{1}{n}G = 0 = E_1E_0,$$

$$E_1E_2 = \frac{t_1t_2}{n^2}C_1C_2 - \frac{t_1}{n^2}C_1G - \frac{t_2}{n^2}GC_2 + (\frac{1}{n}G)^2$$

$$= \frac{1}{n}G - \frac{1}{n}G - \frac{1}{n}G + \frac{1}{n}G = 0 = E_2E_1.$$

Hence

$$I - E_0 - E_1 - E_2$$

is also idempotent and orthogonal to $E_i, i = 0,1,2$, with rank equal to

$$tr(I - E_0 - E_1 - E_2) = t_1t_2 - 1 - (t_1 - 1) - (t_2 - 1)$$

$$= t_1t_2 - t_1 - t_2 + 1 = (t_1 - 1)(t_2 - 1).$$

The quadratic forms

$$Q_1 = \underline{x}'E_1\underline{x} = \frac{t_1}{n}\,\underline{x}'C_1\underline{x} - \frac{1}{n}\,\underline{x}'G\underline{x} = \frac{t_1}{n}\sum_{a_1=1}^{t_1} x_{a_1\cdot}^2 - \frac{1}{n}x_{\cdot\cdot}^2$$

and

$$Q_2 = \underline{x}'E_2\underline{x} = \frac{t_2}{n}\underline{x}'C_2\underline{x} - \frac{1}{n}\,\underline{x}'G\underline{x} = \frac{t_2}{n}\sum_{a_2=1}^{t_2} x_{\cdot a_2}^2 - \frac{1}{n}x_{\cdot\cdot}^2.$$

have respective rank $\alpha_1 = t_1 - 1$ and $\alpha_2 = t_2 - 1$. The difference

$$Q - Q_1 - Q_2 = \sum_{a_1=1}^{t_1}\sum_{a_2=1}^{t_2} x_{a_1a_2}^2 - \frac{t_1}{n}\sum_{a=1}^{t_1} x_{a_1\cdot}^2 - \frac{t_2}{n}\sum_{a=1}^{t_2} x_{\cdot a_2}^2 + \frac{1}{n}x_{\cdot\cdot}^2.$$

has rank $t_1t_2 - 1 - (t_1 - 1) - (t_2 - 1) = (t_1 - 1)(t_2 - 1)$.

Under the model that was adopted in the preceding section, we have

$$x_{a_1a_2} = \mu + \mu(1;\ a_1) + \mu(2;\ a_2) + \mu(1,2;\ a_1,a_2) + e_{a_1a_2},$$

with the $e_{a_1a_2}$ are mutually independent and distributed as $N(0,\ \sigma^2)$, and

$$\sum_{a_1} \mu(1;\ a_1) = 0; \quad \sum_{a_2} \mu(2;\ a_2) = 0;$$

$$\sum_{a_1} \mu(1,2;\ a_1,a_2) = 0 \qquad (a_2 = 1,\ldots,t_2)$$

$$\sum_{a_2} \mu(1,2;\ a_1,a_2) = 0 \qquad (a_1 = 1,\ldots,t_1)$$

Now since

$$E\left[\underline{\zeta}_{a_1}^{(1)'}\underline{x}\right] = \frac{n}{t_1}\left[\mu + \mu(1;\ a_1)\right], \quad E\left[\underline{\zeta}_{a_2}^{(2)'}\underline{x}\right] = \frac{n}{t_2}\left[\mu + \mu(2;\ a_2)\right],$$

the $(t_i - 1)$ mutually orthogonal contrasts of $\underline{x}$ that make up Q_i have as their expectations $(t_i - 1)$ mutually orthogonal contrasts between $\mu(i;\ a_i)$.

Since

$$Q - Q_1 - Q_2 = \underline{x}'(I - \frac{t_1}{n}C_1 - \frac{t_2}{n}C_2 + \frac{1}{n}G)(I - \frac{t_1}{n}C_1 - \frac{t_2}{n}C_2 + \frac{1}{n}G)\underline{x}$$

and

$$(I - \frac{t_1}{n}C_1 - \frac{t_2}{n}C_2 + \frac{1}{n}G)\underline{x} = \underline{x} - \frac{t_1}{n}\Phi_1\Phi_1'\underline{x} - \frac{t_2}{n}\Phi_2\Phi_2'\underline{x} + \frac{1}{n}x_{..}\underline{1},$$

the fth element of this vector is

$$x_{a_1a_2} - \frac{t_1}{n}x_{a_1\cdot} - \frac{t_2}{n}x_{\cdot a_2} + \frac{1}{n}x_{..}$$

if f corresponds to (a_1,a_2). Hence its expectation is given by

$$\mu + \mu(1;a_1) + \mu(2;a_2) + \mu(1,2;a_1,a_2) - \mu - \mu(1;a_1) - \mu - \mu(2;a_2) + \mu$$

$$= \mu(1,2;a_1,a_2).$$

Therefore the $(t_1 - 1)(t_2 - 1)$ mutually orthogonal contrasts of $\underline{x}$ that make up $Q - Q_1 - Q_2$ have as their expectations $(t_1 - 1)(t_2 - 1)$

mutually orthogonal contrasts among $\mu(1,2;\ a_1,a_2)$.

The interactions are, in general, of smaller order compared to the main-effects due to each classification, that is, $\mu(i;\ a_i)$, $i = 1,2$. One usually ignores interactions unless the problem under consideration dictates otherwise, that is $\mu(1,2;\ a_1,a_2) = 0$. Then under our model

$$Q - Q_1 - Q_2 = \sigma^2\chi^2_{(t_1 - 1)(t_2 - 1)},$$

and hence

$$\frac{1}{(t_1 - 1)(t_2 - 1)}\left\{Q - Q_1 - Q_2\right\}$$

gives us an unbiased estimator of the population variance σ^2. Of course

$$Q_1,\ Q_2,\ Q - Q_1 - Q_2$$

are mutually independent. Thus under the hypothesis

$$H_i^0\ ;\ \mu(i;\ a_i) = 0 \qquad (a_i = 1,\ldots,t_i, \quad i = 1,2)$$

the statistics

$$F_i = \frac{(t_1 - 1)(t_2 - 1)}{t_i - 1}\ \frac{Q_i}{Q - Q_1 - Q_2} \qquad (i = 1,2)$$

are distributed as $F^{t_i-1}_{(t_1-1)(t_2-1)}$. This enables us to test the hypothesis H_i^0.

REFERENCES

1. H. Scheffe, *The Analysis of Variance*, Wiley, New York, 1959.

2. O. Carpenter, *Ann. Math. Statist., 21*, 455 (1950).

3. S. Kotz, N. L. Johnson, and D. W. Boyd, *Ann. Math. Statist., 38*, 823 (1967).

4. S. Kotz, N. L. Johnson, and D. W. Boyd, *Ann. Math. Statist., 38*, 838 (1967).

5. J. Ogawa, *Ann. Inst. Stat. Math. Tokyo, 1*, 83 (1949).

6. K. S. Banerjee, *Ann. Math. Statist., 37*, 295 (1966).

7. R. M. Loynes, *Ann. Math. Statist., 37*, 295 (1966).

8. J. Ogawa, *Osaka Math. J., 5(1)*, 1 (1953).

9. H. B. Mann, *Analysis and Design of Experiments*, Dover, New York, 1949.

10. K. Matsushita, On the Independence of Certain Estimates Variances, *Ann. Inst. Stat. Math. Tokyo, 1* (1949).

Chapter 2
DESIGN OF EXPERIMENTS

2.1. General Remarks

One should realize the distinction between two kinds of experimentation. One is the experiment aimed at estimating absolute constants, and the other is the so-called *comparative experiment*. The statistical theory of experimental design is mainly concerned with the latter type, such as comparisons between the effects of different doses of a drug, rather than with the determination of physical constants.

Much of the stimulus to the development of the modern statistical theory of experimental design came originally from agricultural research, and the standard nomenclature used in the theory still bears evidence of it. It should not be mistaken, however, that its applicability is limited only to the agricultural field.

The unit of material to which a *treatment* or *variety* is applied is called a *plot* or an *experimental unit*. A plot may be an area of land on which a crop is grown, in accordance with the original way, but it may be a hospital patient, a piece of animal tissue, *the site* on the body of an animal to receive an injection on a particular occasion, or one of a number of similar machines.

The objective of an experiment is to make comparisons between the effects of different treatments, one of which is applied to one or more plots, in terms of measurement of observations made on the separate plots. Any quantitative measure obtained from a plot may be termed *a yield* or *an observation*.

A group of plots used in the structure of the design because they have certain inherent features in common is *a block* (eg., a more or less homogeneous portion of land or a litter of animals).

Very frequently the treatments tested are composed of several *factors*. The several states of a factor are called *levels* (multi-factor experiment).

The design of an experiment is constituted of the following steps:

1. The set of treatments selected for comparison
2. The specification of the plots to which the treatments are to be applied
3. The rules by which the treatments are to be allocated to plots
4. The specification of the measurements or other records to be made on each plot

The mathematical theory of experimental design is mainly concerned with step 1 (eg. factorial design, fractional factorial, confounding) and step 3 (eg. incomplete block designs). In order to to conduct a good experiment, steps 2 and 4 are also important, but the key is the closest collaboration between experimenter and statistician, so that the nature of the experimental material can be exploited to the full advantage.

2.2. Randomization

The actual procedures of randomization and the mathematical theory of randomization will be presented in later sections. Here some general philosophy is mentioned. Why is any randomization necessary? In the first place, since there is always uncontrollable heterogeneity in experimental units, it is the only way of ensuring that comparisons between treatments are not biased by the fact of one treatment being assigned to inherently better plots than another. However honest the experimenter may be, if he has freedom of choice within the explicit constraints of design, he is very likely to prejudice results for or against one treatment by the plot that he chooses for it, and this should be regarded as due to human nature.

Of course, any blatant favoritism would be recognized and discarded, but there still is *abundant evidence that even small subconscious effects may disturb the true objective character of an experiment.* "It is not merely of some importance but is of fundamental importance that justice should not only be done, but should *manifestly and undoubtedly be seen to be done.*"

Second, as you will see in later sections, the randomization is the only way to let the sampling distribution of the F-statistics appearing in the analysis of variance of the observations obtainable from an arrangement be approximated by the F-distribution.

One should note that there are two kinds of randomness of entirely different nature. One might call the one--randomization--"man-made" and the other--a random sample from some population--"god-given" randomness.

2.3. Randomized Block Design

Any experimenter is well aware of the advantage to be gained by comparing treatments under conditions that are as homogeneous as possible, and it is to this end that much of the effort of experimental design has been directed. In other words, each plot has its own effect, and they are rarely homogeneous. A simple example will make this clear. A shoemaker wishing to compare the durabilities of shoes made from four different types of material might make 10 pairs from each material A, B, C, and D, and arrange that these be worn by volunteers for 6 months, after which they would be collected and the deterioration assessed. A valid procedure would be to obtain 40 volunteers and to allot one pair of shoes to each entirely at random. However, if eight of the wearers of A were farmers, seven of the wearers of B were bank officials, nine of the wearers of C were steelworkers, and eight of the wearers of D were infantrymen in Vietnam, the shoemaker would justifiably distrust a conclusion that shoes D and A were exceptionally liable to wear. One possibility is to restrict the volunteers to a single profession, but this might limit the applicability of the results, and furthermore it would be impossible unless a large number of volunteers from one profession could be secured.

The obvious alternative is to *stratify the available volunteers according to professions or group of professions and then to restrict the randomization by requiring that each type of shoes be allocated to two farmers, three steelworkers, and so on.* All comparisons between shoes are then balanced in repect of professional differences in hardness of wear, yet the investigation still has the merit of being conducted on a wide range of persons.

Suppose that an experiment for comparing v treatments or varieties is to be conducted and that the number of plots available is $n = vb$. Suppose that the n plots are divided into b blocks of size v each in such a way that the variances of plot effects within blocks are as homogeneous as possible. Allocate the treatments to plots at random in each block and independently from block to block. This is called a *randomized complete block design* (see table 2.1.).

TABLE 2.1.

Example of a Randomized Arrangement for Six Blocks of Five Treatments in a Randomized Complete Block Design

Plot	BLOCK					
	I	II	III	IV	V	VI
1	D	D	A	B	E	D
2	E	C	C	A	D	B
3	C	A	B	D	C	E
4	A	B	E	E	A	C
5	B	E	D	C	B	A

We shall consider the analysis of variance of the observations obtainable from this design under the linear model taking the randomization into account.

We number the whole plots from 1 to n in any order. Then we define the incidence vectors and incidence matrices for treatments and blocks.

<u>Blocks:</u>

(2.3.1)
$$\underline{\eta}_i = \begin{pmatrix} \eta_{i1} \\ \vdots \\ \eta_{in} \end{pmatrix}$$

$$n_{if} = \begin{cases} 1, & \text{if the } f\text{th plot belongs to the } i\text{th block} \\ 0, & \text{otherwise} \end{cases} \qquad (i = 1,\ldots,b),$$

(2.3.2)
$$\Psi = \left\| \underline{\eta}_1 \cdots \underline{\eta}_b \right\|$$

(2.3.3)
$$B = \Psi'\Psi, \; B^2 = KB.$$

Hence

(2.3.4)
$$\left(\frac{1}{k} B\right)^2 = \frac{1}{k} B \text{ idempotent } r\left(\frac{B}{k}\right) = \frac{n}{k} = b$$

In our case of *complete blocks*, $v = k$ to which we owe the name "complete block." Let the observation vector be $\underline{x}' = (x_1 \cdots x_n)$; then

(2.3.5)
$$B_i = \underline{\eta}_i' \, \underline{x} \qquad (i = 1,\ldots,b)$$

are block totals. We denote the block-total vector by

(2.3.6)
$$\underline{B}' = (B_1 \cdots B_b).$$

Treatments :

(2.3.7) $$\underline{\zeta}_\alpha = \begin{pmatrix} \zeta_{\alpha 1} \\ \vdots \\ \zeta_{\alpha n} \end{pmatrix},$$

$$\zeta_{\alpha f} = \begin{cases} 1, \text{ if the } f\text{-th plot receives } \alpha\text{-th treatment} \\ 0 \text{ otherwise} \end{cases} \qquad (\alpha = 1,\ldots,v),$$

(2.3.8) $$\Phi = \|\underline{\zeta}_1 \cdots \underline{\zeta}_v\| \qquad \Phi'\Phi = bI_v,$$

(2.3.9) $$T = \Phi\Phi', \qquad T^2 = bT.$$

Hence $(1/b)T$ is idempotent. Treatment totals are given by

(2.3.10) $$T_\alpha = \underline{\zeta}'_\alpha \underline{x} \qquad (\alpha = 1,\ldots,v),$$

(2.3.11) $$\underline{T}' = (T_1 \cdots T_v).$$

The grand total is given by $n\bar{x} = \underline{1}'\underline{x}$, and the corresponding idempotent is $(1/n)G = (1/n)\underline{1}\ \underline{1}'$. Furthermore,

$$\text{total sum of squares} = \sum x_i^2 - n\bar{x}^2 = \underline{x}'\left(I - \frac{1}{n}G\right)\underline{x},$$

$$\text{sum of squares due to treatment} = \frac{1}{b}\left(T_1^2 + \cdots + T_v^2\right) - n\bar{x}^2$$

$$= \underline{x}'\left(\frac{1}{b}T - \frac{1}{n}G\right)\underline{x},$$

and sum of squares due to block $= \frac{1}{k}\left(B_1^2 + \cdots + B_b^2\right) - n\bar{x}^2$

$$= \underline{x}'\left(\frac{1}{k} B - \frac{1}{n} G\right)\underline{x}.$$

Hence by subtraction one obtains

$$\text{sum of squares due to error} = \underline{x}'\left(I - \frac{1}{b} T - \frac{1}{k} B + \frac{1}{n} G\right)\underline{x}.$$

Let $T \equiv \|t_{fg}\|$ and $B \equiv \|b_{fg}\|$. Then

$$t_{fg} = \sum_{\alpha=1}^{v} \zeta_{\alpha f}\zeta_{\alpha g} = \begin{cases} 1 \text{ if the } f\text{th and } g\text{th plots receive a common treatment,} \\ 0 \text{ otherwise.} \end{cases}$$

$$b_{fg} = \sum_{i=1}^{b} n_{if} n_{ig} = \begin{cases} 1 \text{ if the } f\text{th and } g\text{th plots belong to a same block,} \\ 0 \text{ otherwise.} \end{cases}$$

In this sense, T represents a relationship among plots due to treatment, and similarly B represents the relationship among plots due to block; also G represents *the universal relationship* (the identity matrix I is also included). These matrices are called *the relationship matrices* of the randomized block design. One can easily make the following multiplication table :

I	B	T	G
B	kB	G	kG
T	G	bT	bG
G	kG	bG	nG

Hence the linear closure of the four matrices I, G, B, T over the field of all real numbers is a linear associative and commucative algebra of rank 4. Therefore its unit is decomposed into the sum of four mutually orthogonal idempotent matrices. This has been given by the analysis of variance above, that is,

$$I = \frac{1}{n} G + \left(\frac{1}{k} B - \frac{1}{n} G\right) + \left(\frac{1}{b} T - \frac{1}{n} G\right) + \left(I - \frac{1}{b} T - \frac{1}{k} B + \frac{1}{n} G\right) \tag{2.3.12}$$

with the respective ranks

$$\begin{aligned} \alpha_0 &= tr\left(\frac{1}{n} G\right) = 1, \\ \alpha_1 &= tr\left(\frac{1}{k} B - \frac{1}{n} G\right) = b - 1, \\ \alpha_2 &= tr\left(\frac{1}{b} T - \frac{1}{n} G\right) = k - 1, \\ \alpha_3 &= tr\left(I - \frac{1}{b} T - \frac{1}{k} B + \frac{1}{n} G\right) = (b - 1)(k - 1). \end{aligned} \tag{2.3.13}$$

We take the following model :

$$x_f = \mu + \sum_{\alpha=1}^{v} \zeta_{\alpha f}\tau_\alpha + \sum_{i=1}^{b} \eta_{if}\beta_i + \pi_f + e_f \quad (f = 1,\ldots,n), \tag{2.3.14}$$

where x_f is the observation obtainable at the fth plot, μ is the general mean effect, τ_α is the effect of the αth treatment, β_i is the effect of the ith block, π_f is the plot effect or unit error of the f-th plot, and e_f is the error. In vector notation

$$(2.3.15) \qquad \underline{x} = \mu\underline{1} + \phi\underline{\tau} + \Psi\underline{\beta} + \underline{\pi} + \underline{e},$$

the treatment-effects vector $\underline{\tau}' = (\tau_1 \cdots \tau_v)$ and the block-effects vector $\underline{\beta}' = (\beta_1 \cdots \beta_b)$ are subjected to the restrictions $\tau_1 + \cdots + \tau_v = 0$ and $\beta_1 + \cdots + \beta_b = 0$, respectively, and the unit-error vector $\underline{\pi}' = (\pi_1 \cdots \pi_n)$ can be subjected to the restriction $\Psi'\underline{\pi} = 0$, that is, within block totals $\underline{\eta}_i'\underline{\pi} = 0$, $i = 1,\ldots,b$. Finally the error vector $\underline{e}' = (e_1 \cdots e_n)$ is assumed to be distributed as $N(0,\sigma^2 I)$.

For the sake of simplicity in calculation, we shall consider the null distribution of the statistic for testing the null-hypothesis $H_0{:}\underline{\tau} = 0$:

$$(2.3.16) \qquad \frac{1}{b-1} F = \frac{\underline{x}'\left(\frac{1}{b} T - \frac{1}{n} G\right)\underline{x}}{\underline{x}'\left(I - \frac{1}{b} T - \frac{1}{k} B + \frac{1}{n} G\right)\underline{x}}$$

Under the model

$$\underline{x} = \mu\underline{1} + \Phi\underline{\tau} + \Psi\underline{\beta} + \underline{\pi} + \underline{e},$$

$$\left(I - \frac{1}{b} T - \frac{1}{k} B + \frac{1}{n} G\right)\underline{x} = \left(I - \frac{1}{b} T - \frac{1}{k} B + \frac{1}{n} G\right)(\underline{\pi} + \underline{e})$$

$$= \left(I - \frac{1}{b} T\right)\underline{\pi} + \left(I - \frac{1}{b} T - \frac{1}{k} B + \frac{1}{n} G\right)\underline{e}$$

because

$$\left(I - \frac{1}{b} T - \frac{1}{k} B + \frac{1}{n} G\right)\underline{1} = \underline{1} - \frac{1}{b} \Phi\Phi'\underline{1} - \frac{1}{k} \Psi\Psi'\underline{1} + \frac{1}{n} \underline{1}\ \underline{1}'\underline{1}$$

$$= \underline{1} - \underline{1} - \underline{1} + \underline{1} = 0,$$

$$\left(I - \frac{1}{b} T - \frac{1}{k} B + \frac{1}{n} G\right)\Phi = \Phi - \frac{1}{b} \Phi\Phi'\Phi - \frac{1}{k} \Psi\Psi'\Phi + \frac{1}{n} \underline{1}\ \underline{1}'\Phi$$

$$= \Phi - \Phi - \frac{1}{k} \Psi N' + \frac{b}{n} \underline{1}_n \underline{1}_k' = 0.$$

Hence

$$\underline{x}'\left(I - \frac{1}{b} T - \frac{1}{k} B + \frac{1}{n} G\right)\underline{x} = \underline{e}'\left(I - \frac{1}{b} T - \frac{1}{k} B + \frac{1}{n} G\right)\underline{e} + 2\underline{e}'\left(I - \frac{1}{b} T\right)\underline{\pi} + \underline{\pi}'\left(I - \frac{1}{b} T\right)\underline{\pi} \tag{2.3.17}$$

irrespective of the null hypothesis H_0: $\underline{\tau} = 0$ and under the null hypothesis H_0: $\underline{\tau} = 0$ the treatment sum part becomes

$$\left(\frac{1}{b} T - \frac{1}{n} G\right)\underline{x} = \mu\left(\frac{1}{b} T - \frac{1}{n} G\right)\underline{1} + \left(\frac{1}{b} T - \frac{1}{n} G\right)\Phi\underline{\tau}$$

$$+ \left(\frac{1}{b} T - \frac{1}{n} G\right)\Psi\underline{\beta} + \left(\frac{1}{b} T - \frac{1}{n} G\right)\underline{\pi} + \left(\frac{1}{b} T - \frac{1}{n} G\right)\underline{e}$$

$$= \Phi\underline{\tau} + \frac{1}{n} T\underline{\pi} + \left(\frac{1}{b} T - \frac{1}{n} G\right)\underline{e}$$

$$= \frac{1}{b} T\underline{\pi} + \left(\frac{1}{b} T - \frac{1}{n} G\right)\underline{e}$$

hence

(2.3.18)
$$\underline{x}'\left(\frac{1}{b} T - \frac{1}{n} G\right)\underline{x} = \underline{e}'\left(\frac{1}{b} T - \frac{1}{n} G\right)\underline{e} + 2\underline{e}'\left(\frac{1}{b} T - \frac{1}{n}\right)G \frac{1}{n} T\underline{\pi} + \frac{1}{b} \underline{\pi}'T\underline{\pi},$$

one can see that they are mutually independent. Let

$$\sigma^2\chi_1^2 = \underline{x}'\left(\frac{1}{b} T - \frac{1}{n} G\right)\underline{x}'.$$

Then χ_1^2 is the noncentral χ^2 with $v - 1$ degrees of freedom and with the noncentrality parameter

(2.3.19)
$$\lambda_1 = \frac{1}{\sigma^2 b} \underline{\pi}'T\underline{\pi}.$$

Its probability element is given by

(2.3.20)
$$exp\left(-\frac{\lambda_1}{2}\right)\sum_{\mu=0}^{\infty} \frac{\left(\frac{\lambda_1}{2}\right)^{\mu}}{\mu!} \frac{\left(\frac{\chi_1^2}{2}\right)^{[(k-1)/2] + \mu - 1}}{\Gamma\left(\frac{k-1}{2} + \mu\right)} exp\left(-\frac{\chi_1^2}{2}\right) d\left(\frac{\chi_1^2}{2}\right).$$

Let

$$\sigma^2\chi_2^2 = \underline{x}'\left(1 - \frac{1}{b}T - \frac{1}{k}B + \frac{1}{n}G\right)\underline{x}.$$

Then χ_2^2 is the noncentral χ^2 with $(b - 1)(k - 1)$ degrees of freedom and with the noncentrality parameter

(2.3.21)
$$\lambda_2 = \frac{\underline{\pi}'\underline{\pi} - (1/b)\,\underline{\pi}'T\underline{\pi}}{\sigma^2}$$

Its probability element is given by

(2.3.22)

$$exp\left(-\frac{\lambda_2}{2}\right)\sum_{\nu=0}^{\infty}\frac{\left(\frac{\lambda_2}{2}\right)^{\nu}}{\nu!}\frac{\left(\frac{\chi_2^2}{2}\right)^{[(b-1)(k-1)]/2+\nu-1}}{\Gamma\left(\frac{(b-1)(k-1)}{2}+\nu\right)}exp\left(-\frac{\chi_2^2}{2}\right)d\left(\frac{\chi_2^2}{2}\right).$$

The probability element of the joint distribution of χ_1^2 and χ_2^2 is

(2.3.23)

$$exp\left(-\frac{\lambda_1+\lambda_2}{2}\right)\sum_{\mu=0}^{\infty}\sum_{\nu=0}^{\infty}\frac{\left(\frac{\lambda_1}{2}\right)^{\mu}\left(\frac{\lambda_2}{2}\right)^{\nu}}{\mu!\nu!}\frac{\left(\frac{\chi_1^2}{2}\right)^{[(k-1)/2]+\mu-1}\left(\frac{\chi_2^2}{2}\right)^{\{[(b-1)(k-1)]/2\}+\nu-1}}{\Gamma\left(\frac{k-1}{2}+\mu\right)\Gamma\left(\frac{(b-1)(k-1)}{2}+\nu\right)}$$

$$\times\, exp\left(-\frac{\chi_1^2}{2}-\frac{\chi_2^2}{2}\right)d\left(\frac{\chi_1^2}{2}\right)d\left(\frac{\chi_2^2}{2}\right)$$

$$= exp\left(-\frac{\pi'\pi}{2\sigma^2}\right)\sum_{\mu=0}^{\infty}\sum_{\nu=0}^{\infty}\frac{\left(\frac{\lambda_1}{2}\right)^{\mu}\left(\frac{\lambda_2}{2}\right)^{\nu}}{\mu!\nu!}\left(\frac{F}{b-1}\right)^{[(k-1)/2]+\mu-1}$$

$$\times\frac{\left(\frac{\chi_2^2}{2}\right)^{\{[b(k-1)]/2\}+\mu+\nu-1}}{\Gamma\left(\frac{k-1}{2}+\mu\right)\Gamma\left(\frac{(b-1)(k-1)}{2}+\nu\right)}\,exp\left|-\frac{\chi_2^2}{2}\left(1+\frac{F}{b-1}\right)\right|\,d\left(\frac{\chi_2^2}{2}\right)d\left(\frac{F}{b-1}\right).$$

Since

$$\frac{\chi_1^2}{2}=\frac{\chi_2^2}{2}\,\frac{F}{b-1},$$

integrating out $\chi_1^2/2$, one obtains the probability element for $F/(b-1)$ as follows :

$$exp\left(-\frac{\pi'\pi}{2\sigma^2}\right)\sum_{\mu}\sum_{\nu}\frac{\left(\frac{\lambda_1}{2}\right)^{\mu}\left(\frac{\lambda_2}{2}\right)^{\nu}}{\mu!\nu!}\left(\frac{F}{b-1}\right)^{[(k-1)/2]+\mu-1}\left(1+\frac{F}{b-1}\right)^{-\{[b(k-1)/2]+\mu+\nu\}}$$

$$\times\frac{\Gamma\left(\frac{b(k-1)}{2}+\mu+\nu\right)}{\Gamma\left(\frac{k-1}{2}+\mu\right)\Gamma\left(\frac{(b-1)(k-1)}{2}+\nu\right)}\,d\left(\frac{F}{b-1}\right)$$

$$=\frac{\Gamma\left(\frac{b(k-1)}{2}\right)}{\Gamma\left(\frac{k-1}{2}\right)\Gamma\left(\frac{(b-1)(k-1)}{2}\right)}\left(\frac{F}{b-1}\right)^{[(k-1)/2]-1}$$

$$\times \left(1 + \frac{F}{b-1}\right)^{-\{[b(k-1)/2]\}} d\left(\frac{F}{b-1}\right)$$

$$exp\left(-\frac{\underline{\pi}'\underline{\pi}}{2\sigma^2}\right) \sum_{l=0}^{\infty} \frac{\left(\frac{\underline{\pi}'\underline{\pi}}{2\sigma^2}\right)^l}{l!} \left(1 + \frac{F}{b-1}\right)^{-l} \sum_{\mu=0}^{l} \frac{l!}{\mu!\,(l-\mu)!} \theta^\mu (1-\theta)^{l-\mu}$$

$$\times \left(\frac{F}{b-1}\right)^\mu \frac{\Gamma\left(\frac{b(k-1)}{2} + l\right) \Gamma\left(\frac{k-1}{2}\right) \Gamma\left(\frac{(b-1)(k-1)}{2}\right)}{\Gamma\left(\frac{b(k-1)}{2}\right) \Gamma\left(\frac{k-1}{2} + \mu\right) \Gamma\left(\frac{(b-1)(k-1)}{2} + l - \mu\right)},$$

(2.3.24)

where we have put

$$\theta \equiv \frac{1}{b} \frac{\underline{\pi}'T\underline{\pi}}{\underline{\pi}'\underline{\pi}} = \frac{1}{b} \frac{\Pi_1^2 + \cdots + \Pi_v^2}{\underline{\pi}'\underline{\pi}}. \qquad (2.3.25)$$

Even in the normal theory this is the noncentral F-distribution depending on the parameter θ. *The randomization procedure gives us a way of disposing of this nuisance parameter.* We adopt a spcial numbering system of plots, that is, $f = (i-1)k + j$ if the plot f is the jth plot in the ith block and

$$\pi_f \equiv \pi_j^{(i)}, \quad \sum_{j=1}^{k} \pi_j^{(i)2} = \Delta_i, \quad \sum_{i=1}^{b} \Delta_i = \Delta = \underline{\pi}'\underline{\pi}.$$

Let the permutation matrix corresponding to permutation

$$\sigma = \begin{pmatrix} 1 & 2 & \cdots & k \\ \\ \sigma(1) & \sigma(2) & \cdots & \sigma(k) \end{pmatrix}$$

be S_σ; that is,

$$S_\sigma \begin{bmatrix} 1 \\ 2 \\ \vdots \\ k \end{bmatrix} = \begin{bmatrix} \sigma(1) \\ \sigma(2) \\ \vdots \\ \sigma(k) \end{bmatrix}.$$

Also we are using the following notations:

$$\zeta_{\alpha f} = \zeta_{\alpha j}^{(i)} \quad \text{for} \quad f = (i - 1)k + j$$

and

$$\underline{\zeta}_\alpha^{(i)\prime} = \left[\zeta_{\alpha 1}^{(i)} \;\cdots\; \zeta_{\alpha k}^{(i)}\right],$$

$$\text{(2.3.26)} \qquad \Phi^{(i)} = \left\| \underline{\zeta}_1^{(i)} \;\cdots\; \underline{\zeta}_k^{(i)} \right\|.$$

The randomization in the ith block means that we take a permutation σ with probability $1/k!$ and permute the treatment (i.e., $S_\sigma \underline{\tau}$) and so our model becomes for the j-th plot in the i-th block

$$x_{(i-1)k+j} = \gamma + \sum_{\alpha=1}^{k} \zeta_{\sigma(\alpha)j}^{(i)} \tau_{\sigma(\alpha)} + \beta_j + \pi_j^{(i)} + e_{(i-1)k+j}$$
$$= \gamma + \sum_{\alpha=1}^{k} \zeta_{\alpha\sigma'(j)}^{(i)} \tau_\alpha + \beta_j + \pi_j^{(i)} + e_{(i-1)k+j}$$

and hence

$$\text{(2.3.27)} \qquad \Phi^{(i)} = S_{\sigma'} = S_{\sigma'} \Phi_0^{(i)}.$$

This means that $\Phi^{(i)}$ should be replaced by $S_{\sigma'} \Phi_0^{(i)}$ or $S_{\sigma'}$, if especially $\Phi_0^{(i)} = I_k$. Since σ' runs all over $\mathfrak{S}_k$ with σ, there will be no harm to use σ instead of σ'.

Now let

$$U_{\underline{\sigma}} = \left\| \begin{matrix} S_{\sigma_1} & & & \\ & S_{\sigma_2} & & \\ & & \ddots & \\ & & & S_{\sigma_b} \end{matrix} \right\|$$

then the incidence matrix Φ becomes a random (discrete) variable by the randomization procedure such that

$$\text{(2.3.28)} \qquad \mathrm{P}\left(\Phi = U_{\underline{\sigma}}\Phi_0\right) = \frac{1}{(k!)^b},$$

where $\sigma_1,\ldots,\sigma_b$ range independently over the whole of $\mathfrak{S}_k$ and Φ_0 is any fixed incidence matrix. For convenience we take

$$\Phi_0\Phi_0' = I_k \times G_b.$$

We calculate in the first place

$$E\left(\underline{\pi}'\Phi\Phi'\underline{\pi}\right) = (k')^{-b} \sum_{\sigma_1,\ldots,\sigma_b \in \mathfrak{S}_k} \underline{\pi}'U_{\sigma}\Phi_0\Phi_0'U'\underline{\pi}$$

$$= (k!)^{-b} \sum_{\sigma_1,\ldots,\sigma_b \in \mathfrak{S}_k} \left[\underline{\pi}^{(1)\prime}S_{\sigma_1} + \cdots + \underline{\pi}^{(b)\prime}S_{\sigma_b}\right]$$

$$\times \left[S_{\sigma_1}'\underline{\pi}^{(1)} + \cdots + S_{\sigma_b}'\underline{\pi}^{(b)}\right]$$

$$= \underline{\pi}^{(1)\prime}\,\underline{\pi}^{(1)} + \cdots + \underline{\pi}^{(b)\prime}\,\underline{\pi}^{(b)} + \frac{1}{(k!)^2} \sum_{\substack{\sigma_l,\sigma_m \in \mathfrak{S}_k \\ (l \neq m)}} \times\ \underline{\pi}^{(l)\prime}\, S_{\sigma_l} S'_{\sigma_m} \underline{\pi}^{(m)}$$

$$= \Delta + \frac{1}{(k!)^2} \sum_{\substack{\sigma_l,\sigma_m \in \mathfrak{S}_k \\ (l \neq m)}} \underline{\pi}^{(l)\prime}\, S_{\sigma_l} S'_{\sigma_m} \underline{\pi}^{(m)}$$

$$E\left[\underline{\pi}^{(l)\prime}\, S'_{\sigma_l} S_{\sigma_m} \underline{\pi}^{(m)}\right] = (k!)^{-2} \sum_{\sigma,\tau \in \mathfrak{S}_k} \underline{\pi}^{(l)\prime}\, S_{\sigma} S'_{\tau} \underline{\pi}^{(m)}$$

$$= (k!)^{-2} \sum_{\sigma \in \mathfrak{S}_k} \underline{\pi}^{(l)\prime}\, S_{\sigma} \underline{\pi}^{(m)} = 0$$

$$E(\theta) = \frac{1}{b\Delta} E(\underline{\pi}' T \underline{\pi}) = \frac{1}{b} \tag{2.3.29}$$

$$E(\underline{\pi}' T \underline{\pi})^2 = E\left[\Delta + \sum_{l \neq m} \underline{\pi}^{(l)\prime}\, S_{\sigma_l} S'_{\sigma_m} \underline{\pi}^{(m)}\right]^2$$

$$= \Delta^2 + 2\Delta E\left[\sum_{l \neq m} \underline{\pi}^{(l)\prime}\, S_{\sigma_l} S'_{\sigma_m} \underline{\pi}^{(m)}\right]$$

$$+ \sum_{l \neq m} E\left[\underline{\pi}^{(l)\prime} S_{\sigma_l} S'_{\sigma_m} \underline{\pi}^{(m)} \underline{\pi}^{(l)\prime} S_{\sigma_l} S'_{\sigma_m} \underline{\pi}^{(m)} + \underline{\pi}^{(l)\prime} \right.$$

$$\left. \times S_{\sigma_l} S'_{\sigma_m} \underline{\pi}^{(m)} \underline{\pi}^{(m)\prime} S_{\sigma_m} S'_{\sigma_l} \underline{\pi}^{(l)} \right]$$

(2.3.30)

$$+ \sum_{l \neq m \neq p} E\left\{\underline{\pi}^{(l)\prime} S_{\sigma_l} S'_{\sigma_m} \underline{\pi}^{(m)} \left[\underline{\pi}^{(l)\prime} S_{\sigma_l} S'_{\sigma_p} \underline{\pi}^{(p)} + \underline{\pi}^{(m)\prime} \right.\right.$$

$$\left.\left. \times S_{\sigma_l} S'_{\sigma_p} \underline{\pi}^{(p)} + \underline{\pi}^{(p)\prime} S_{\sigma_p} S'_{\sigma_l} \underline{\pi}^{(l)} + \underline{\pi}^{(p)\prime} S_{\sigma_p} S'_{\sigma_m} \underline{\pi}^{(m)} \right]\right\}$$

$$+ E\left\{ \sum_{l \neq m \neq p \neq q} \underline{\pi}^{(l)\prime} S_{\sigma_l} S'_{\sigma_m} \underline{\pi}^{(m)} \underline{\pi}^{(p)\prime} S_{\sigma_p} S'_{\sigma_q} \underline{\pi}^{(q)} \right\}.$$

It is easy to see that for $l \neq m \neq p \neq q$

$$E\left[\underline{\pi}^{(l)\prime} S_{\sigma_l} S'_{\sigma_m} \underline{\pi}^{(m)} \underline{\pi}^{(p)\prime} S_{\sigma_p} S'_{\sigma_q} \underline{\pi}^{(q)}\right] = 0 \tag{2.3.31}$$

and for $l \neq m$, substituting σ, τ in σ_l, σ_m respectively

$$E\left[\underline{\pi}^{(l)\prime} S_{\sigma_l} S'_{\sigma_m} \underline{\pi}^{(m)}\right]^2$$

$$= (k!)^{-2} \sum_{\sigma, \tau \in \mathfrak{S}_k} \left[\underline{\pi}^{(l)\prime} S_\sigma S'_\tau \underline{\pi}^{(m)}\right]^2$$

$$= (k!)^{-1} \sum_{\sigma\in\mathfrak{S}_k} \left[\underline{\pi}^{(l)'} S_\sigma \underline{\pi}^{(m)}\right]^2$$

$$= (k!)^{-1} \sum_{\sigma\in\mathfrak{S}_k} \left[\pi_1^{(l)}\pi_{\sigma(k)}^{(m)} + \cdots + \pi_k^{(l)}\pi_{\sigma(k)}^{(m)}\right]^2$$

(2.3.32)

$$= \sum_{i=1}^{k} \pi_i^{(l)2} \frac{1}{k!} \sum_{\sigma\in\mathfrak{S}_k} \pi_{\sigma(i)}^{(m)2} + \sum_{l\neq m} \pi_i^{(l)}\pi_j^{(l)} \frac{1}{k!} \sum_{\sigma\in\mathfrak{S}_k} \pi_{\sigma(i)}^{(m)}\pi_{\sigma(j)}^{(m)}$$

$$= \sum_i \pi_i^{(l)2} \frac{1}{k} \Delta_m + \sum_{i\neq j} \pi_i^{(l)}\pi_j^{(m)} \frac{1}{k(k-1)} \sum_{p\neq q} \pi_p^{(m)}\pi_q^{(m)}$$

$$= \frac{1}{k} \Delta_l\Delta_m + \frac{1}{k(k-1)} \left[\sum_{i\neq j} \pi_i^{(l)}\pi_j^{(l)}\right]\left[\sum_{p\neq q} \pi_p^{(m)}\pi_q^{(m)}\right] = \frac{1}{k-1} \Delta_l\Delta_m,$$

whence

$$E\left[\sum_{l\neq m} \underline{\pi}^{(l)'} S_{\sigma_l} S'_{\sigma_m} \underline{\pi}^{(m)}\right]^2 = \frac{1}{k-1} \sum_{l\neq m} \Delta_l\Delta_m$$

$$= \frac{1}{k-1}\left(\Delta^2 - \sum_{i=1}^{b} \Delta_i^2\right).$$

Similarly

$$E\left[\sum_{l\neq m} \underline{\pi}^{(l)'} S_{\sigma_l} S'_{\sigma_m} \underline{\pi}^{(m)} \underline{\pi}^{(m)'} S_{\sigma_m} S'_{\sigma_l} \underline{\pi}^{(l)}\right]$$

$$= \frac{1}{k-1}\left(\Delta^2 - \sum_i \Delta_i^2\right).$$

Consequently one obtains

(2.3.33)

$$Var\ \theta = \frac{1}{b^2\Delta^2} E(\underline{\pi}'T\underline{\pi})^2 - E^2(\theta) = \frac{1}{b^2}\left[\frac{2}{(k-1)\Delta^2}\left(\Delta^2 - \sum \Delta_i^2\right)\right].$$

Since $\sum \Delta_i^2 = \sum(\Delta_i - \bar{\Delta})^2 + b\bar{\Delta}^2$, $\quad \Delta = b\bar{\Delta}$.

$$\frac{\Delta^2 - \sum\Delta_i^2}{\Delta^2} = 1 - \left[\sum(\Delta_i - \bar{\Delta})^2/b^2\bar{\Delta}^2\right] - \frac{1}{b},$$

$$V \equiv \frac{1}{b-1}\frac{\sum(\Delta_i - \bar{\Delta})^2}{\bar{\Delta}^2},$$

where $\sqrt{V}$ is the coefficient of variation of unit errors of block,

$$Var\ \theta = \frac{1}{b^2}\left[\frac{2}{k-1}\left(1 - \frac{b-1}{b}V - \frac{1}{b}\right)\right]$$

(2.3.34)

$$= \frac{1}{b^2}\left[\frac{2(k-1)}{(k-1)} - \frac{2(b-1)}{(k-1)b^2}V\right] = \frac{2(b-1)}{b^3(k-1)}\left(1 - \frac{V}{b}\right).$$

Now we calculate the mean and variance of the β-distribution :

(2.3.35)

$$\frac{\Gamma[(f_1 + f_2)/2]}{\Gamma(f_1/2)\ \Gamma(f_2/2)} t^{(f_1/2)-1} (1-t)^{(f_2/2)-1},$$

getting

$$\frac{f_1}{f_1 + f_2} \quad \text{and} \quad \frac{2f_1f_2}{(f_1 + f_2)^2(f_1 + f_2 + 2)}$$

Using the moment method, we put

(2.3.36)

$$\frac{f_1}{f_1 + f_2} = \frac{1}{b}; \qquad \frac{2f_1f_2}{(f_1 + f_2)(f_1 + f_2 + 2)} = \frac{2(b - 1)}{b^2(k - 1)}\left(1 - \frac{V}{b}\right).$$

Then

(2.3.37)

$$\frac{f_1}{f_1 + f_2} = \frac{1}{b}; \qquad \frac{f_2}{(f_1 + f_2)(f_1 + f_2 + 2)} = \frac{b - 1}{b^2(k - 1)} \quad 1 - \frac{V}{b}.$$

Hence

$$\frac{f_2}{f_1 + f_2} = \frac{b - 1}{b}, \qquad \frac{1}{f_1 + f_2 + 2} = \frac{1}{b(k - 1)}\left(1 - \frac{V}{b}\right),$$

$$f_1 + f_2 = bf_1 \cdot \rightarrow f_2 = (b - 1)f_1,$$

$$f_1 + f_2 + 2 = (k - 1)b\left(1 - \frac{V}{b}\right)^{-1},$$

$$f_1 + f_2 = (k - 1)b\left(1 - \frac{V}{b}\right)^{-1} - 2$$

$$= (k - 1)b\left[\frac{1}{1 - (V/b)} - \frac{2}{b(k - 1)}\right].$$

Consequently we obtain

$$f_1 = (k - 1)\left[\frac{1}{1 - (V/b)} - \frac{2}{b(k - 1)}\right],$$

(2.3.38)

$$f_2 = (b - 1)(k - 1)\left[\frac{1}{1 - (V/b)} - \frac{2}{b(k - 1)}\right].$$

Since V is the square of the coefficient of variation of within block variances of plot effects, if this can be assumed to be very small for b being fairly large, we can put

(2.3.39) $$\frac{V}{b} \to 0.$$

Hence

$$\left(1 - \frac{V}{b}\right)^{-1} - \frac{2}{b(k - 1)} \doteqdot 1.$$

Thus, from the point of view of the moment method, the distribution of θ due to the permutation distribution induced by randomization can be approximated by the β-distribution :

(2.3.40)

$$\frac{\Gamma\{[b(k-1)]/2\}}{\Gamma[(k-1)/2]\Gamma\{[(b-1)(k-1)]/2\}}\,\theta^{[(k-1)/2]-1}(1-\theta)^{\{[(b-1)(k-1)]/2\}-1}.$$

Thus

$$E\left[\theta^{\mu}(1-\theta)^{l-\mu}\right] = \frac{\Gamma\left(\frac{b(k-1)}{2}\right)}{\Gamma\left(\frac{k-1}{2}\right)\Gamma\left(\frac{(b-1)(k-1)}{2}\right)}\int_0^1 \theta^{[(k-1)/2]+\mu-1}$$

$$\times (1 - \theta)^{[(b-1)(k-1)]/2+l-\mu-1}\, d\theta$$

(2.3.41)

$$= \frac{\Gamma\left(\frac{b(k-1)}{2}\right)\Gamma\left(\frac{k-1}{2}+\mu\right)\Gamma\left(\frac{(b-1)(k-1)}{2}+l-\mu\right)}{\Gamma\left(\frac{k-1}{2}\right)\Gamma\left(\frac{(b-1)(k-1)}{2}\right)\Gamma\left(\frac{b(k-1)}{2}+l\right)}.$$

The distribution of $F/(b-1)$ after the randomization becomes

$$\frac{\Gamma\left(\frac{b(k-1)}{2}\right)}{\Gamma\left(\frac{k-1}{2}\right)\Gamma\left(\frac{(b-1)(k-1)}{2}\right)} \left(\frac{F}{b-1}\right)^{[(k-1)/2]-1}\left(1+\frac{F}{b-1}\right)^{-[b(k-1)/2]}$$

$$\times\, d\left(\frac{F}{b-1}\right) exp\left(-\frac{\Delta}{2\sigma^2}\right) \sum_{l=0}^{\infty} \frac{\left(\frac{\Delta}{2\sigma^2}\right)^l}{l!}\left(1+\frac{F}{b-1}\right)^{-l} \sum_{\mu=0}^{l} \frac{l!}{\mu!(l-\mu)!}$$

(2.3.42)

$$\times\, E\left(\theta^{\mu}(1-\theta)^{l-\mu}\right)\frac{\Gamma\left(\frac{k-1}{2}\right)\Gamma\left(\frac{(b-1)(k-1)}{2}\right)\Gamma\left(\frac{b(k-1)}{2}+l\right)}{\Gamma\left(\frac{b(k-1)}{2}\right)\Gamma\left(\frac{k-1}{2}+\mu\right)\Gamma\left(\frac{(b-1)(k-1)}{2}+l-\mu\right)}$$

$$\times \left(\frac{F}{b-1}\right)^{\mu}$$

$$= \frac{\Gamma\left(\frac{b(k-1)}{2}\right)}{\Gamma\left(\frac{k-1}{2}\right)\Gamma\left(\frac{(b-1)(k-1)}{2}\right)}$$

$$\times \left(\frac{F}{b-1}\right)^{[(k-1)/2]-1}\left(1+\frac{F}{b-1}\right)^{-[b(k-1)/2]} d\left(\frac{F}{b-1}\right)$$

This is the central F-distribution. Although the above argument is quite heuristic, one may think that this gives us an outline of the mathematical theory of the randomization of the block design.

So far as the estimation is concerned, the randomized block design is valid when the experimenter chooses to assign his k variates to k plots in each block at random, as long as this is done before the experiment begins. However, if we are concerned with the test, intrablock variances should be homogeneous over b blocks in the sense that the coefficient of variation of the block variances of the plot effects is sufficiently small for large values of b. Leaves on one plant for virus inoculation, animals in one litter for nutritional trials, blood samples from one subject for comparison of alternative techniques for cell counting, wheels on one vehicle for comparison of tire durability, batches of insects tested on one day for effect of inserticides, and so forth, are commonly used as a block. *Any property of the plots that can be determined before an experiment begins may be used to group them into blocks, the experience of both experimenter and statiscian will suggest properties likely to be sufficiently closely associated with "yields" for their use to reduce or homogenize intra-block variances.*

2.4 Latin-Square Design, Greco-Latin Square Design

The *Latin square* is an arrangement for permitting two sets of block constraints, usually termed *rows* and *columns*, to be used simultaneously. Table 2.3 is an example of Latin-square arrangement for comparing five virus inoculations of plants, the plot being a single leaf. The Table shows how leaves on one plant and leaves of approximately the same size were made to act as blocks simultaneously.

TABLE 2.2

Analysis of Variance for a Randomized Block Design

Source of variation	Degrees of freedom	Sum of squares	Expectation of mean sum of squares
Block	$b - 1$	$\frac{1}{k} \sum B_{i}^2 - n\bar{x}^2$	$\sigma^2 + \frac{k \sum \beta_{i}^2}{b - 1}$
Treatment	$k - 1$	$\frac{1}{b} \sum T_{\alpha}^2 - n\bar{x}^2$	$\sigma^2 + \frac{b \sum \tau_{\alpha}^2}{k - 1}$
Residual or error	$(b - 1)(k - 1)$	By subtraction	σ^2
Total	$bk - 1$	$\sum x_{i}^2 - n\bar{x}^2$	

TABLE 2.3

Randomized Allocation of Treatments A, B, C, D, E in a Plant Virus Experiment of Latin-Square Design

		Leaf size				
		1	2	3	4	5
Plant No.	I	A	E	D	C	B
	II	C	D	A	B	E
	III	B	C	E	A	D
	IV	E	A	B	D	C
	V	D	B	C	E	A

Five plants with five leaves each were required. Each inoculation occurred once in each plant, once on each leaf size, and treatments give sets of totals (rows = plant, column = leaf size, treatment = letters) such that the corresponding contrasts are mutually orthogonal. Hence this is also an orthogonal design.

Latin squares exist for every value of t. The total numbers of squares of different sizes are

2×2	2
3×3	12
4×4	576
5×5	161,280
6×6	812,851,200
7×7	61,479,419,904,000

in which permutation of letters are counted separately. For instance,

$$\begin{matrix} A & B \\ B & A \end{matrix} \quad \text{and} \quad \begin{matrix} B & A \\ A & B \end{matrix}$$

are the two 2×2 Latin squares. The former goes into the latter either by the interchange of A and B or by the interchange of rows or columns.

Two Latin squares are said to belong to a *transformation set* if one goes into the other by permutation of letters or rows or columns, or by permutation of the roles of row, column, and letter. From one square one can derive $3!(t!)^3$ squares but they are not all distinct. The whole set of squares of *degree* t are divided into a certain number of transformation sets.

The transpose of a Latin square is also a Latin square that is called *the conjugate square*. A symmetric square is called a *self-conjugate square*.

For example, for $t = 3$, one can write down a Latin square

A	B	C
B	C	A
C	A	B

This is called *a standard square* and is self-conjugate. One can show that there is no other standard square for $t = 3$. Since any Latin square can be brought into a standard form first by a permutation of t columns and then fixing the first row and by a permutation of $(t - 1)$ rows, there are $t!(t - 1)!$ Latin squares derivable from a standard square by column and row permutations.

When $t = 4$, there are two transoformation sets; their standard squares are listed below :

First transformation set :

Three self-conjugate standard squares

A	B	C	D		A	B	C	D		A	B	C	D
B	A	D	C		B	C	D	A		B	D	A	C
C	D	B	A		C	D	A	B		C	A	D	B
D	C	A	B		D	A	B	C		D	C	B	A

Second transformation set :

One self-conjugate standard square

A	B	C	D
B	A	D	C
C	D	A	B
D	C	B	A

Therefore the total number is 4 × 4!3! = 4 × 24 × 6 = 576.

When $t = 5$, there are also two transformation sets. The first set contains 25 standard dquares and their conjugates (hence 50 standard squares), and the second set contains 6 self-conjugate standard squares. The total number of squares is 56 × 5! × 4! = 161,280.

The 5 × 5 Latin Squares

First transformation set : 25 standard squares and their conjugates

A B C D E	A B C D E	A B C D E	A B C D E	A B C D E
B A E C D	B A D E C	B A E C D	B A D E C	B C D E A
C D A E B	C E B A D	C E D A B	C D E A B	C E B A D
D E B A C	D C E B A	D C B E A	D E B C A	D A E B C
E C D B A	E D A C B	E D A B C	E C A B D	E D A C B

A B C D E	A B C D E	A B C D E	A B C D E	A B C D E
B C D E A	B C E A D	B C E A D	B C D E A	B C A E D
C E A B D	C D B E A	C A D E R	C A E B D	C E D A B
D A E C B	D E A C B	D E B C A	D E A C B	D A E B C
E D B A C	E A D B C	E D A B C	E D B A C	E D B C A

A B C D E	A B C D E	A B C D E	A B C D E	A B C D E
B C A E D	B D E C A	B D A E C	B D E C A	B D A E C
C D E B A	C A B E D	C E D B A	C A D E B	C E D B A
D E B A C	D E A B C	D C E A B	D E A B C	D A E C B
E A D C B	E C D A B	E A B C D	E C B A D	E C B A D

A B C D E	A B C D E	A B C D E	A B C D E	A B C D E
B D E A C	B D A E C	B D E A C	B E D A C	B E D A C
C E D B A	C E B A D	C E A B D	C A B E D	C A E B D
D C A E B	D A E C B	D C B E A	D C E B A	D C A E B
E A B C D	E C D B A	E A D C B	E D A C B	E D B C A

A B C D E	*A B C D E*	*A B C D E*	*A B C D E*	*A B C D E*
B E A C D	*B E D C A*	*B E D C A*	*B E A C D*	*B E D A C*
C D E B A	*C A E B D*	*C D E A B*	*C D B E A*	*C D A E B*
D A B E C	*D C A E B*	*D A B E C*	*D C E A B*	*D C E B A*
E C D A B	*E D B A C*	*E C A B D*	*E A D B C*	*E A B C D*

Second transformation set : Six self-conjugate standard squares

A B C D E	*A B C D E*	*A B C D E*	*A B C D E*	*A B C D E*	*A B C D E*
B C E A D	*B C D E A*	*B D E C A*	*B D A E C*	*B E D A C*	*B E A C D*
C E D B A	*C D E A B*	*C E B A D*	*C A E B D*	*C D B E A*	*C A D E B*
D A B E C	*D E A B C*	*D C A E B*	*D E B C A*	*D A E C B*	*D C E B A*
E D A C B	*E A B C D*	*E A D B C*	*E C D A B*	*E C A B D*	*E D B A C*

A complete collection of standard squares up to $t = 7$ is presented by Fisher and Yates [1]. The analysis of variance table for the Latin square design becomes as follows :

TABLE 2.4

Analysis of Variance of Latin Square Design

Source of variation	Degrees of Freedom	Sum of squares	Expectation of mean sum of squares
Rows	$t - 1$	$\frac{1}{t} \sum R_{i}^{2} - n\bar{x}^{2}$	$\sigma^{2} + \frac{t}{t-1} \sum \pi_{i}^{2}$
Columns	$t - 1$	$\frac{1}{t} \sum C_{j}^{2} - n\bar{x}^{2}$	$\sigma^{2} + \frac{t}{t-1} \sum \gamma_{j}^{2}$
Treatments	$t - 1$	$\frac{1}{t} \sum T_{\alpha}^{2} - n\bar{x}^{2}$	$\sigma^{2} + \frac{t}{t-1} \sum \tau_{\alpha}^{2}$
Errors	$(t - 1)(t - 2)$	By subtraction	σ^{2}
Total	$t^{2} - 1$	$\sum x_{f}^{2} - n\bar{x}^{2}$	

Given two Latin squares of the same degree, if one is superimposed on the other, all combinations of letters -- taking the order into account -- occur exactly once, and then the two squares are said to be *orthogonal to each other*. If one of the two orthogonal Latin squares is superimposed on the other, this is sometimes called *a Greco-Latin square*.

5 × 5 Greco-Latin Square

$A\ \beta$	$E\ \gamma$	$D\ \alpha$	$C\ \delta$	$B\ \varepsilon$
$C\ \alpha$	$D\ \varepsilon$	$A\ \gamma$	$B\ \beta$	$E\ \delta$
$B\ \gamma$	$A\ \delta$	$C\ \varepsilon$	$E\ \alpha$	$D\ \beta$
$D\ \delta$	$B\ \alpha$	$E\ \beta$	$A\ \varepsilon$	$C\ \gamma$
$E\ \varepsilon$	$C\ \beta$	$B\ \delta$	$D\ \gamma$	$A\ \alpha$

One can consider *super Greco-Latin squares*.

A Completely Orthogonalized 4 × 4 Latin Square

111	222	333	444
234	143	412	321
342	431	124	213
423	314	241	132

One can write this down as follows :

11111 12222 13333 14444 21234 22143 23412 24321

31342 32431 33124 34213 41423 42314 43241 44132

A complete set of orthogonal Latin squares of order 5 is as follows :

1111	2222	3333	4444	5555
2345	3451	4512	5123	1234
3524	4135	5241	1352	2413
4253	5314	1425	2531	3142
5432	1543	2154	3215	4321

There are no orthogonal Latin squares for $t = 6$ (Euler's 36 officers problem).

A complete set of orthogonal Latin squares of order 7 is as follows:

111111	222222	333333	444444	555555	666666	777777
234567	345671	456712	567123	671234	712345	123456
357246	461357	572461	613572	724613	135724	246135
473625	514736	625147	736251	147362	251473	362514
526374	637415	741526	152637	263741	374152	415263
642753	753164	164275	275316	316427	427531	531642
765432	176543	217654	321765	432176	543217	654321

The analysis of variance table for a $t \times t$ Greco-Latin square design is presented in the following Table 2.5.

One can show that there are $n - 1$ mutually orthogonal Latin squares of order n if $n = p^m$ (prime power). This is called *a complete set of orthogonal Latin squares*.

2.5 Randomization in Latin-Square Design

There are k^2 experimental units or plots that are arranged in a $k \times k$ square. Suppose that we are given k treatments to be compared with each other by experimentation. An arrangement is said to be a *Latin-square design of order* k if each treatment occurs in

TABLE 2.5

Analysis of Variance for a $t \times t$ Greco-Latin Square

Source of variation	Degrees of Freedom	Sum of Squares
Rows	$t - 1$	$\frac{1}{t} \sum R_{i}^{2} - n\bar{x}^2$
Columns	$t - 1$	$\frac{1}{t} \sum C_{j}^{2} - n\bar{x}^2$
Latin letters	$t - 1$	$\frac{1}{t} \sum T_{\alpha}^{2} - n\bar{x}^2$
Greek letters	$t - 1$	$\frac{1}{t} \sum S_{\alpha}^{2} - n\bar{x}^2$
Error	$(t - 1)(t - 3)$	By subtraction
Total	$t^2 - 1$	$\sum x_{f}^{2} - n\bar{x}^2$

each row and in each column exactly once. We number the whole experimental units from 1 to k^2 in some way. A special numbering system [i.e., for the fth experimental unit, which is located at the intersection of the ith row and the pth column, $f = (i - 1)k + p$], will be used later.

We assume that there are no interactions between treatments and the experimental units at all. Let the fertility of the fth plot be π_{f}^{*}, and let

$$k^2\bar{\pi}^* \equiv \sum_{f=1}^{k^2} \pi_{f}^{*},$$

$$k\, r_i \equiv \sum_{f \text{ in the } i\text{th row}} \pi_f^* - k\, \bar{\pi}^* \qquad (i = 1, 2, \ldots, k),$$

$$k\, c_p \equiv \sum_{f \text{ in the } p\text{th column}} \pi_f^* - k\, \bar{\pi}^* \qquad (p = 1, 2, \ldots, k),$$

and finally let

$$\pi_f \equiv \pi_f^* - \bar{\pi}^* - r_i - c_p, \quad \text{if } f = (i - 1)k + p.$$

The terms $\bar{\pi}^*$, r_i, and c_p are called *the general mean, the row effect* of the ith row and *the column effect* of the pth column of fertility respectively. The term π_f is called *the plot effect* of the fth plot, and that may be considered as an interaction between row and column. It is clear that

$$\sum_{f \text{ in the } i\text{th row}} \pi_f = \sum_{f \text{ in the } p\text{th column}} \pi_f = 0$$

and

$$\sum_{i=1}^{k} r_i = \sum_{p=1}^{k} c_p = 0.$$

The observation at the fth plot is denoted by x_f and the n-dimensional vector $\underline{x}$ whose fth component is x_f is called *the observation vector*. The n-dimensional vector $\underline{\pi}$ whose fth component is π_f is called *the plot-effect vector*. The k-dimensional vectors

$$\underline{r} = \begin{bmatrix} r_1 \\ r_2 \\ \vdots \\ r_k \end{bmatrix} \quad \text{and} \quad \underline{c} = \begin{bmatrix} c_1 \\ c_2 \\ \vdots \\ c_k \end{bmatrix}$$

are called the row- and column-effect vectors, respectively.

The theoretical model underlying our analysis of Latin-square design of order k will be

$$(2.5.1) \qquad \underline{x} = g\,\underline{1} + \underline{\bar{\phi}}\,\underline{t} + \Psi_1\underline{r} + \Psi_2\underline{c} + \underline{\pi} + \underline{e},$$

where g is the general mean, $\underline{t}$ is the treatment-effect vector satisfying the relation $\sum_{\alpha=1}^{k^2} = 0$, $\underline{e}' = (e_1, e_2, \ldots, e_{k^2})$ is *the error vector* distributed as $N(\underline{0}, \sigma^2 I_{k^2})$, and $\underline{1}' = (1,1,\ldots,1)$; $\underline{\bar{\phi}}$ is the incidence matrix for treatments :

$$\underline{\bar{\phi}} = \|\underline{\zeta}_1, \underline{\zeta}_2, \ldots, \underline{\zeta}_k\|,$$

where

$$\underline{\zeta}_\alpha = \begin{bmatrix} \zeta_{\alpha 1} \\ \zeta_{\alpha 2} \\ \vdots \\ \zeta_{\alpha k^2} \end{bmatrix},$$

$$\underline{\zeta}_{\alpha f} = \begin{cases} 1, & \text{if the } \alpha\text{th treatment occurs at the } f\text{th plot,} \\ 0, & \text{otherwise,} \end{cases}$$

In (2.5.1) $\underline{\Psi}_1$ and $\underline{\Psi}_2$ are incidence matrices for rows and columns, respectively :

$$\Psi_1 = \left\| \underline{\eta}_1^{(1)}, \underline{\eta}_2^{(1)}, \ldots, \underline{\eta}_k^{(1)} \right\|,$$

$$\Psi_2 = \left\| \underline{\eta}_1^{(2)}, \underline{\eta}_2^{(2)}, \ldots, \underline{\eta}_k^{(2)} \right\|,$$

where

$$\underline{\eta}_i^{(1)} = \begin{bmatrix} \eta_{i1}^{(1)} \\ \eta_{i2}^{(1)} \\ \vdots \\ \eta_{ik^2}^{(1)} \end{bmatrix}$$

$$\eta_{if}^{(1)} = \begin{cases} 1 \text{ if the } f\text{th plot belongs to the } i\text{th row,} \\ 0 \text{ otherwise,} \end{cases}$$

and

$$\underline{\eta}_p^{(2)} = \begin{bmatrix} \eta_{p1}^{(2)} \\ \eta_{p2}^{(2)} \\ \vdots \\ \eta_{pk^2}^{(2)} \end{bmatrix}$$

$$\eta_{pf}^{(2)} = \begin{cases} 1 \text{ if the } f\text{th plot belongs to the } p\text{th column,} \\ 0 \text{ otherwise.} \end{cases}$$

Let us denote the treatment sums, row sums, and column sums by

$$T_\alpha = \underline{\zeta}'_\alpha \underline{x} \qquad (\alpha = 1, 2, \ldots, k),$$

$$R_i = \underline{\eta}_i^{(1)'} \underline{x} \qquad (i = 1, 2, \ldots, k),$$

and

$$C_p = \underline{\eta}_p^{(2)'} \underline{x} \qquad (p = 1, 2, \ldots, k),$$

respectively. The sum of squares due to treatment $S_t{}^2$ is given by

$$S_t{}^2 = \frac{1}{k} \sum_{\alpha=1}^{k} T_\alpha{}^2 - k^2 \bar{x}^2,$$

and the sum of squares due to error $s_e{}^2$ is given by

$$S_e{}^2 = \sum_{f=1}^{k^2} x_f{}^2 - k^2 \bar{x}^2 - \left(\frac{1}{k} \sum_{\alpha=1}^{k} T\alpha^2 - k^2 - x^2\right)$$

$$- \left(\frac{1}{k} \sum_{i=1}^{k} R_i{}^2 - k^2 \bar{x}^2\right) - \left(\frac{1}{k} \sum_{p=1}^{k} C_p{}^2 - k^2 \bar{x}^2\right),$$

where

$$\bar{x} = \frac{1}{k^2} \sum_{f=1}^{k^2} x_f.$$

As is well known, if there are no interactions between the row effects and the column effects (i.e. $\underline{\pi} = \underline{0}$), then the sampling distribution of the statistic

$$F = \frac{k^2 - 3k + 2}{k - 1} \frac{S_t^2}{S_e^2} = (k - 2)\frac{S_t^2}{S_e^2}$$

under the null hypothesis $H_0 : \underline{t} = \underline{0}$ is the F-distribution with $(k - 1, k^2 - 3k + 2)$ degrees of freedom. However, in most practical cases there do exist interactions or plot effects. In such cases the sampling distribution of the statistic F is no longer the central F-distribution even under the null hypothesis H_0. It will be shown that the null distribution of the statistic is the noncentral F-distribution whose noncentrality parameter depends upon the specific Latin square under consideration and the plot effect vector $\underline{\pi}$.

The randomization procedure has been recommended in order to get rid of the unknown plot effect vector $\underline{\pi}$; in other words take a Latin square of order k at random from the whole set of all possible Latin square of order k and apply the experimentation. It should be noted that due to the randomization procedure the incidence matrix for treatments Φ becomes random variable (discrete).

The author has shown that the null distribution of the statistic that is a multiple of the ratio of the sum of squares due to treatments to the sum of squares due to error is the central F-distribution in a good approximation in the case of randomized block design [2], and a balanced incomplete block design [3], provided that the variances of the plot effects within blocks are nearly uniform.

Our purpose is to examine the same question in the case of a Latin square design under model (2.5.1), the Neyman model [4]. This problem was treated by Welch [5] under the Fisher model [i.e., the Fisher model amounts to $\underline{e} = \underline{0}$ in (2.5.1)].

From the definitions of incidence matrices it follows that

$$(2.5.2) \quad \underline{\bar{\Phi}}'\underline{\bar{\Phi}} = kI_k, \quad \Psi_1'\Psi_1 = kI_k, \quad \Psi_2'\Psi_2 = kI_k.$$

Due to the property of the Latin square arrangement, we have

$$(2.5.3) \quad \underline{\bar{\Phi}}'\Psi_1 = \underline{\bar{\Phi}}'\Psi_2 = \Psi_1'\Psi_2 = \Psi_2'\Psi_1 = G_k,$$

where G_k denotes the $k \times k$ matrix whose elements are all unity.

Let

$$T \equiv \underline{\bar{\Phi}}\,\underline{\bar{\Phi}}', \quad R \equiv \Psi_1 \Psi_1', \quad G \equiv \Psi_2 \Psi_2'.$$

Then it will be seen from (2.5.3) that

$$T^2 = kT, \quad R^2 = kR, \quad G^2 = kG.$$

Further we have

$$TR = RT = G, \quad TC = CT = G, \quad RC = CR = G$$

due to (2.5.3), where G is the $k^2 \times k^2$ matrix whose elements are all unity.

It is easily confirmed that

$$S_t^2 = \frac{1}{k}\sum_{\alpha=1}^{k} T_\alpha^2 - k^2\bar{x}^2 = \underline{x}'\left(\frac{1}{k}T - \frac{1}{k^2}G\right)\underline{x},$$

$$S_r^{\,2} = \frac{1}{k}\sum_{i=1}^{k} R_{i}^{\,2} - k^2\bar{x}^2 = \underline{x}'\left(\frac{1}{k}R - \frac{1}{k^2}G\right)\underline{x},$$

$$S_c^{\,2} = \frac{1}{k}\sum_{p=1}^{k} C_p^{\,2} - k^2\bar{x}^2 = \underline{x}'\left(\frac{1}{k}C - \frac{1}{k^2}G\right)\underline{x},$$

and hence it follows by subtraction that

$$S_e^{\,2} = \underline{x}'\left(I_{k^2} - \frac{1}{k}T - \frac{1}{k}R - \frac{1}{k}C + \frac{2}{k^2}G\right)\underline{x}.$$

The preceding four sums of squares are all quadratic forms with respect to the observations, and it will be seen that their coefficient matrices are mutually orthogonal idempotents.

Now, under the null hypothesis $H_0 : \underline{t} = \underline{0}$, we have

$$S_t^{\,2} = \underline{e}'\left(\frac{1}{k}T - \frac{1}{k^2}G\right)\underline{e} + 2\underline{e}'\left(\frac{1}{k}T - \frac{1}{k^2}G\right)\underline{\pi} + \underline{\pi}'\left(\frac{1}{k}T - \frac{1}{k^2}G\right)\underline{\pi}, \tag{2.5.4}$$

and

$$\begin{aligned} S_e^{\,2} &= \underline{e}'\left(I - \frac{1}{k}T - \frac{1}{k}R - \frac{1}{k}C + \frac{2}{k^2}G\right)\underline{e} \\ &\quad + 2\underline{e}'\left(I - \frac{1}{k}T - \frac{1}{k}R - \frac{1}{k}C + \frac{2}{k^2}G\right)\underline{\pi} \\ &\quad + \underline{\pi}'\left(I - \frac{1}{k}T - \frac{1}{k}R - \frac{1}{k}C + \frac{2}{k^2}G\right)\underline{\pi}. \end{aligned} \tag{2.5.5}$$

It should be remarked that (2.5.5) is true even in non-null case.

For a specific Latin square design of order k, S_t^2 and S_e^2 are mutually independent in the stochastic sense [6]. The distribution of the variate

$$\chi_1^2 = \frac{S_t^2}{\sigma^2}$$

is the noncentral χ^2-distribution with $(k - 1)$ degrees of freedom the noncentrality parameter

$$\lambda_1 = \frac{1}{2\sigma^2} \frac{1}{k} \underline{\pi}' T \underline{\pi} = \frac{\Delta}{2\sigma^2} \theta,$$

say, where $\Delta = \underline{\pi}'\underline{\pi}$, that is,

$$exp\left(-\frac{\lambda_1}{2}\right) \sum_{\mu=0}^{\infty} \frac{\left(\frac{\lambda_1}{2}\right)^{\mu}}{\mu!} \frac{\left(\frac{\chi_1^2}{2}\right)^{[(k-1)/2]+\mu-1}}{\Gamma\left(\frac{k-1}{2} + \mu\right)} exp\left(-\frac{\chi_1^2}{2}\right) d\left(\frac{\chi_1^2}{2}\right) .$$

The distribution of the variate

$$\chi_2^2 = \frac{S_e^2}{\sigma^2}$$

is the noncentral χ^2-distribution with $(k^2 - 3k + 2)$ degrees of freedom and with the noncentrality parameter

$$\lambda_2 = \frac{\Delta}{\sigma^2} (1 - \theta),$$

that is,

$$\exp\left(-\frac{\lambda_2}{2}\right) \sum_{\nu=0}^{\infty} \frac{\left(\frac{\lambda_2}{2}\right)^{\nu}}{\nu!} \frac{\left(\frac{\chi_2^2}{2}\right)^{[(k^2-3k+2)/2]+\nu-1}}{\Gamma\left(\frac{k -3k+2}{2}+\nu\right)} \exp\left(-\frac{\chi_2^2}{2}\right) d\left(\frac{\chi_2^2}{2}\right)$$

Thus the null distribution of the F-statistic for a specific Latin square of order k turns out to be the so-called noncentral F-distribution;

$$\frac{\Gamma\left(\frac{k^2-2k+1}{2}\right)}{\Gamma\left(\frac{k-1}{2}\right)\Gamma\left(\frac{k^2-3k+2}{2}\right)} \left(\frac{F}{k-2}\right)^{[(k-1)/2]-1}\left(1+\frac{F}{k-2}\right)^{-[(k^2-2k+1)/2]}$$

$$\times\, d\left(\frac{F}{k-2}\right) \exp\left(-\frac{\Delta}{2\sigma^2}\right) \sum_{\tau=0}^{\infty} \frac{\left[\frac{\Delta}{2\sigma^2}\right]^{\tau}}{\tau!} \left(1+\frac{F}{k-2}\right)^{-\tau} \sum_{\mu+\nu=\tau} \frac{\tau!}{\mu!\,\nu!} \left(\frac{F}{k-2}\right)^{\mu}$$

$$\times\, \theta^{\mu}(1-\theta)^{\nu} \frac{\Gamma\left(\frac{k^2-2k+1}{2}+\tau\right)\Gamma\left(\frac{k-1}{2}\right)\Gamma\left(\frac{k^2-3k+2}{2}\right)}{\Gamma\left(\frac{k-1}{2}+\mu\right)\Gamma\left(\frac{k^2-3k+2}{2}+\nu\right)\Gamma\left(\frac{k^2-2k+1}{2}\right)}.$$

If the incidence matrix Φ and therefore $T = \Phi\,\Phi'$ become random variable due to randomization, then

$$\theta = \frac{1}{k\Delta}\,\underline{\pi}'T\underline{\pi}$$

is a discrete random variable in the range $0 \leqq \theta \leqq 1$. If the permutation distribution of θ due to randomization can be approximated by the following β-distribution

(2.5.7)

$$\frac{\Gamma\left(\frac{k^2 - 2k + 1}{2}\right)}{\Gamma\left(\frac{k - 1}{2}\right)\Gamma\left(\frac{k^2 - 3k + 2}{2}\right)} \theta^{[(k-1)/2]-1} (1 - \theta)^{[(k^2-3k+2)/2]-1} d\theta,$$

then by averaging (2.5.6) with respect to θ, we get the central F-distribution with degrees of freedom $(k - 1, k^2 - 3k + 2)$, that is

$$\frac{\Gamma\left(\frac{k^2 - 2k + 1}{2}\right)}{\Gamma\left(\frac{k - 1}{2}\right)\Gamma\left(\frac{k^2 - 3k + 2}{2}\right)} \left(\frac{F}{k - 2}\right)^{[(k-1)/2]-1} \left(1 + \frac{F}{k - 2}\right)^{-[(k^2-2k+1)/2]} \times d\left(\frac{F}{k - 2}\right),$$

which is commonly met in practice.

We shall examine the meaning of the approximation (2.5.7) to the permutation distribution of θ due to randomization.

An example of a Latin square of order 4 is as follows :

A	*B*	*C*	*D*
B	*C*	*D*	*A*
C	*D*	*A*	*B*
D	*A*	*B*	*C*

Four Latin letters A, B, C, D instead of four treatments are so arranged in a 4×4 square that each Latin letter appears in each row and in each column exactly once. Furthermore, the above special Latin square is constructed by the cyclic permutations of the first row. Any permutation of the rows, columns, letters and the roles of these three items of a Latin square leaves it still a Latin square. Therefore there are $3!(k!)^3$ Latin squares derivable by permutation from a given Latin square of order k. However, these $3!(k!)^3$ Latin squares are not necessarily different. The set of all Latin squares obtained by a permutation of the rows, columns, letters and the roles of these three items of a Latin square from one square is called a *transformation set*. In general the set of all Latin squares of order k is divided into several transformation sets. For example, the set of all Latin squares of order 4 or 5 is divided into two transformation sets.

Suppose that a Latin square of order k is given. Then by a suitable permutation of columns the k letters in the first row are arranged in their natural order $A\ B\ C \cdots K$. Then by a suitable permutation of $k - 1$ rows (except the first) the k letters in the first column can also be arranged in their natural order $A\ B\ C \cdots K$. The Latin square thus obtained is called *a standard square*. For instance, the above-mentioned Latin square of order 4 is a standard square. Conversely, we get $k!(k - 1)!$ different Latin squares of order k by permuting the rows and columns of a standard square. Thus the total number of different Latin squares in a transformation set is $k!(k - 1)!$ times the number of different standard squares contained in that transformation set.

Now we shall calculate the mean value of θ with respect to the permutation distribution due to randomization.

Let $$\pi_f = \pi_p^{(i)} \quad \text{if} \quad f = (i - 1)k + p$$

and

$$\underline{\pi} = \begin{bmatrix} \underline{\pi}^{(1)} \\ \underline{\pi}^{(2)} \\ \vdots \\ \underline{\pi}^{(k)} \end{bmatrix}, \qquad \underline{\pi}^{(i)} = \begin{bmatrix} \pi_1^{(i)} \\ \pi_2^{(i)} \\ \vdots \\ \pi_k^{(i)} \end{bmatrix}.$$

Let us denote the variance of the plot effects in the ith row by

$$\Delta^{(i)} = \underline{\pi}^{(i)\prime}\, \underline{\pi}^{(i)}.$$

Let us define the plot-effect vector of the pth column by

$$\underline{\pi}_p = \begin{bmatrix} \pi_p^{(1)} \\ \pi_p^{(2)} \\ \vdots \\ \pi_p^{(k)} \end{bmatrix},$$

and let us denote the variance of the plot effects in the pth column by

$$\Delta_p = \underline{\pi}_p^{\prime}\, \underline{\pi}_p.$$

It is clear that

$$\Delta = \underline{\pi}'\, \underline{\pi} = \sum_{i=1}^{k} \Delta^{(i)} = \sum_{p=1}^{k} \Delta_p.$$

Let the $k \times k$ matrix whose elements are 0 or 1 corresponding to a permutation

$$\sigma = \begin{pmatrix} 1 & 2 & \cdots & k \\ \sigma(1) & \sigma(2) & \cdots & \sigma(k) \end{pmatrix}$$

be S_σ;

$$S_\sigma \begin{bmatrix} 1 \\ 2 \\ \vdots \\ k \end{bmatrix} = \begin{bmatrix} \sigma(1) \\ \sigma(2) \\ \vdots \\ \sigma(k) \end{bmatrix}, \qquad S'_\sigma \begin{bmatrix} \sigma(1) \\ \sigma(2) \\ \vdots \\ \sigma(k) \end{bmatrix} = \begin{bmatrix} 1 \\ 2 \\ \vdots \\ k \end{bmatrix}.$$

Since an application of a permutation σ to k treatments is equivalent to the transformation

$$\underline{t}^* = S_\sigma \underline{t}, \qquad S_\sigma = \left\| s_{ij}^\sigma \right\| \qquad (i,j = 1,\ldots,k),$$

the model to be considered should be

$$\underline{x} = g\,\underline{1} + \overline{\underline{\phi}}\, S'_\sigma \underline{t}^* + \Psi_1 \underline{r} + \Psi_2 \underline{c} + \underline{\pi} + \underline{e}.$$

In this case the quantity corresponding to $\underline{\pi}' T\, \underline{\pi}$ is

$$\underline{\pi}' T^* \underline{\pi} = \underline{\pi}' \Phi\, S'_\sigma\, S_\sigma \Phi' \underline{\pi} = \underline{\pi}' \Phi\, \Phi' \underline{\pi} = \underline{\pi}' T\, \underline{\pi},$$

which is the same as before. Thus in order to calculate the mean of $\underline{\pi}' T\, \underline{\pi}$ in a transformation set, it is sufficient to consider the Latin squares transformable with each other by a permutation of treatments as identical.

Let the transformation sets of Latin squares of order k be $\mathcal{T}_1, \mathcal{T}_2, \ldots, \mathcal{T}_k$, containing $n_1, n_2, \ldots, n_k$ $(n = \sum_{\nu=1}^{k} n_\nu)$ different standard squares, respectively. Then

$$E(\underline{\pi}'T\,\underline{\pi}) = \frac{1}{n}\sum_{\nu=1}^{k} n_\nu\, E_\nu(\underline{\pi}'T\,\underline{\pi}),$$

where E_ν denotes the averaging operator in the transformation set $\mathcal{T}_\nu$.

Now we shall start from a Latin square transformable by the permutations of rows and columns from a specific standard square belonging to the transformation set $\mathcal{T}_\nu$. Let

$$U_\sigma = \begin{Vmatrix} S_\sigma & & & 0 \\ & S_\sigma & & \\ & & \ddots & \\ 0 & & & S_\sigma \end{Vmatrix} \quad \text{and} \quad V_\tau = \begin{Vmatrix} s_{11}^\tau & 0 & s_{1k}^\tau & 0 \\ \ddots & s_{11}^\tau & \cdots & \ddots\; s_{1k}^\tau \\ \vdots & & & \vdots \\ s_{k1}^\tau & 0 & s_{kk}^\tau & 0 \\ \ddots & \cdots & & \ddots \\ 0 & s_{k1}^\tau & 0 & s_{kk}^\tau \end{Vmatrix} = I_k \times S_\tau',$$

where σ runs over $\mathfrak{S}_{k-1}$ consisting of all permutations of $2,3,\ldots,k$ and τ runs over $\mathfrak{S}_k$ consisting of all permutations of $1,2,\ldots,k$, and further let

$$t_{gf} = t_{pq}^{ij} \quad \text{if} \quad f = (i-1)k + p, \quad g = (j-1)k + q.$$

We have the average for this subset of $\mathcal{T}_\nu$

$$E_{\nu}^{(\alpha)}\ (\underline{\pi}'T\ \underline{\pi}) = \frac{1}{k!(k-1)!} \sum_{\alpha\in\mathfrak{S}_{k-1}} \sum_{\tau\in\mathfrak{S}_k} \underline{\pi}'U_{\sigma}'V_{\tau}'TV_{\tau}U_{\sigma}\underline{\pi}$$

and

$$E_{\nu}(\underline{\pi}'T\ \underline{\pi}) = \frac{1}{n_{\nu}} \sum_{\alpha=1}^{n_{\nu}} E_{\nu}^{(\alpha)}\ (\underline{\pi}'T\ \underline{\pi}) .$$

First of all we shall consider the average with respect to τ. Since

$$\frac{1}{k!} \sum_{\tau\in\mathfrak{S}_k} \underline{\pi}'V_{\tau}'TV_{\tau}\underline{\pi} = \frac{1}{k!} \sum_{\tau\in\mathfrak{S}_k} (V_{\tau}\ \underline{\pi})'\ T(V_{\tau}\ \underline{\pi})$$

$$= \frac{1}{k!} \sum_{p,q=1}^{k} \sum_{i,j=1}^{k} t_{pq}^{ij} \sum_{\tau\in\mathfrak{S}_k} \pi_p^{\tau(i)}\pi_q^{\tau(i)}$$

and

$$\frac{1}{k!} \sum_{\tau\in\mathfrak{S}_k} \pi_p^{\tau(i)}\pi_q^{\tau(j)} = \begin{cases} \dfrac{1}{k} \displaystyle\sum_{l=1}^{k} \pi_p^{(l)}\pi_q^{(l)} & \text{if } i = j, \\[2ex] \dfrac{1}{k(k-1)} \displaystyle\sum_{l\neq m} \pi_p^{(l)}\pi_q^{(m)} & \text{if } i \neq j, \end{cases}$$

It follows that

$$\frac{1}{k!} \sum_{\tau \in \mathfrak{S}_k} (V_\tau \underline{\pi})' T (V_\tau \underline{\pi}) = \frac{1}{k} \sum_{i=1}^{k} \sum_{p,q=1}^{k} t_{pq}^{ii} \sum_{l=1}^{k} \pi_p^{(l)} \pi_q^{(m)}$$

$$+ \frac{1}{k(k-1)} \sum_{i \neq j} \sum_{p,q=1}^{k} t_{pq}^{ij} \sum_{l \neq m} \pi_p^{(l)} \pi_q^{(m)}.$$

By the very nature of the Latin square, it is seen that

$$t_{pq}^{ii} = \delta_{pq}, \qquad \sum_{i \neq j} t_{pq}^{ij} = k(1 - \delta_{pq}).$$

Hence we have

$$\frac{1}{k!} \sum_{\tau \in \mathfrak{S}_k} (V_\tau \underline{\pi})' T (V_\tau \underline{\pi}) = \frac{k}{k-1} \Delta,$$

and therefore

$$E^{(\alpha)} (\underline{\pi}' T \, \underline{\pi}) = \frac{k}{k-1} \Delta.$$

Consequently

$$E_\nu (\underline{\pi}' T \, \underline{\pi}) = \frac{1}{n_\nu} \sum_{\alpha=1}^{n_\nu} E_\nu^{(\alpha)} (\underline{\pi}' T \, \underline{\pi}) = \frac{k}{k-1} \Delta$$

or

$$E_\nu(\theta) = \frac{1}{k-1} .$$

Finally we get

$$E(\theta) = \frac{1}{k - 1}.$$

This coincides with the first moment of the β-distribution given by (2.5.7). Thus in this sense (2.5.7) is an approximation to the permutation distribution of θ due to randomization.

In the case of Latin-square design the variance of θ with respect to its permutation distribution is quite complicated, as will be shown in appendix to this chapter, and it seems hard to give any clear-cut statistical interpretation of the variance of θ in the situation under consideration.

2.6. Appendix : Calculation of the Variance of $\underline{\pi}'T\,\underline{\pi}$ with Respect to Its Permutation Distribution Due to Randomization

In calculating the average

$$E_{\nu}^{(\alpha)}(\underline{\pi}'T\,\underline{\pi})^2 = \frac{1}{k!(k-1)!} \sum_{\tau \in \mathfrak{S}_{k-1}} \sum_{\sigma \in \mathfrak{S}_k} (\underline{\pi}'U_\sigma' V_\tau' T V_\tau U_\sigma \underline{\pi})^2$$

there will be no harm to consider that τ and σ run independently over all the $\mathfrak{S}_k$. Thus

$$E_{\nu}^{(\alpha)}(\underline{\pi}'T\,\underline{\pi})^2 = \frac{1}{(k!)^2} \sum_{\tau \in \mathfrak{S}_k} \sum_{\sigma \in \mathfrak{S}_k} (\underline{\pi}'U_\sigma' V_\tau' T V_\tau U_\sigma \underline{\pi})^2.$$

Now,

$$(V_\tau U_\sigma \underline{\pi})' T (V_\tau U_\sigma \underline{\pi}) = \sum_{p,q} \sum_{i,j} t_{pq}^{ij} \pi_{\sigma(p)}^{\tau(i)} \pi_{\sigma(q)}^{\tau(j)}$$

$$= \sum_{p} \left[\sum_{i} t_{pp}^{ii} \pi_{\sigma(p)}^{\tau(i)^2} + \sum_{i \neq j} t_{pp}^{ij} \pi_{\sigma(p)}^{\tau(i)} \pi_{\sigma(p)}^{\tau(j)} \right]$$

$$+ \sum_{p \neq q} \left[\sum_{i} t_{pq}^{ii} \pi_{\sigma(p)}^{\tau(i)} \pi_{\sigma(q)}^{\tau(i)} + \sum_{i \neq j} t_{pq}^{ij} \pi_{\sigma(p)}^{\tau(i)} \pi_{\sigma(q)}^{\tau(j)} \right]$$

$$= \sum_{p} \sum_{i} \pi_{\sigma(p)}^{\tau(i)^2} + \sum_{p \neq q} \sum_{i \neq j} t_{pq}^{ij} \pi_{\sigma(p)}^{\tau(i)} \pi_{\sigma(q)}^{\tau(j)}.$$

Hence we have

$$[(V_\tau U_\sigma \underline{\pi})' T (V_\tau U_\sigma \underline{\pi})]^2 = \left[\sum_{p} \sum_{i} \pi_{\sigma(p)}^{\tau(i)^2} \right]^2$$

$$+ 2 \left[\sum_{p} \sum_{i} \pi_{\sigma(p)}^{\tau(i)^2} \right] \left[\sum_{p \neq q} \sum_{i \neq j} t_{pq}^{ij} \pi_{\sigma(p)}^{\tau(i)} \pi_{\sigma(q)}^{\tau(j)} \right]$$

$$+ \left[\sum_{p \neq q} \sum_{i \neq j} t_{pq}^{ij} \pi_{\sigma(p)}^{\tau(i)} \pi_{\sigma(p)}^{\tau(j)} \right]^2.$$

I. Calculation of $E_\nu^{(\alpha)} \left(\sum_p \sum_i \pi_{\sigma(p)}^{\tau(i)^2} \right)$:

$$\left[\sum_p \sum_i \pi_{\sigma(p)}^{\tau(i)^2}\right]^2 = \sum_{p,q} \sum_{i,j} \pi_{\sigma(p)}^{\tau(i)^2} \pi_{\sigma(q)}^{\tau(j)^2}$$

$$= \sum_p \left[\sum_i \pi_{\sigma(p)}^{\tau(i)^4} + \sum_{i \neq j} \pi_{\sigma(p)}^{\tau(i)^2} \pi_{\sigma(p)}^{\tau(j)^2}\right]$$

$$+ \sum_{p \neq q} \left[\sum_i \pi_{\sigma(p)}^{\tau(i)^2} \pi_{\sigma(q)}^{\tau(i)^2} + \sum_{i \neq j} \pi_{\sigma(p)}^{\tau(i)^2} \pi_{\sigma(q)}^{\tau(j)^2}\right].$$

Here we shall introduce the following notations after Welch [5] :

$$D \equiv \sum_p \sum_i \pi_p^{(i)^4},$$

$$F \equiv \Delta^2 = \left(\sum_p \sum_i \pi_p^{(i)^2}\right)^2,$$

$$G \equiv \sum_i \left(\sum_p \pi_p^{(i)^2}\right)^2 + \sum_p \left(\sum_i \pi_p^{(i)^2}\right)^2 = \sum_i \Delta^{(i)^2} + \sum_p \Delta_p^2 ,$$

$$H \equiv \sum_{i,j} \left(\sum_p \pi_p^{(i)} \pi_p^{(i)}\right)^2 = \sum_{i \neq j} \left(\sum_p \pi_p^{(i)} \pi_p^{(i)}\right)^2 + \sum_i \Delta^{(i)^2} .$$

Then since

$$E_\nu^{(\alpha)}\left[\pi_{\sigma(p)}^{\tau(i)^4}\right] = \frac{1}{k^2} \sum_p \sum_i \pi_p^{(i)^4} = \frac{1}{k^2} D ,$$

$$E_\nu^{(\alpha)}\left[\pi_{\sigma(p)}^{\tau(i)2}\pi_{\sigma(p)}^{\tau(j)2}\right] = \frac{1}{k^2(k-1)}\sum_p\sum_{i\neq j}\pi_p^{(i)2}\pi_p^{(j)2} = \frac{1}{k^2(k-1)}\left(\sum_p \Delta_p^2 - D\right),$$

$$E_\nu^{(\alpha)}\left[\pi_{\sigma(p)}^{\tau(i)2}\pi_{\sigma(q)}^{\tau(i)2}\right] = \frac{1}{k^2(k-1)}\sum_{p\neq q}\sum_i \pi_p^{(i)2}\pi_q^{(i)2} = \frac{1}{k^2(k-1)}$$

$$\times \left[\sum_i \Delta^{(i)2} - D\right],$$

$$E_\nu^{(\alpha)}\left[\pi_{\sigma(p)}^{\tau(i)2}\pi_{\sigma(q)}^{\tau(j)2}\right] = \frac{1}{k^2(k-1)^2}\sum_{p\neq q}\sum_{i\neq j}\pi_p^{(i)2}\pi_q^{(j)2}$$

$$= \frac{1}{k^2(k-1)^2}(F - G + D),$$

we have

$$E_\nu^{(\alpha)}\left[\sum_p\sum_i \pi_{\sigma(p)}^{\tau(i)2}\right]^2 = F = \Delta^2.$$

II. Calculation of $E_\nu^{(\alpha)}\left[\sum_p\sum_i\pi_{\sigma(p)}^{\tau(i)2}\right]\left[\sum_{p\neq q}\sum_{i\neq j} t_{pq}^{ij}\right.$

$$\left.\times\ \pi_{\sigma(p)}^{\tau(i)}\ \pi_{\sigma(q)}^{\tau(j)}\right],$$

The general term in the expansion of

$$\left[\sum_p\sum_i \pi_{\sigma(p)}^{\tau(i)2}\right]\left[\sum_{p\neq q}\sum_{i\neq j} t_{pq}^{ij}\ \pi_{\sigma(p)}^{\tau(i)}\ \pi_{\sigma(q)}^{\tau(j)}\right]$$

is of the type

$$t_{pq}^{ij}\,\pi_{\sigma(p)}^{\tau(i)}\,\pi_{\sigma(q)}^{\tau(j)}\,\pi_{\sigma(r)}^{\tau(l)^2} \qquad (i \neq j,\quad p \neq q).$$

Therefore there are terms of the following nine types :

(i) $r = p,\quad l = i;$

(ii) $r = p,\quad l = j;$

(iii) $r = p,\quad l \neq i \neq j;$

(iv) $r = q,\quad l = i;$

(v) $r = q,\quad l = j;$

(vi) $r = q,\quad l \neq i \neq j;$

(vii) $r \neq p \neq q,\quad l = i;$

(viii) $r \neq p \neq q,\quad l = j;$

(ix) $r \neq p \neq q,\quad l \neq i \neq j.$

We calculate the averages of terms of those nine types as follows :

(i) $$E_{\nu}^{(\alpha)}\left[\pi_{\sigma(p)}^{\tau(i)^3}\,\pi_{\sigma(q)}^{\tau(j)}\right] = \frac{1}{k^2(k-1)^2}\sum_{p\neq q}\sum_{i\neq j}\pi_p^{(i)^3}\pi_q^{(j)}$$

$$= \frac{1}{k^2(k-1)^2}\sum_p\sum_i \pi_p^{(i)^4} = \frac{1}{k^2(k-1)^2}D,$$

(ii)
$$E_\nu^{(\alpha)}\left[\pi_{\sigma(p)}^{\tau(i)}\,\pi_{\sigma(q)}^{\tau(j)^2}\,\pi_{\sigma(p)}^{\tau(j)}\right] = \frac{1}{k^2(k-1)^2}\sum_{p\neq q}\sum_{i\neq j}\pi_p^{(i)}\pi_p^{(j)^2}\pi_q^{(j)}$$

$$= \frac{1}{k^2(k-1)^2}\sum_p\sum_i \pi_p^{(i)^4} = \frac{1}{k^2(k-1)^2}D,$$

(iii)
$$E_\nu^{(\alpha)}\left[\pi_{\sigma(p)}^{\tau(i)}\,\pi_{\sigma(q)}^{\tau(j)}\,\pi_{\sigma(p)}^{\tau(l)^2}\right] = \frac{1}{k^2(k-1)^2(k-2)}$$

$$\times \sum_{p\neq q}\sum_{i\neq j\neq l}\pi_p^{(i)}\pi_q^{(j)}\pi_p^{(l)^2} = \frac{1}{k^2(k-1)^2(k-2)}\left[\sum_p \Delta_p^2 - 2D\right],$$

(iv)
$$E_\nu^{(\alpha)}\left[\pi_{\sigma(p)}^{\tau(i)}\,\pi_{\sigma(q)}^{\tau(j)}\,\pi_{\sigma(q)}^{\tau(i)^2}\right] = \frac{1}{k^2(k-1)^2}\sum_{p\neq q}\sum_{i\neq j}\pi_p^{(i)}\pi_q^{(j)}\pi_q^{(i)^2}$$

$$= \frac{1}{k^2(k-1)^2}D,$$

(v)
$$E_\nu^{(\alpha)}\left[\pi_{\sigma(p)}^{\tau(i)}\,\pi_{\sigma(q)}^{\tau(j)^3}\right] = \frac{1}{k^2(k-1)^2}\sum_{p\neq q}\sum_{i\neq j}\pi_p^{(i)}\pi_q^{(j)^3}$$

$$= \frac{1}{k^2(k-1)^2}D,$$

(vi) $$E_\nu^{(\alpha)}\left[\pi_{\sigma(p)}^{\tau(i)}\pi_{\sigma(q)}^{\tau(j)}\pi_{\sigma(q)}^{\tau(l)^2}\right] = \frac{1}{k^2(k-1)^2(k-2)}$$

$$\times \sum_{p\neq q}\sum_{i\neq j\neq l} \pi_p^{(i)}\pi_q^{(j)}\pi_q^{(l)^2} = \frac{1}{k^2(k-1)^2(k-2)}\left[\sum_p \Delta_p^2 - 2D\right],$$

(vii) $$E_\nu^{(\alpha)}\left[\pi_{\sigma(p)}^{\tau(i)}\pi_{\sigma(q)}^{\tau(j)}\pi_{\sigma(r)}^{\tau(i)^2}\right] = \frac{1}{k^2(k-1)^2(k-2)}$$

$$\times \sum_{p\neq q\neq r}\sum_{i\neq j} \pi_p^{(i)}\pi_q^{(j)}\pi_r^{(i)^2} = \frac{1}{k^2(k-1)^2(k-2)}\left[\sum_i \Delta^{(i)^2} - 2D\right],$$

(viii) $$E_\nu^{(\alpha)}\left[\pi_{\sigma(p)}^{\tau(i)}\pi_{\sigma(q)}^{\tau(j)}\pi_{\sigma(r)}^{\tau(j)^2}\right] = \frac{1}{k^2(k-1)^2(k-2)}$$

$$\times \sum_{p\neq q\neq r}\sum_{i\neq j} \pi_p^{(i)}\pi_q^{(j)}\pi_r^{(j)^2} = \frac{1}{k^2(k-1)^2(k-2)}\left[\sum_i \Delta^{(i)^2} - 2D\right],$$

(ix) $$E_\nu^{(\alpha)}\left[\pi_{\sigma(p)}^{\tau(i)}\pi_{\sigma(q)}^{\tau(j)}\pi_{\sigma(r)}^{\tau(l)^2}\right] = \frac{1}{k^2(k-1)^2(k-2)^2}$$

$$\times \sum_{p\neq q\neq r}\sum_{i\neq j\neq l} \pi_p^{(i)}\pi_q^{(j)}\pi_r^{(l)^2} = \frac{1}{k^2(k-1)^2(k-2)^2}(F-2G+4D).$$

Since we have

$$\sum_{p \neq q} \sum_{i \neq j} t_{pq}^{ij} = k^2(k - 1),$$

it follows that

$$E_{\nu}^{(\alpha)}\left[\sum_{p} \sum_{i} \pi_{\sigma(p)}^{\tau(i)^2}\right]\left[\sum_{p \neq q} \sum_{i \neq j} t_{pq}^{ij} \pi_{\sigma(p)}^{\tau(i)} \pi_{\sigma(q)}^{\tau(j)}\right] = \frac{1}{k - 1} F.$$

III. Calculation of $E_{\nu}^{(\alpha)}\left[\sum_{p \neq q} \sum_{i \neq j} t_{pq}^{ij} \pi_{\sigma(p)}^{\tau(i)} \pi_{\sigma(q)}^{\tau(j)}\right]^2$

$$= E_{\nu}^{(\alpha)}\left[\sum_{p \neq q} \sum_{p' \neq q'} \sum_{i \neq j} \sum_{i' \neq j'} t_{pq}^{ij} t_{p'q'}^{i'j'} \pi_{\sigma(p)}^{\tau(i)} \pi_{\sigma(q)}^{\tau(j)} \pi_{\sigma(p')}^{\tau(i')} \pi_{\sigma(q')}^{\tau(j')}\right]$$

Since there are seven different types of the pairing of (p,q) and (p',q'), and also seven different types of the pairing of (i,j) and (i',j'), there are 49 terms of different types on the whole. First,

$$\left[\sum_{p \neq q} \sum_{i \neq j} t_{pq}^{ij} \pi_{\sigma(p)}^{\tau(i)} \pi_{\sigma(q)}^{\tau(j)}\right]\left[\sum_{p' \neq q'} \sum_{i' \neq j'} t_{p'q'}^{i'j'} \pi_{\sigma(p')}^{\tau(i')} \pi_{\sigma(q')}^{\tau(j')}\right]$$

$$= \sum_{p \neq q}\left[\sum_{i \neq j} t_{pq}^{ij} \pi_{\sigma(p)}^{\tau(i)} \pi_{\sigma(q)}^{\tau(j)}\right]\left[\sum_{i' \neq j'} t_{pq}^{i'j'} \pi_{\sigma(p)}^{\tau(i')} \pi_{\sigma(q)}^{\tau(j')}\right] \cdots\cdots\cdots\cdots (P)$$

$$+ \sum_{p \neq q}\left[\sum_{i \neq j} t_{pq}^{ij} \pi_{\sigma(p)}^{\tau(i)} \pi_{\sigma(q)}^{\tau(j)}\right]\left[\sum_{i' \neq j'} t_{qp}^{i'j'} \pi_{\sigma(q)}^{\tau(i')} \pi_{\sigma(p)}^{\tau(j')}\right] \cdots\cdots\cdots\cdots (Q)$$

$$+ \sum_{p\neq q\neq p'} \left[\sum_{i\neq j} t_{pq}^{ij} \pi_{\sigma(p)}^{\tau(i)} \pi_{\sigma(q)}^{\tau(j)} \right] \left[\sum_{i'\neq j'} t_{p'q}^{i'j'} \pi_{\sigma(p')}^{\tau(i')} \pi_{\sigma(q)}^{\tau(j')} \right] \cdots\cdots\cdots (R)$$

$$+ \sum_{p\neq q\neq p'} \left[\sum_{i\neq j} t_{pq}^{ij} \pi_{\sigma(p)}^{\tau(i)} \pi_{\sigma(q)}^{\tau(j)} \right] \left[\sum_{i'\neq j'} t_{p'p}^{i'j'} \pi_{\sigma(p')}^{\tau(i')} \pi_{\sigma(p)}^{\tau(j')} \right] \cdots\cdots\cdots (S)$$

$$+ \sum_{p\neq q\neq q'} \left[\sum_{i\neq j} t_{pq}^{ij} \pi_{\sigma(p)}^{\tau(i)} \pi_{\sigma(q)}^{\tau(j)} \right] \left[\sum_{i'\neq j'} t_{pq'}^{i'j'} \pi_{\sigma(p)}^{\tau(i')} \pi_{\sigma(q')}^{\tau(j')} \right] \cdots\cdots\cdots (T)$$

$$+ \sum_{p\neq q\neq q'} \left[\sum_{i\neq j} t_{pq}^{ij} \pi_{\sigma(p)}^{\tau(i)} \pi_{\sigma(q)}^{\tau(j)} \right] \left[\sum_{i'\neq j'} t_{qq'}^{i'j'} \pi_{\sigma(q)}^{\tau(i')} \pi_{\sigma(q')}^{\tau(j')} \right] \cdots\cdots\cdots (U)$$

$$+ \sum_{p\neq q\neq p'\neq q'} \left[\sum_{i\neq j} t_{pq}^{ij} \pi_{\sigma(p)}^{\tau(i)} \pi_{\sigma(q)}^{\tau(j)} \right] \left[\sum_{i'\neq j'} t_{p'q'}^{i'j'} \pi_{\sigma(p')}^{\tau(i')} \pi_{\sigma(q')}^{\tau(j')} \right] \cdots\cdots (V)$$

Then (P), (Q), $\cdots$, (V) are expanded into sums of seven terms that are of different types of pairing of (i,j) and (i',j').

$$P = \sum_{p\neq q} \sum_{i\neq j} t_{pq}^{ij^2} \pi_{\sigma(p)}^{\tau(i)^2} \pi_{\sigma(q)}^{\tau(j)^2} \cdots\cdots\cdots\cdots\cdots\cdots\cdots\cdots\cdots\cdots\cdots (P_1)$$

$$+ \sum_{p\neq q} \sum_{i\neq j} t_{pq}^{ij} t_{pq}^{ji} \pi_{\sigma(p)}^{\tau(i)} \pi_{\sigma(q)}^{\tau(j)} \pi_{\sigma(p)}^{\tau(j)} \pi_{\sigma(q)}^{\tau(i)} \cdots\cdots\cdots\cdots\cdots\cdots\cdots\cdots (P_2)$$

$$+ \sum_{p\neq q} \sum_{i\neq j\neq i'} t_{pq}^{ij} t_{pq}^{i'j} \pi_{\sigma(p)}^{\tau(i)} \pi_{\sigma(q)}^{\tau(j)} \pi_{\sigma(p)}^{\tau(i')} \pi_{\sigma(q)}^{\tau(j)} \cdots\cdots\cdots\cdots\cdots\cdots (P_3)$$

$$+ \sum_{p \neq q} \sum_{i \neq j \neq i'} t_{pq}^{ij} t_{pq}^{i'i} \pi_{\sigma(p)}^{\tau(i)} \pi_{\sigma(q)}^{\tau(j)} \pi_{\sigma(p)}^{\tau(i')} \pi_{\sigma(q)}^{\tau(i)} \quad \ldots\ldots\ldots\ldots\ldots\ldots (P_4)$$

$$+ \sum_{p \neq q} \sum_{i \neq j \neq j'} t_{pq}^{ij} t_{pq}^{ij'} \pi_{\sigma(p)}^{\tau(i)} \pi_{\sigma(q)}^{\tau(j)} \pi_{\sigma(p)}^{\tau(i)} \pi_{\sigma(q)}^{\tau(j')} \quad \ldots\ldots\ldots\ldots\ldots\ldots (P_5)$$

$$+ \sum_{p \neq q} \sum_{i \neq j \neq j'} t_{pq}^{ij} t_{pq}^{jj'} \pi_{\sigma(p)}^{\tau(i)} \pi_{\sigma(q)}^{\tau(j)} \pi_{\sigma(p)}^{\tau(j)} \pi_{\sigma(q)}^{\tau(j')} \quad \ldots\ldots\ldots\ldots\ldots\ldots (P_6)$$

$$+ \sum_{p \neq q} \sum_{i \neq j \neq i' \neq j'} t_{pq}^{ij} t_{pq}^{i'j'} \pi_{\sigma(p)}^{\tau(i)} \pi_{\sigma(q)}^{\tau(j)} \pi_{\sigma(p)}^{\tau(i')} \pi_{\sigma(q)}^{\tau(j')} \quad \ldots\ldots\ldots\ldots\ldots (P_7).$$

Since $t_{pq}^{ij} = 0$ or 1 and hence

$$\sum_{p \neq q} \sum_{i \neq j} t_{pq}^{ij^2} = \sum_{p \neq q} \sum_{i \neq j} t_{pq}^{ij} = k^2(k - 1),$$

we have

$$E_\nu^{(\alpha)}(P_1) = \sum_{p \neq q} \sum_{i \neq j} t_{pq}^{ij^2} E_\nu^{(\alpha)} \left[\pi_{\sigma(p)}^{\tau(i)^2} \pi_{\sigma(q)}^{\tau(j)^2} \right]$$

$$= \frac{1}{k - 1} \sum_{p \neq q} \sum_{i \neq j} \pi_p^{(i)^2} \pi_q^{(j)^2}$$

$$= \frac{1}{k - 1} \sum_{p \neq q} \sum_{i} \pi_p^{(i)^2} \left[\Delta_p - \pi_q^{(i)^2} \right]$$

$$= \frac{1}{k - 1} \left\{ \sum_p \Delta_p (\Delta - \Delta_p) - \sum_p \sum_i \pi_p^{(i)^2} \left[\Delta^{(i)} - \pi_p^{(i)^2} \right] \right\}$$

$$= \frac{1}{k-1}\left[\Delta^2 - \sum_p \Delta_p^2 - \sum_i \Delta^{(i)^2} + \sum_p \sum_i \pi_p^{(i)^4}\right]$$

$$= \frac{1}{k-1}(F - G + D).$$

Now, $t_{pq}^{ij} t_{pq}^{ji} = 1$ if and only if $t_{pq}^{ij} = t_{pq}^{ji} = 1$. In other words, $t_{pq}^{ij} t_{pq}^{ji} = 1$ if and only if a treatment α occurs in the $[(i - 1)k + p]$th and $[(j - 1)k + q]$th plots and a treatment β occurs in the $[(j - 1)k + p]$th and $[(i - 1)k + q]$th plots. We shall say that two treatments α and β are *reciprocal* with respect to the pth and qth columns in such a case. Thus the sum $\sum_{i \neq j} t_{pq}^{ij} t_{pq}^{ji}$ gives the number of pairs of two treatments that are reciprocal with respect to columns p and q. Therefore

$$\rho_\nu = \sum_{p \neq q} \sum_{i \neq j} t_{pq}^{ij} t_{pq}^{ji}$$

is the total number of pairs of two treatments that are reciprocal with respect to some pairs of columns. Since

$$\sum_{p \neq q} \sum_{i \neq j} t_{\sigma(p)\sigma(q)}^{\tau(i)\tau(j)} t_{\sigma(p)\sigma(q)}^{\tau(j)\tau(i)} = \sum_{p \neq q} \sum_{i \neq j} t_{pq}^{ij} t_{pq}^{ji},$$

it is clear that ρ_ν is invariant in $\mathcal{T}_\nu$.

$$E_\nu^{(\alpha)}(P_2) = \sum_{p \neq q} \sum_{i \neq j} t_{pq}^{ij} t_{pq}^{ji} E_\nu^{(\alpha)}\left(\pi_{\sigma(p)}^{\tau(i)} \pi_{\sigma(q)}^{\tau(j)} \pi_{\sigma(p)}^{\tau(j)} \pi_{\sigma(q)}^{\tau(i)}\right)$$

$$= \frac{\rho_\nu}{k^2(k-1)^2} \sum_{p \neq q} \sum_{i \neq j} \pi_p^{(i)} \pi_q^{(j)} \pi_p^{(j)} \pi_q^{(i)}$$

$$= \frac{\rho_\nu}{k^2(k-1)^2} \quad (H - G + D).$$

Since

$$\sum_{i \neq j \neq i'} t_{pq}^{ij} t_{pq}^{i'j} = \sum_{i \neq j} t_{pq}^{ij} \left(1 - t_{pq}^{ij} - t_{pq}^{ij}\right) = 0$$

and similarly

$$\sum_{i \neq j \neq j'} t_{pq}^{ij} t_{pq}^{ij'} = \sum_{i \neq j} t_{pq}^{ij} \left(1 - t_{pq}^{ii} - t_{pq}^{ij}\right) = 0,$$

we have

$$E_\nu^{(\alpha)}(P_3) = \sum_{p \neq q} \sum_{i \neq j \neq i'} t_{pq}^{ij} t_{pq}^{i'j} E_\nu^{(\alpha)} \left[\pi_{\sigma(p)}^{\tau(i)} \pi_{\sigma(q)}^{\tau(j)} \pi_{\sigma(p)}^{\tau(i')} \pi_{\sigma(q)}^{\tau(j)} \right] = 0$$

and

$$E_\nu^{(\alpha)}(P_5) = \sum_{p \neq q} \sum_{i \neq j \neq j'} t_{pq}^{ij} t_{pq}^{ij'} E_\nu^{(\alpha)} \left[\pi_{\sigma(p)}^{\tau(i)} \pi_{\sigma(q)}^{\tau(j)} \pi_{\sigma(p)}^{\tau(i)} \pi_{\sigma(q)}^{\tau(j')} \right] = 0.$$

Since

$$\sum_{p \neq q} \sum_{i \neq j \neq i'} t_{pq}^{ij} t_{pq}^{i'i} = \sum_{p \neq q} \sum_{i \neq j} t_{pq}^{ij} \left(1 - t_{pq}^{ii} - t_{pq}^{ji}\right) = k^2(k-1) - \rho_\nu$$

and similarly

$$\sum_{p \neq q} \sum_{i \neq j \neq j'} t_{pq}^{ij} t_{pq}^{jj'} = \sum_{p \neq q} \sum_{i \neq j} t_{pq}^{ij} \left(1 - t_{pq}^{ji} - t_{pq}^{ij}\right) = k^2(k-1) - \rho_\nu$$

we have

$$E_\nu^{(\alpha)}(P_4) = \frac{k^2(k-1) - \rho_\nu}{k^2(k-1)^2(k-2)} \sum_{p \neq q} \sum_{i \neq j \neq i'} \pi_p^{(i)} \pi_q^{(j)} \pi_p^{(i')} \pi_q^{(i)}$$

$$= \frac{k^2(k-1) - \rho_\nu}{k^2(k-1)^2(k-2)} \sum_{p \neq q} \left[\sum_i \pi_p^{(i)2} \pi_q^{(i)2} - \sum_{i \neq j} \pi_p^{(i)} \pi_p^{(j)} \pi_q^{(i)} \pi_q^{(j)} \right]$$

$$= \frac{k^2(k-1) - \rho_\nu}{k^2(k-1)^2(k-2)} \left(\sum_i \Delta^{(i)2} - H + G - 2D \right),$$

and in a similar manner we get

$$E_\nu^{(\alpha)}(P_6) = \frac{k^2(k-1) - \rho_\nu}{k^2(k-1)^2(k-2)} \left[\sum_i \Delta^{(i)2} - H + G - 2D \right] .$$

Finally since

$$\sum_{p \neq q} \sum_{i \neq j \neq i' \neq j} t_{pq}^{ij} t_{pq}^{i'j'} = \sum_{p \neq q} \sum_{i \neq j \neq i'} t_{pq}^{ij} \left[1 - t_{pq}^{i'i} - t_{pq}^{i'j} - t_{pq}^{i'i'} \right]$$

$$= \sum_{p \neq q} \sum_{i \neq j} t_{pq}^{ij} \left[k - 2 - \left(1 - t_{pq}^{ii} - t_{pq}^{ji} \right) - \left(1 - t_{pq}^{ij} - t_{pq}^{jj} \right) \right]$$

$$= k^2(k-1)(k-3) + \rho_\nu,$$

we have

$$E_\nu^{(\alpha)}(P_7) = \frac{k^2(k-1)(k-3) + \rho_\nu}{k^2(k-1)^2(k-2)^2(k-3)} \sum_{p \neq q} \sum_{i \neq j \neq i' \neq j'} \pi_p^{(i)} \pi_q^{(j)} \pi_p^{(i')} \pi_q^{(j')}$$

$$= \frac{k^2(k-1)(k-3) + \rho_\nu}{k^2(k-1)^2(k-2)^2(k-3)} \sum_{p \neq q} \left[2 \sum_{i \neq j} \pi_p^{(i)} \pi_p^{(j)} \pi_q^{(i)} \pi_q^{(j)} + \sum_{i \neq j} \pi_p^{(i)2} \pi_q^{(j)2} - 3 \sum_i \pi_p^{(i)2} \pi_q^{(i)2} \right]$$

$$= \frac{k^2(k-1)(k-3) + \rho_\nu}{k^2(k-1)^2(k-2)(k-3)} \left[6D + F - 3G + 2H - 3 \sum_i \Delta^{(i)^2} \right].$$

In a completely similar argument we get the following :

$$Q = \sum_{p \neq q} \sum_{i \neq j} t_{pq}^{ij} t_{qp}^{ij} \pi_{\sigma(p)}^{\tau(i)} \pi_{\sigma(q)}^{\tau(j)} \pi_{\sigma(q)}^{\tau(i)} \pi_{\sigma(p)}^{\tau(j)} \qquad \ldots\ldots\ldots\ldots (Q_1)$$

$$+ \sum_{p \neq q} \sum_{i \neq j} t_{pq}^{ij} t_{qp}^{ji} \pi_{\sigma(p)}^{\tau(i)} \pi_{\sigma(q)}^{\tau(j)} \pi_{\sigma(q)}^{\tau(j)} \pi_{\sigma(p)}^{\tau(i)} \qquad \ldots\ldots\ldots\ldots (Q_2)$$

$$+ \sum_{p \neq q} \sum_{i \neq j \neq i'} t_{pq}^{ij} t_{qp}^{i'j} \pi_{\sigma(p)}^{\tau(i)} \pi_{\sigma(q)}^{\tau(j)} \pi_{\sigma(q)}^{\tau(i')} \pi_{\sigma(p)}^{\tau(j)} \qquad \ldots\ldots\ldots\ldots (Q_3)$$

$$+ \sum_{p \neq q} \sum_{i \neq j \neq i'} t_{pq}^{ij} t_{qp}^{i'i} \pi_{\sigma(p)}^{\tau(i)} \pi_{\sigma(q)}^{\tau(j)} \pi_{\sigma(q)}^{\tau(i')} \pi_{\sigma(p)}^{\tau(i)} \qquad \ldots\ldots\ldots\ldots (Q_4)$$

$$+ \sum_{p \neq q} \sum_{i \neq j \neq j'} t_{pq}^{ij} t_{qp}^{ij'} \pi_{\sigma(p)}^{\tau(i)} \pi_{\sigma(q)}^{\tau(j)} \pi_{\sigma(q)}^{\tau(i)} \pi_{\sigma(p)}^{\tau(j')} \qquad \ldots\ldots\ldots\ldots (Q_5)$$

$$+ \sum_{p \neq q} \sum_{i \neq j \neq j'} t_{pq}^{ij} t_{qp}^{jj'} \pi_{\sigma(p)}^{\tau(i)} \pi_{\sigma(q)}^{\tau(j)} \pi_{\sigma(q)}^{\tau(j)} \pi_{\sigma(p)}^{\tau(j')} \qquad \ldots\ldots\ldots\ldots (Q_6)$$

$$+ \sum_{p \neq q} \sum_{i \neq j \neq i' \neq j'} t_{pq}^{ij} t_{qp}^{i'j'} \pi_{\sigma(p)}^{\tau(i)} \pi_{\sigma(q)}^{\tau(j)} \pi_{\sigma(q)}^{\tau(i')} \pi_{\sigma(p)}^{\tau(j')} \ldots\ldots\ldots\ldots (Q_7).$$

Expectations of seven Q_i are given as follows :

$$E_\nu^{(\alpha)}(Q_1) = \frac{\rho_\nu}{k^2(k-1)^2}(D - G + H),$$

$$E_\nu^{(\alpha)}(Q_2) = \frac{1}{k-1}(D - G + H),$$

$$E_\nu^{(\alpha)}(Q_3) = E_\nu^{(\alpha)}(Q_5) = \frac{k^2(k-1) - \rho_\nu}{k^2(k-1)^2(k-2)}\left[-2D + G - H + \sum_i \Delta^{(i)^2}\right],$$

$$E_\nu^{(\alpha)}(Q_4) = E_\nu^{(\alpha)}(Q_6) = 0,$$

$$E_\nu^{(\alpha)}(Q_7) = \frac{k^2(k-1)(k-3) + \rho_\nu}{k^2(k-1)^2(k-2)(k-3)}\left[6D + F - 3G + 2H - 3\sum_i \Delta^{(i)^2}\right].$$

R can be similarly expanded as the sum of seven terms R_i as follows :

$$R = \sum_{p\neq q\neq p'} \sum_{i\neq j} t_{pq}^{ij} t_{p'q}^{ij} \pi_{\sigma(p)}^{\tau(i)} \pi_{\sigma(q)}^{\tau(j)} \pi_{\sigma(p')}^{\tau(i)} \pi_{\sigma(q)}^{\tau(j)} \quad \ldots\ldots\ldots\ldots\ldots\ldots (R_1)$$

$$+ \sum_{p\neq q\neq p'} \sum_{i\neq j} t_{pq}^{ij} t_{p'q}^{ji} \pi_{\sigma(p)}^{\tau(i)} \pi_{\sigma(q)}^{\tau(j)} \pi_{\sigma(p')}^{\tau(j)} \pi_{\sigma(q)}^{\tau(i)} \quad \ldots\ldots\ldots\ldots\ldots\ldots (R_2)$$

$$+ \sum_{p\neq q\neq p'} \sum_{i\neq j\neq i'} t_{pq}^{ij} t_{p'q}^{i'j} \pi_{\sigma(p)}^{\tau(i)} \pi_{\sigma(q)}^{\tau(j)} \pi_{\sigma(p')}^{\tau(i')} \pi_{\sigma(q)}^{\tau(j)} \quad \ldots\ldots\ldots\ldots\ldots (R_3)$$

$$+ \sum_{p\neq q\neq p'} \sum_{i\neq j\neq j'} t_{pq}^{ij} t_{p'q}^{i'i} \pi_{\sigma(p)}^{\tau(i)} \pi_{\sigma(q)}^{\tau(j)} \pi_{\sigma(p')}^{\tau(i')} \pi_{\sigma(q)}^{\tau(i)} \quad \ldots\ldots\ldots\ldots\ldots (R_4)$$

$$+ \sum_{p \neq q \neq p'} \sum_{i \neq j \neq j'} t_{pq}^{ij} t_{p'q}^{ij'} \pi_{\sigma(p)}^{\tau(i)} \pi_{\sigma(q)}^{\tau(j)} \pi_{\sigma(p')}^{\tau(i)} \pi_{\sigma(q)}^{\tau(j')} \ldots\ldots\ldots\ldots (R_5)$$

$$+ \sum_{p \neq q \neq p'} \sum_{i \neq j \neq j'} t_{pq}^{ij} t_{p'q}^{jj'} \pi_{\sigma(p)}^{\tau(i)} \pi_{\sigma(q)}^{\tau(i)} \pi_{\sigma(p')}^{\tau(j)} \pi_{\sigma(q)}^{\tau(j')} \ldots\ldots\ldots\ldots (R_6)$$

$$+ \sum_{p \neq q \neq p'} \sum_{i \neq j \neq i' \neq j'} t_{pq}^{ij} t_{p'q}^{i'j'} \pi_{\sigma(p)}^{\tau(i)} \pi_{\sigma(q)}^{\tau(j)} \pi_{\sigma(p')}^{\tau(i')} \pi_{\sigma(q)}^{\tau(j')} \ldots\ldots\ldots\ldots (R_7).$$

Expectations of the seven terms of R_i are

$$E_\nu^{(\alpha)}(R_1) = 0,$$

$$E_\nu^{(\alpha)}(R_2) = \frac{k^2(k-1) - \rho_\nu}{k^2(k-1)^2(k-2)} \left[-2D + G - H + \sum_p \Delta_p^2 \right] ,$$

$$E_\nu^{(\alpha)}(R_3) = E_\nu^{(\alpha)}(R_5) = \frac{1}{(k-1)(k-2)}(4D + F - 2G) ,$$

$$E_\nu^{(\alpha)}(R_4) = E_\nu^{(\alpha)}(R_6) = \frac{k^2(k-1)(k-3) + \rho_\nu}{k^2(k-1)^2(k-2)^2}(4D - 2G + H) ,$$

$$E_\nu^{(\alpha)}(R_7) = \frac{k^2(k-1)(k-3)^2 - \rho_\nu}{k^2(k-1)^2(k-2)^2(k-3)}(-12D - F + 6G - 2H) .$$

Similarly for S, we have

$$S = \sum_{p \neq q \neq p'} \sum_{i \neq j} t_{pq}^{ij} t_{p'p}^{ij} \pi_{\sigma(p)}^{\tau(i)} \pi_{\sigma(q)}^{\tau(j)} \pi_{\sigma(p')}^{\tau(i)} \pi_{\sigma(p)}^{\tau(i)} \ldots\ldots\ldots\ldots (S_1)$$

$$+ \sum_{p \neq q \neq p'} \sum_{i \neq j} t_{pq}^{ij} t_{p'p}^{ji} \pi_{\sigma(p)}^{\tau(i)} \pi_{\sigma(q)}^{\tau(j)} \pi_{\sigma(p')}^{\tau(j)} \pi_{\sigma(p)}^{\tau(i)} \ldots\ldots\ldots\ldots (S_2)$$

$$+ \sum_{p \neq q \neq p'} \sum_{i \neq j \neq i'} t_{pq}^{ij} t_{p'p}^{i'j} \pi_{\sigma(p)}^{\tau(i)} \pi_{\sigma(q)}^{\tau(j)} \pi_{\sigma(p')}^{\tau(i')} \pi_{\sigma(p)}^{\tau(j)} \cdots\cdots (S_3)$$

$$+ \sum_{p \neq q \neq p'} \sum_{i \neq j \neq i'} t_{pq}^{ij} t_{p'p}^{i'i} \pi_{\sigma(p)}^{\tau(i)} \pi_{\sigma(q)}^{\tau(j)} \pi_{\sigma(p')}^{\tau(i')} \pi_{\sigma(p)}^{\tau(i)} \cdots\cdots (S_4)$$

$$+ \sum_{p \neq q \neq p'} \sum_{i \neq j \neq j'} t_{pq}^{ij} t_{p'p}^{ij'} \pi_{\sigma(p)}^{\tau(i)} \pi_{\sigma(q)}^{\tau(j)} \pi_{\sigma(p')}^{\tau(i)} \pi_{\sigma(p)}^{\tau(j')} \cdots\cdots (S_5)$$

$$+ \sum_{p \neq q \neq p'} \sum_{i \neq j \neq j'} t_{pq}^{ij} t_{p'p}^{jj'} \pi_{\sigma(p)}^{\tau(i)} \pi_{\sigma(q)}^{\tau(j)} \pi_{\sigma(p')}^{\tau(j)} \pi_{\sigma(p)}^{\tau(j')} \cdots\cdots (S_6)$$

$$+ \sum_{p \neq q \neq p'} \sum_{i \neq j \neq i' \neq j'} t_{pq}^{ij} t_{p'p}^{i'j'} \pi_{\sigma(p)}^{\tau(i)} \pi_{\sigma(q)}^{\tau(j)} \pi_{\sigma(p')}^{\tau(i')} \pi_{\sigma(p)}^{\tau(j')} \cdots\cdots (S_7).$$

Expectations of the seven terms of S_i are

$$E_\nu^{(\alpha)}(S_1) = \frac{k^2(k-1) - \rho_\nu}{k^2(k-1)^2(k-2)} \left[-2D + G - H + \sum_p \Delta_p^2 \right] ,$$

$$E_\nu^{(\alpha)}(S_2) = 0 ,$$

$$E_\nu^{(\alpha)}(S_3) = E_\nu^{(\alpha)}(S_5) = \frac{k^2(k-1)(k-3) + \rho_\nu}{k^2(k-1)^2(k-2)^2}(4D - 2G + H) ,$$

$$E_\nu^{(\alpha)}(S_4) = E_\nu^{(\alpha)}(S_4) = \frac{1}{(k-1)(k-2)}(4D + F - 2G) ,$$

$$E_\nu^{(\alpha)}(S_7) = \frac{k^2(k-1)(k-3)^2 - \rho_\nu}{k^2(k-1)^2(k-2)^2(k-3)}(-12D - F + 6G - 2H) .$$

Similarly for T, we have

$$T = \sum_{p \neq q \neq q'} \sum_{i \neq j} t_{pq}^{ij} t_{pq'}^{ij} \pi_{\sigma(p)}^{\tau(i)} \pi_{\sigma(q)}^{\tau(j)} \pi_{\sigma(p)}^{\tau(i)} \pi_{\sigma(q')}^{\tau(j)} \cdots\cdots (T_1)$$

$$+ \sum_{p \neq q \neq q'} \sum_{i \neq j} t_{pq}^{ij} t_{pq'}^{ij} \pi_{\sigma(p)}^{\tau(i)} \pi_{\sigma(q)}^{\tau(j)} \pi_{\sigma(p)}^{\tau(j)} \pi_{\sigma(q')}^{\tau(i)} \cdots\cdots\cdots (T_2)$$

$$+ \sum_{p \neq q \neq q'} \sum_{i \neq j \neq i'} t_{pq}^{ij} t_{pq'}^{i'j} \pi_{\sigma(p)}^{\tau(i)} \pi_{\sigma(q)}^{\tau(j)} \pi_{\sigma(p)}^{\tau(i')} \pi_{\sigma(q')}^{\tau(j)} \cdots\cdots\cdots (T_3)$$

$$+ \sum_{p \neq q \neq q'} \sum_{i \neq j \neq i'} t_{pq}^{ij} t_{pq'}^{i'i} \pi_{\sigma(p)}^{\tau(i)} \pi_{\sigma(q)}^{\tau(j)} \pi_{\sigma(p)}^{\tau(i')} \pi_{\sigma(q')}^{\tau(i)} \cdots\cdots\cdots (T_4)$$

$$+ \sum_{p \neq q \neq q'} \sum_{i \neq j \neq j'} t_{pq}^{ij} t_{pq'}^{ij'} \pi_{\sigma(p)}^{\tau(i)} \pi_{\sigma(q)}^{\tau(j)} \pi_{\sigma(p)}^{\tau(i)} \pi_{\sigma(q')}^{\tau(j')} \cdots\cdots\cdots (T_5)$$

$$+ \sum_{p \neq q \neq q'} \sum_{i \neq j \neq j'} t_{pq}^{ij} t_{pq'}^{jj'} \pi_{\sigma(p)}^{\tau(i)} \pi_{\sigma(q)}^{\tau(j)} \pi_{\sigma(p)}^{\tau(j)} \pi_{\sigma(q')}^{\tau(j')} \cdots\cdots\cdots (T_6)$$

$$+ \sum_{p \neq q \neq q'} \sum_{i \neq j \neq i' \neq j'} t_{pq}^{ij} t_{pq'}^{i'j'} \pi_{\sigma(p)}^{\tau(i)} \pi_{\sigma(q)}^{\tau(j)} \pi_{\sigma(p)}^{\tau(i')} \pi_{\sigma(q')}^{\tau(j')} \cdots\cdots\cdots (T_7).$$

Expectations of the seven terms of T_i are

$$E_\nu^{(\alpha)}(T_1) = 0 \ ,$$

$$E_\nu^{(\alpha)}(T_2) = \frac{k^2(k-1) - \rho_\nu}{k^2(k-1)^2(k-2)} \left[-2D + G - H + \sum_p \Delta_p^2 \right] \ ,$$

$$E_\nu^{(\alpha)}(T_3) = E_\nu^{(\alpha)}(T_5) = \frac{k^2(k-1)(k-3) + \rho_\nu}{k^2(k-1)^2(k-2)^2}(4D - 2G + H) \ ,$$

$$E_\nu^{(\alpha)}(T_4) = E_\nu^{(\alpha)}(T_6) = \frac{1}{(k-1)(k-2)}(4D + F - 2G) \ ,$$

$$E_\nu^{(\alpha)}(T_7) = \frac{k^2(k-1)(k-3)^2 - \rho_\nu}{k^2(k-1)^2(k-2)^2(k-3)} (-12D - F + 6G - 2H) .$$

Similarly for U, we have

$$U = \sum_{p\neq q\neq q'} \sum_{i\neq j} t_{pq}^{ij} t_{qq'}^{ij} \pi_{\sigma(p)}^{\tau(i)} \pi_{\sigma(q)}^{\tau(j)} \pi_{\sigma(q)}^{\tau(i)} \pi_{\sigma(q')}^{\tau(j)} \cdots\cdots (U_1)$$

$$+ \sum_{p\neq q\neq q'} \sum_{i\neq j} t_{pq}^{ij} t_{qq'}^{ji} \pi_{\sigma(p)}^{\tau(i)} \pi_{\sigma(q)}^{\tau(j)} \pi_{\sigma(q)}^{\tau(j)} \pi_{\sigma(q')}^{\tau(i)} \cdots\cdots (U_2)$$

$$+ \sum_{p\neq q\neq q'} \sum_{i\neq j\neq i'} t_{pq}^{ij} t_{qq'}^{i'j} \pi_{\sigma(p)}^{\tau(i)} \pi_{\sigma(q)}^{\tau(j)} \pi_{\sigma(q)}^{\tau(i')} \pi_{\sigma(q')}^{\tau(j)} \cdots\cdots (U_3)$$

$$+ \sum_{p\neq q\neq q'} \sum_{i\neq j\neq i'} t_{pq}^{ij} t_{qq'}^{i'i} \pi_{\sigma(p)}^{\tau(i)} \pi_{\sigma(q)}^{\tau(j)} \pi_{\sigma(q)}^{\tau(i')} \pi_{\sigma(q')}^{\tau(i)} \cdots\cdots (U_4)$$

$$+ \sum_{p\neq q\neq q'} \sum_{i\neq j\neq j'} t_{pq}^{ij} t_{qq'}^{ij'} \pi_{\sigma(p)}^{\tau(i)} \pi_{\sigma(q)}^{\tau(j)} \pi_{\sigma(q)}^{\tau(i)} \pi_{\sigma(q')}^{\tau(j')} \cdots\cdots (U_5)$$

$$+ \sum_{p\neq q\neq q'} \sum_{i\neq j\neq j'} t_{pq}^{ij} t_{qq'}^{jj'} \pi_{\sigma(p)}^{\tau(i)} \pi_{\sigma(q)}^{\tau(j)} \pi_{\sigma(q)}^{\tau(j)} \pi_{\sigma(q')}^{\tau(j')} \cdots\cdots (U_6)$$

$$+ \sum_{p\neq q\neq q'} \sum_{i\neq j\neq i'\neq j'} t_{pq}^{ij} t_{qq'}^{i'j'} \pi_{\sigma(p)}^{\tau(i)} \pi_{\sigma(q)}^{\tau(j)} \pi_{\sigma(q)}^{\tau(i')} \pi_{\sigma(q')}^{\tau(j')} \cdots\cdots (U_7).$$

Expectations of the seven terms of U_i are

$$E_\nu^{(\alpha)}(U_1) = \frac{k^2(k-1) - \rho_\nu}{k^2(k-1)^2(k-2)} \left[-2D + G - H + \sum_p \Delta_p^2\right] ,$$

$$E_\nu^{(\alpha)}(U_2) = 0,$$

$$E_\nu^{(\alpha)}(U_3) = E_\nu^{(\alpha)}(U_5) = \frac{k^2(k-1)(k-3) + \rho_\nu}{k^2(k-1)^2(k-2)^2}(4D - 2G + H),$$

$$E_\nu^{(\alpha)}(U_4) = E_\nu^{(\alpha)}(U_6) = \frac{1}{(k-1)(k-2)}(4D - 2G + F),$$

$$E_\nu^{(\alpha)}(U_7) = \frac{k^2(k-1)\ (k-3)^2 - \rho_\nu}{k^2(k-1)^2(k-2)^2(k-3)}(-12D - F + 6G - 2H).$$

Finally one has the similar expansion for V as the sum of seven V_i,

$$V = \sum_{p\neq q\neq p'\neq q'} \sum_{i\neq j} t_{pq}^{ij} t_{p'q'}^{ij}\ \pi_{\sigma(p)}^{\tau(i)} \pi_{\sigma(q)}^{\tau(j)} \pi_{\sigma(p')}^{\tau(i)} \pi_{\sigma(q')}^{\tau(j)} \cdots\cdots\cdots (V_1)$$

$$+ \sum_{p\neq q\neq p'\neq q'} \sum_{i\neq j} t_{pq}^{ij} t_{p'q'}^{ji}\ \pi_{\sigma(p)}^{\tau(i)} \pi_{\sigma(q)}^{\tau(j)} \pi_{\sigma(p')}^{\tau(j)} \pi_{\sigma(q')}^{\tau(i)} \cdots\cdots\cdots (V_2)$$

$$+ \sum_{p\neq q\neq p'\neq q'} \sum_{i\neq j\neq i'} t_{pq}^{ij} t_{p'q'}^{i'j}\ \pi_{\sigma(p)}^{\tau(i)} \pi_{\sigma(q)}^{\tau(j)} \pi_{\sigma(p')}^{\tau(i')} \pi_{\sigma(q')}^{\tau(j)} \cdots\cdots\cdots (V_3)$$

$$+ \sum_{p\neq q\neq p'\neq q'} \sum_{i\neq j\neq i'} t_{pq}^{ij} t_{p'q'}^{i'i}\ \pi_{\sigma(p)}^{\tau(i)} \pi_{\sigma(q)}^{\tau(j)} \pi_{\sigma(p')}^{\tau(i')} \pi_{\sigma(q')}^{\tau(i)} \cdots\cdots\cdots (V_4)$$

$$+ \sum_{p\neq q\neq p'\neq q'} \sum_{i\neq j\neq j'} t_{pq}^{ij} t_{p'q'}^{ij'}\ \pi_{\sigma(p)}^{\tau(i)} \pi_{\sigma(q)}^{\tau(j)} \pi_{\sigma(p')}^{\tau(i)} \pi_{\sigma(q')}^{\tau(j')} \cdots\cdots\cdots (V_5)$$

$$+ \sum_{p\neq q\neq p'\neq q'} \sum_{i\neq j\neq j'} t_{pq}^{ij} t_{p'q'}^{jj'}\ \pi_{\sigma(p)}^{\tau(i)} \pi_{\sigma(q)}^{\tau(j)} \pi_{\sigma(p')}^{\tau(j)} \pi_{\sigma(q')}^{\tau(j')} \cdots\cdots\cdots (V_6)$$

$$+ \sum_{p\neq q\neq p'\neq q'} \sum_{i\neq j\neq i'\neq j'} t_{pq}^{ij} t_{p'q'}^{i'j'}\ \pi_{\sigma(p)}^{\tau(i)} \pi_{\sigma(q)}^{\tau(j)} \pi_{\sigma(p')}^{\tau(i')} \pi_{\sigma(q')}^{\tau(j')} \cdots\cdots (V_7).$$

Expectations of the seven terms of V_i are

$$E_\nu^{(\alpha)}(V_1) = E_\nu^{(\alpha)}(V_2) = \frac{k^2(k-1)(k-3) + \rho_\nu}{k^2(k-1)^2(k-2)(k-3)}$$

$$\times \left(6D + F - 6G + 2H - 3\sum_p \Delta_p^2\right),$$

$$E_\nu^{(\alpha)}(V_3) = E_\nu^{(\alpha)}(V_4) = E_\nu^{(\alpha)}(V_5) = E_\nu^{(\alpha)}(V_6)$$

$$= \frac{k^2(k-1)\ (k-3)^2 - \rho_\nu}{k^2(k-1)^2(k-2)^2(k-3)}$$

$$\times\ (-12D - F + 6G - 2H),$$

$$E_\nu^{(\alpha)}(V_7) = \frac{k^2(k-1)(k-3)(k^2 - 6k + 10) + 2\rho_\nu}{k^2(k-1)^2(k-2)^2(k-3)^2}$$

$$\times (36D + 3F - 18G + 6H).$$

We shall summarize these results in Table 2.6.

TABLE 2.6.

	Terms being independent of ρ_ν	Coefficients of ρ_ν
P_1	$\frac{1}{k-1}(D + F - G)$	0
P_2	0	$\frac{1}{k^2(k-1)^2}(D - G + H)$
P_3	0	0

TABLE 2.6. (Cont'd)

	Terms being independent of ρ_ν	Coefficient of ρ_ν
(P) P_4	$\frac{1}{(k-1)(k-2)} \times \left[-2D + G - H + \sum_i \Delta^{(i)2}\right]$	$\frac{-1}{k^2(k-1)^2(k-2)} \times \left[-2D + G - H + \sum_i \Delta^{(i)2}\right]$
P_5	0	0
P_6	$\frac{1}{(k-1)(k-2)} \times \left[-2D + G - H + \sum_i \Delta^{(i)2}\right]$	$\frac{-1}{k^2(k-1)^2(k-2)} \times \left[-2D + G - H + \sum_i \Delta^{(i)2}\right]$
P_7	$\frac{1}{(k-1)(k-2)} \times \left[6D + F - 3G + 2H - 3 \sum_i \Delta^{(i)2}\right]$	$\frac{1}{k^2(k-1)^2(k-2)(k-3)} \times \left[6D + F - 3G + 2H - 3 \sum_i \Delta^{(i)2}\right]$
Q_1	0	$\frac{1}{k^2(k-1)^2} (D - G + H)$
Q_2	$\frac{1}{k-1} (D - G + H)$	0

TABLE 2.6 (Cont'd)

		Terms being independent of ρ_ν	Coefficients of ρ_ν
	Q_3	$\frac{1}{(k-1)(k-2)} \times \left[-2D + G - H + \sum_i \Delta^{(i)2}\right]$	$\frac{-1}{k^2(k-1)^2(k-2)} \times \left[-2D + G - H + \sum_i \Delta^{(i)2}\right]$
(Q)	Q_4	0	0
	Q_5	$\frac{1}{(k-1)(k-2)} \times \left[2D + G - H + \sum_i \Delta^{(i)2}\right]$	$\frac{-1}{k^2(k-1)^2(k-2)} \times \left[-2D + G - H + \sum_i \Delta^{(i)2}\right]$
	Q_6	0	0
	Q_7	$\frac{1}{(k-1)(k-2)} \times \left[6D + F - 3G + 2H - 3 \times \sum_i \Delta^{(i)2}\right]$	$\frac{1}{k^2(k-1)^2(k-2)(k-3)} \times \left[6D + F - 3G + 2H - 3 \sum_i \Delta^{(i)2}\right]$
	R_1	0	0
	R_2	$\frac{1}{(k-1)(k-2)}$	$\frac{-1}{k^2(k-1)^2(k-2)}$

TABLE 2.6. (Cont'd)

		Terms being independent of ρ_ν	Coefficients of ρ_ν
		$\times\left[-2D + G - H + \sum_p \Delta_p^2\right]$	$\times\left[-2D + G - H + \sum_p \Delta_p^2\right]$
(R)	R_3	$\frac{1}{(k-1)(k-2)}(4D + F - 2G)$	0
	R_4	$\frac{k-3}{(k-1)(k-2)^2}(4D - 2G + H)$	$\frac{1}{k^2(k-1)^2(k-2)^2}(4D - 2G + H)$
	R_5	$\frac{1}{(k-1)(k-2)}(4D + F - 2G)$	0
	R_6	$\frac{k-3}{(k-1)(k-2)^2}(4D - 2G + H)$	$\frac{1}{k^2(k-1)^2(k-2)^2}(4D - 2G + H)$
	R_7	$\frac{k-3}{(k-1)(k-2)^2}$ $\times(-12D - F + 6G - 2H)$	$\frac{-1}{k^2(k-1)^2(k-2)^2(k-3)}$ $\times(-12D - F + 6G - 2H)$
	S_1	$\frac{1}{(k-1)(k-2)}$ $\times\left[-2D + G - H + \sum_p \Delta_p^2\right]$	$\frac{-1}{k^2(k-1)^2(k-2)^2}$ $\times\left[-2D + G - H + \sum_p \Delta_p^2\right]$
	S_2	0	0

TABLE 2.6. (Cont'd)

	Terms being independent of ρ_ν	Coefficients of ρ_ν
(S) S_3	$\frac{k-3}{(k-1)(k-2)^2}(4D - 2G + H)$	$\frac{1}{k^2(k-1)^2(k-2)^2}(4D - 2G + H)$
S_4	$\frac{1}{(k-1)(k-2)}(4D + F - 2G)$	0
S_5	$\frac{k-3}{(k-1)(k-3)^2}(4D - 2G + H)$	$\frac{1}{k^2(k-1)^2(k-2)^2}(4D - 2G + H)$
S_6	$\frac{1}{(k-1)(k-2)}(4D + F - 2G)$	0
S_7	$\frac{k-3}{(k-1)(k-2)^2}$ $\times(-12D - F + 6G - 2H)$	$\frac{-1}{k^2(k-1)^2(k-2)^2(k-3)}$ $\times(-12D - F + 6G - 2H)$
T_1	0	0
T_2	$\frac{1}{(k-1)(k-2)}$ $\times\left[-2D + G - H + \sum_p \Delta_p^2\right]$	$\frac{-1}{k^2(k-1)^2(k-2)}$ $\times\left[-2D + G - H + \sum_p \Delta_p^2\right]$

TABLE 2.6. (Cont'd)

		Terms being independent of ρ_ν	Coefficients of ρ_ν
	T_3	$\frac{k-3}{(k-1)(k-2)^2}(4D - 2G + H)$	$\frac{1}{k^2(k-1)^2(k-2)^2}(4D - 2G + H)$
(T)	T_4	$\frac{1}{(k-1)(k-2)}(4D + F - 2G)$	0
	T_5	$\frac{k-3}{(k-1)(k-2)^2}(4D - 2G + H)$	$\frac{1}{k^2(k-1)^2(k-2)^2}(4D - 2G + H)$
	T_6	$\frac{1}{(k-1)(k-2)}(4D + F - 2G)$	0
	T_7	$\frac{k-3}{(k-1)(k-2)^2} \times(-12D - F + 6G - 2H)$	$\frac{-1}{k^2(k-1)^2(k-2)^2(k-3)} \times(-12D - F + 6G - 2H)$
	U_1	$\frac{1}{(k-1)(k-2)} \times\left[-2D + G - H + \sum_p \Delta_p^2\right]$	$\frac{-1}{k^2(k-1)^2(k-2)} \times\left[-2D + G - H + \sum_p \Delta_p^2\right]$
(U)	U_2	0	0
	U_3	$\frac{k-3}{(k-1)(k-2)^2}(4D - 2G + H)$	$\frac{1}{k^2(k-1)^2(k-2)^2}(4D - 2G + H)$
	U_4	$\frac{1}{(k-1)(k-2)}(4D + F - 2G)$	0

TABLE 2.6. (Cont'd)

	Terms being independent of ρ_ν	Coefficients of ρ_ν
U_5	$\frac{k-3}{(k-1)(k-2)^2}(4D - 2G + H)$	$\frac{1}{k^2(k-1)^2(k-2)^2}(4D - 2G + H)$
U_6	$\frac{1}{(k-1)(k-2)}(4D + F - 2G)$	0
U_7	$\frac{k-3}{(k-1)(k-2)^2}$ $\times(-12D - F + 6G - 2H)$	$\frac{-1}{k^2(k-1)^2(k-2)^2(k-3)}$ $\times(-12D - F + 6G - 2H)$
V_1	$\frac{1}{(k-1)(k-2)}$ $\times\left[6D + F - 6G + 2H - 3\sum_p \Delta_p^2\right]$	$\frac{1}{k^2(k-1)^2(k-2)(k-3)}$ $\times\left[6D + F - 6G + 2H - 3\sum_p \Delta_p^2\right]$
V_2 (V)	$\frac{1}{(k-1)(k-2)}$ $\times\left[6D + F - 6G + 2H - 3\sum_p \Delta_p^2\right]$	$\frac{1}{k^2(k-1)^2(k-2)(k-3)}$ $\times\left[6D + F - 6G + 2H - 3\sum_p \Delta_p^2\right]$
V_3	$\frac{k-3}{(k-1)(k-2)^2}$ $\times(-12D - F + 6G - 2H)$	$\frac{-1}{k^2(k-1)^2(k-2)^2(k-3)}$ $\times(-12D - F + 6G - 2H)$

TABLE 2.6. (Cont'd)

	Terms being independent of ρ_ν	Coefficients of ρ_ν
V_4	$\frac{k-3}{(k-1)(k-2)^2}$ $\times(-12D - F + 6G - 2H)$	$\frac{-1}{k^2(k-1)^2(k-2)^2(k-3)}$ $\times(-12D - F + 6G - 2H)$
V_5	$\frac{k-3}{(k-1)(k-2)^2}$ $\times(-12D - F + 6G - 2H)$	$\frac{-1}{k^2(k-1)^2(k-2)^2(k-3)}$ $\times(-12D - F + 6G - 2H)$
V_6	$\frac{k-3}{(k-1)(k-2)^2}$ $\times(-12D - F + 6G - 2H)$	$\frac{-1}{k^2(k-1)^2(k-2)^2(k-3)}$ $\times(-12D - F + 6G - 2H)$
V_7	$\frac{k^2-6k+10}{(k-1)(k-2)^2(k-3)}$ $\times(36D + 3F - 18G + 6H)$	$\frac{2}{k^2(k-1)^2(k-2)^2(k-3)^2}$ $\times(36D + 3F - 18G + 6H)$

Using the results obtained, we get

$$E_\nu^{(\alpha)}(\underline{\pi}'T\,\underline{\pi})^2$$

$$= \frac{1}{(k-1)(k-2)^2(k-3)}\left[2k^2(k-1)D + (k^4 - 4k^3 + 2k^2 + 6k - 6)F - 2k(k^2 - 3k + 3)G - 2(k^2 - 6k + 6)H\right]$$

$$+ \frac{\rho_\nu}{k^2(k-1)^2(k-2)^2(k-3)} [2k^2(k-1)^2 D + 2(2k^2 - 6k + 3)F$$
$$- 2k(k-1)(k^2 - 3k + 3)G$$
$$+ 2(k^4 - 6k^3 + 13k^2 - 12k + 6)H],$$

and therefore $\mathcal{E}_\nu(\underline{\pi}'T\underline{\pi})^2$ is the same as $\mathcal{E}_\nu^{(\alpha)}(\underline{\pi}'T\underline{\pi})^2$.

Let

$$\bar{\rho} = \frac{1}{n} \sum_\nu n_\nu \rho_\nu.$$

Then we finally get

$$E(\underline{\pi}'T\ \underline{\pi}) = \frac{1}{n} \sum_{\nu=1}^{k} n_\nu E_\nu(\underline{\pi}'T\ \underline{\pi})^2$$

$$= \frac{1}{(k-1)(k-2)^2(k-3)} [2k^2(k-1)D + (k^4 - 4k^3 + 2k^2 + 6k - 6)F$$
$$- 2k(k^2 - 3k + 3)G - 2(k^2 - 6k + 6)H]$$
$$+ \frac{\bar{\rho}}{k^2(k-1)^2(k-2)^2(k-3)^2} [2k^2(k-1)^2 D + 2(2k^2 - 6k + 3)F$$
$$- 2k(k-1)(k^2 - 3k + 3)G$$
$$+ 2(k^4 - 6k^3 + 13k^2 - 12k + 6)H].$$

This coincides with the result obtained by Welch [5] in a sense. Thus

$$\mathrm{Var}(\theta) = \frac{2}{k^2(k-1)(k-2)^2(k-3)} [k^2(k-1)\frac{D}{F} + (k^4 - 5k^3 + 8k^2$$
$$- 6k + 3) - k(k^2 - 3k + 3)\frac{G}{F}$$
$$- (k^2 - 6k + 6)\frac{H}{F}]$$

$$+ \frac{2\bar{P}}{k^2(k-1)^2(k-2)^2(k-3)^2} \Big[k^2(k-1)^2 \frac{D}{F} + (2k^2 - 6k + 3) - k(k-1)(k^2 - 3k + 3) \frac{G}{F} + (k^4 - 6k^3 + 13k^2 - 12k + 6) \frac{H}{F} \Big].$$

REFERENCES

1. R. A. Fisher and F. Yates, *Statistical Tables for Biological, Agricultural, and Medical Research,* 4th Edition, Oliver and Boyd, Edinburgh, 1953.

2. J. Ogawa, *Ann. Inst. Stat. Math. Tokyo, 13,* 105 (1961).

3. J. Ogawa, *Ann. Math. Statist., 34,* 1558 (1963).

4. J. Neyman, K. Iwaszkiewicz and St. Kolodzieczyk, *J. Roy. Stat. Soc. Suppl. 2,* 107 (1935).

5. B. L. Welch, *Biometrika, 29,* 21 (1937).

6. J. Ogawa, *Ann. Inst. Stat. Math. Tokyo, 1,* 83 (1949).

Chapter 3

FACTORIAL DESIGN

3.1 Need for the Factorial Principle

In a comparative experiment one deals with "multifactor experiment." In some circumstances the aim of an experiment may be to find the combination of a number of factors that maximizes a yield. For example, research on deep-sea fisheries might be directed to determining the optimum storage conditions for a certain species of fish. Among the factors that could be varied would be storage temperature at sea between catching and landing, duration of storage at sea, storage temperature during transport from port to wholesale market or retailer, and methods of packing (type of container, proportion of ice to fish, etc.). The "yield" will be a measure of the quality of the fish at the end of the journey, at the consumer's table.

An experimenter who is seeking *the optimal combination* of all these factors might begin by experimenting on factor A alone -- say storage temperature at sea; to do so, he must run trial batches of fish under specified conditions in respect of the other factors B, C,... . Having found an optimal temperature, he might then choose a second factor B -- say method of packing -- and compare alternatives in respect of this factor, using of course the optimal temperature suggested by his first experiment. However, if the best method of packing appears to be any other than that used as a standard in the first experiment, there must be some doubt whether the findings on factor A are now appropriate to the new recommendation on factor B. A third experiment on factor C alone will further confuse the issue unless the first two experiments happened to have been conducted at the level of C that now appears to be the best. Unless the experimenter is very lucky, he may have to reexamine each factor several times in new experiments before he can reach any conclusion.

An example of a hypothetical two-factor experiment is given in Table 3.1.

TABLE 3.1.

Example of Hypothetical Two-Factor Experiment

		Temperature		
		A_1 (100° C)	A_2 (120° C)	A_3 (140° C)
Percentage of Catalyzer	B_1 (4%)	40	60	70
	B_2 (5%)	60	85	80
	B_3 (6%)	70	90	70
	B_4 (7%)	85	80	55

In laboratory research of a kind that permits observations to be obtained quickly the consequent delay due to the reexaminations may not be important, but even so the procedure may be uneconomic. In research of a kind that requires a long period for the results of an experiment to be obtained the progress of the research would be intolerably slow if the optimal could be investigated only by one factor at a time and if the results from one experiment might make necessary repetitions of trials of factors already studied. This is certainly true of agricultural research, in which normally one experiment must take a whole year, and it is not surprising that the impetus to devise experimental designs appropriate to the study of multifactors simultaneously came from agricultural researches.

- Quoted from D. J. Finney: *An Introduction to The Theory of Experimental Design*, 1963.

3.2 2^n Factorial Designs

First we consider the case $n = 2$, that is, two factors at two levels each. Given two factors A and B, each at two levels, we denote the treatment combinations by the symbols $a^{\alpha}b^{\beta}$, $\alpha, \beta = 0,1$, with the conventions $a^0 = b^0 = 1$ and $a1 = a$, $1b = b$. Then the whole treatment combinations are ab, a, b, 1, where ab means that both factors A and B are at higher levels, a means that factor A is at its higher level whereas factor B is at its lower level, and 1 means that both factors A and B are at their lower levels. The corresponding observations are denoted by x_{ab}, x_a, x_b, x_1, respectively. The total sum of squares corrected by the mean

$$Q = x_{ab}^2 + x_a^2 + x_b^2 + x_1^2 - 4\bar{x}^2 \tag{3.2.1}$$

can be decomposed into the sum of three squares of orthonormal contrasts :

$$\begin{aligned} Q &= \frac{1}{4}(x_{ab} + x_a - x_b - x_1)^2 \\ &+ \frac{1}{4}(x_{ab} - x_a + x_b - x_1)^2 \\ &+ \frac{1}{4}(x_{ab} - x_a - x_b + x_1)^2. \end{aligned} \tag{3.2.2}$$

The meanings of the three contrasts can be interpreted as follows :

$$\begin{aligned} &x_{ab} + x_a - x_b - x_1 \\ &\quad = x_{ab} - x_b + x_a - x_1 \\ &\quad = \text{effect of factor } A \text{ at the higher level of factor } B \\ &\quad\quad + \text{effect of factor } A \text{ at the lower level of factor } B \\ &\quad = 2A : \text{main effect of factor } A. \end{aligned}$$

$$x_{ab} - x_a + x_b - x_1$$

= effect of factor B at the higher level of factor A
+ effect of factor B at the lower level of factor A

= $2B$: main effect of factor B.

$$x_{ab} - x_a - x_b + x_1$$

$= x_{ab} - x_a - (x_b - x_1)$

= effect of factor B at the higher level of factor A
- effect of factor B at the lower level of factor A

= change of the factor B due to the levels of factor A

= $2AB$: interaction of the factors A and B.

Hence we have the unbiased linear estimators of the effects :

$$L_A = \frac{1}{2}(x_{ab} + x_a - x_b - x_1),$$

$$L_B = \frac{1}{2}(x_{ab} - x_a + x_b - x_1),$$

$$L_{AB} = \frac{1}{2}(x_{ab} - x_a - x_b + x_1),$$

$$L_0 = \frac{1}{4}(x_{ab} + x_a + x_b + x_1) = \bar{x}.$$

Now it should be noted that L_A, L_B, L_{AB} are normalized orthogonal contrasts. In vector notation we can write

$$
\begin{bmatrix} L_A \\ L_B \\ L_{AB} \\ 2L_0 \end{bmatrix} = \begin{bmatrix} \frac{1}{2} & \frac{1}{2} & -\frac{1}{2} & -\frac{1}{2} \\ \frac{1}{2} & -\frac{1}{2} & \frac{1}{2} & -\frac{1}{2} \\ \frac{1}{2} & -\frac{1}{2} & -\frac{1}{2} & \frac{1}{2} \\ \frac{1}{2} & \frac{1}{2} & \frac{1}{2} & \frac{1}{2} \end{bmatrix} \begin{bmatrix} x_{ab} \\ x_a \\ x_b \\ x_1 \end{bmatrix}. \tag{3.2.3}
$$

Hence we get

$$
\begin{bmatrix} x_{ab} \\ x_a \\ x_b \\ x_1 \end{bmatrix} = \begin{bmatrix} \frac{1}{2} & \frac{1}{2} & \frac{1}{2} & \frac{1}{2} \\ \frac{1}{2} & -\frac{1}{2} & -\frac{1}{2} & \frac{1}{2} \\ -\frac{1}{2} & \frac{1}{2} & -\frac{1}{2} & \frac{1}{2} \\ -\frac{1}{2} & -\frac{1}{2} & \frac{1}{2} & \frac{1}{2} \end{bmatrix} \begin{bmatrix} L_A \\ L_B \\ L_{AB} \\ 2L_0 \end{bmatrix}
$$

or

$$
\begin{aligned}
x_{ab} &= \tfrac{1}{2}(L_A + L_B + L_{AB} + 2L_0), \\
x_a &= \tfrac{1}{2}(L_A - L_B - L_{AB} + 2L_0), \\
x_b &= \tfrac{1}{2}(- L_A + L_B - L_{AB} + 2L_0), \\
x_1 &= \tfrac{1}{2}(- L_A - L_B + L_{AB} + 2L_0).
\end{aligned} \tag{3.2.4}
$$

Hence, taking the expectations, we get

$$
(3.2.5)\quad
\begin{aligned}
E(x_{ab}) &= \frac{1}{2}(A + B + AB + 2\mu) = \mu + \frac{1}{2}A + \frac{1}{2}B + \frac{1}{2}AB \\
E(x_a) &= \frac{1}{2}(A - B - AB + 2\mu) = \mu + \frac{1}{2}A - \frac{1}{2}B - \frac{1}{2}AB \\
E(x_b) &= \frac{1}{2}(-A + B - AB + 2\mu) = \mu - \frac{1}{2}A + \frac{1}{2}B - \frac{1}{2}AB \\
E(x_1) &= \frac{1}{2}(-A - B + AB + 2\mu) = \mu - \frac{1}{2}A - \frac{1}{2}B + \frac{1}{2}AB
\end{aligned}
$$

Next we consider the three factors of two levels each. Let the observations be

$$x_{abc},\ x_{ab},\ x_{ac},\ x_{bc},\ x_a,\ x_b,\ x_c,\ x_1.$$

One can write down the linear estimators of interactions and main effects as follows

$$
(3.2.6)\quad
\begin{aligned}
L_{ABC} &= \frac{1}{4}(x_{abc} - x_{ab} - x_{ac} - x_{bc} + x_a + x_b + x_c - x_1), \\
L_{AB} &= \frac{1}{4}(x_{abc} + x_{ab} - x_{ac} - x_{bc} - x_a - x_b + x_c + x_1), \\
L_{AC} &= \frac{1}{4}(x_{abc} - x_{ab} + x_{ac} - x_{bc} - x_a + x_b - x_c + x_1), \\
L_{BC} &= \frac{1}{4}(x_{abc} - x_{ab} - x_{ac} + x_{bc} + x_a - x_b - x_c + x_1), \\
L_A &= \frac{1}{4}(x_{abc} + x_{ab} + x_{ac} - x_{bc} + x_a - x_b - x_c - x_1), \\
L_B &= \frac{1}{4}(x_{abc} + x_{ab} - x_{ac} + x_{bc} - x_a + x_b - x_c - x_1), \\
L_C &= \frac{1}{4}(x_{abc} - x_{ab} + x_{ac} + x_{bc} - x_a - x_b + x_c - x_1), \\
L_0 &= \frac{1}{8}(x_{abc} + x_{ab} + x_{ac} + x_{bc} + x_a + x_b + x_c + x_1) = \bar{x}.
\end{aligned}
$$

Although L_{ABC}, L_{AB}, L_{AC}, L_{BC}, L_A, L_B, L_C are all contrasts, they are not normalized. In order to normalize them one has to multiply them with $\sqrt{2}$ and multiply L_0 by $2\sqrt{2}$.

$$
\begin{aligned}
x_{abc} &= \tfrac{1}{2}(L_{ABC} + L_{AB} + L_{AC} + L_{BC} + L_A + L_B + L_C + 2L_0),\\
x_{ab} &= \tfrac{1}{2}(-L_{ABC} + L_{AB} - L_{AC} - L_{BC} + L_A + L_B - L_C + 2L_0),\\
x_{ac} &= \tfrac{1}{2}(-L_{ABC} - L_{AB} + L_{AC} - L_{BC} + L_A - L_B + L_C + 2L_0),\\
x_{bc} &= \tfrac{1}{2}(-L_{ABC} - L_{AB} - L_{AC} + L_{BC} - L_A + L_B + L_C + 2L_0),\\
x_{a} &= \tfrac{1}{2}(L_{ABC} - L_{AB} - L_{AC} + L_{BC} + L_A - L_B - L_C + 2L_0),\\
x_{b} &= \tfrac{1}{2}(L_{ABC} - L_{AB} + L_{AC} - L_{BC} - L_A + L_B - L_C + 2L_0),\\
x_{c} &= \tfrac{1}{2}(L_{ABC} + L_{AB} - L_{AC} - L_{BC} - L_A - L_B + L_C + 2L_0),\\
x_{1} &= \tfrac{1}{2}(-L_{ABC} + L_{AB} + L_{AC} + L_{BC} - L_A - L_B - L_C + 2L_0).
\end{aligned}
\tag{3.2.7}
$$

The partition of the sum of squares is given as follows :

$$
\begin{aligned}
&x^2_{abc} + x^2_{ab} + x^2_{ac} + x^2_{bc} + x^2_{a} + x^2_{b} + x^2_{c} + x^2_{1} - 8\bar{x}^2\\
&\quad = 2(L^2_{ABC} + L^2_{AB} + L^2_{AC} + L^2_{BC} + L^2_{A} + L^2_{B} + L^2_{C}).
\end{aligned}
\tag{3.2.8}
$$

An easy way to write down the contrasts corresponding to main effects and interactions is as follows :

	Effects of factors							
Treatment	*A*	*B*	*C*	*AB*	*AC*	*BC*	*ABC*	μ
1	−	−	−	+	+	+	−	+
a	+	−	−	−	−	+	+	+
b	−	+	−	−	+	−	+	+
ab	+	+	−	+	−	−	−	+
c	−	−	+	+	−	−	+	+
ac	+	−	+	−	+	−	−	+
bc	−	+	+	−	−	+	−	+
abc	+	+	+	+	+	+	+	+
Divisor	4	4	4	4	4	4	4	8

In general one can write symbolically

$$2^n\mu = (a + 1)(b + 1)(c + 1)\ldots,$$

$$2^{n-1}A = (a - 1)(b - 1)(c + 1)\ldots,$$

$$2^{n-1}B = (a + 1)(b - 1)(c + 1)\ldots,$$

$$\ldots$$

$$2^{n-1}AB = (a - 1)(b - 1)(c + 1)\ldots,$$

$$\ldots$$

$$2^{n-1}ABCDEF = (a - 1)(b - 1)(c - 1)(d - 1)(e - 1)(f - 1)\ldots .$$

For instance, $2^{3-1}B = (a + 1)(b - 1)(c + 1)$ means that if one expands the right-hand side, one gets the suffices of the observations in the linear form

$$abc + ab - ac + bc - a + b - c - 1;$$

This should be interpreted as

$$x_{abc} + x_{ab} - x_{ac} + x_{bc} - x_a + x_b - x_c - x_1,$$

and the symbolic equation should be interpreted as

$$2^2 L_B = x_{abc} + x_{ab} - x_{ac} + x_{bc} - x_a + x_b - x_c - x_1.$$

There is a famous algorithm, called *Yates' Algorithm* by which one can calculate the contrasts and their squares quite mechanically. That will be shown in the following Table 3.2:

TABLE 3.2

Yates' Algorithm

	(I) Observations	(II)	(III)	(IV)	(V) Sum of squares	
(1)	12	34	82	196	M	$\frac{1}{8}M^2 = 4862$
a	22	48	114	48	A	288
b	18	50	22	28	B	98
ab	30	64	26	8	AB	8
c	20	10	14	32	C	128
ac	30	12	14	4	AC	2
bc	24	10	2	0	BC	0
abc	40	16	6	4	ABC	2

Explanation of the Yates' Algorithm is as follows :

(I)	(II)	(III)	(IV)	
x_1	x_1+x_a	$x_1+x_a+x_b+x_{ab}$	$x_1+x_a+x_b+x_{ab}+x_c+x_{ac}+x_{bc}+x_{abc}$	$\to \mu$
x_a	x_b+x_{ab}	$x_c+x_{ac}+x_{bc}+x_{abc}$	$-x_1+x_a-x_b+x_{ab}-x_c+x_{ac}-x_{bc}+x_{abc}$	$\to A$
x_b	x_c+x_{ac}	$-x_1+x_a-x_b+x_{ab}$	$-x_1-x_a+x_b+x_{ab}-x_c-x_{ac}+x_{bc}+x_{abc}$	$\to B$
x_{ab}	$x_{bc}+x_{abc}$	$-x_c+x_{ac}-x_{bc}+x_{abc}$	$x_1-x_a-x_b+x_{ab}+x_c-x_{ac}-x_{bc}+x_{abc}$	$\to AB$
x_c	$-x_1+x_a$	$-x_1-x_a+x_b+x_{ab}$	$-x_1-x_a-x_b-x_{ab}+x_c+x_{ac}+x_{bc}+x_{abc}$	$\to C$
x_{ac}	$-x_b+x_{ab}$	$-x_c-x_{ac}+x_{bc}+x_{abc}$	$x_1-x_a+x_b-x_{ab}-x_c+x_{ac}-x_{bc}+x_{abc}$	$\to AC$
x_{bc}	$-x_c+x_{ac}$	$x_1-x_a-x_b+x_{ab}$	$x_1+x_a-x_b-x_{ab}-x_c-x_{ac}+x_{bc}+x_{abc}$	$\to BC$
x_{abc}	$-x_{bc}+x_{abc}$	$x_c-x_{ac}-x_{bc}+x_{abc}$	$-x_1+x_a+x_b-x_{ab}+x_c-x_{ac}-x_{bc}+x_{abc}$	$\to ABC$

Here the general definitions of contrasts belonging to the interactions are repeated again. Given n factors $F_1, F_2, \ldots, F_n$ and given that F_i has s_i levels, we have $s = s_1 s_2 \cdots s_n$ treatment combinations $f_1^{a_1} f_2^{a_2} \cdots f_n^{a_n}$, and the corresponding observation is denoted by $x_{a_1 a_2 \cdots a_n}$. Now we are going to use the same symbols $f_1^{a_1} f_2^{a_2} \cdots f_n^{a_n} = \tau(a_1, a_2, \cdots, a_n)$ as the mean values of $x_{a_1 a_2 \cdots a_n}$.

$$E(x_{a_1 a_2 \cdots a_n}) = \tau(a_1, a_2, \cdots, a_n) = f_1^{a_1} f_2^{a_2} \cdots f_n^{a_n}.$$

A linear function

$$\sum C_{a_1 a_2 \cdots a_n} \tau(a_1, a_2, \ldots, a_n)$$

is called a contrast if

$$\sum c_{a_1 a_2 \cdots a_n} = 0.$$

A treatment contrast is said to belong to the (k - 1)*th-order interaction between* $F_{i_1}, F_{i_2}, \ldots, F_{i_k}$ *if (i)* $c_{a_1 \ldots a_n}$ *depends only on* $a_{i_1}, a_{i_2}, \ldots, a_{i_k}$ *and (ii) the sum of* $c_{a_1 \ldots a_n}$ *over each one of the suffices* $a_{i_1}, \ldots, a_{i_k}$ *is zero.* The zero-order interactions are called *main effects*. If one writes

$$\phi_i = f_i^0 + f_i^1 + \cdots + f_i^{s_i - 1} \qquad (i = 1, 2, \ldots, n),$$

then a typical linear function of treatments belonging to the $(k - 1)$th-order interaction between the factors $F_{i_1}, \ldots, F_{i_k}$ is

$$\frac{\phi_1 \phi_2 \cdots \phi_n}{\phi_{i_1} \cdots \phi_{i_k}} \sum_{x_{i_1}, \ldots, x_{i_k}} l(x_{i_1}, \ldots, x_{i_k}) f_{i_1}^{x_{i_1}} \cdots f_{i_k}^{x_{i_k}}$$

$$\sum_{x_{i_1}} l(x_{i_1}, \ldots, x_{i_k}) = \cdots = \sum_{x_{i_k}} l(x_{i_1}, \ldots, x_{i_k}) = 0.$$

One way of obtaining these linear functions is to choose

$$l(x_{i_1}, \ldots, x_{i_k}) = l_{1x_{i_1}} \cdots l_{jx_{i_j}} \cdots l_{kx_{i_k}}, \quad \sum_{x_{i_j}} l_{jx_{i_j}} = 0.$$

Thus there are $(s_{i_1} - 1)(s_{i_2} - 1) \cdots (s_{i_k} - 1)$ independent contrasts

$$\frac{\phi_1\phi_2\cdots\phi_n}{\phi_{i_1}\cdots\phi_{i_k}} \sum_{x_{i_1}} l_{1x_{i_1}} f_{i_1}^{x_{i_1}} \sum_{x_{i_2}} l_{2x_{i_2}} f_{i_2}^{x_{i_2}} \cdots \sum_{x_{i_k}} l_{kx_{i_k}} f_{i_k}^{x_{i_k}}$$

belonging to the $(k - 1)$th-order interaction between $F_{i_1},\ldots,F_{i_k}$.

In the case of $s_1 = \cdots = s_n = s$ this is called *a symmetrical factorial design.*

3.3 3^n Factorial Designs

In the case of 3^2 factorial design these are nine treatment combinations :

$$a^2b^2,\ a^2b\ ,\ a^2,\ ab^2,\ ab,\ a,\ b^2,\ b,\ 1.$$

We arrange the observations corresponding to these treatment combinations into a square.

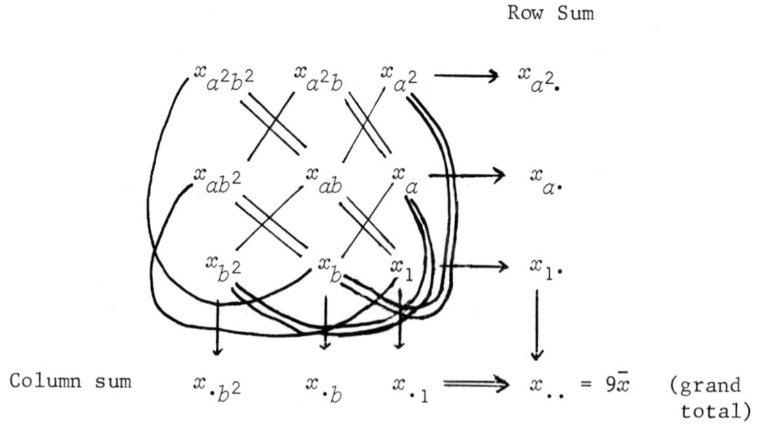

$$Q = x^2_{a^2b^2} + x^2_{a^2b} + \cdots + x^2_1 - \frac{x^2_{\cdot\cdot}}{9}$$ total sum of squares with eight degrees of freedom

$Q_A = \frac{1}{3}(x^2_{a^2\cdot} + x^2_{a\cdot} + x^2_{1\cdot}) - 9\bar{x}^2$ sum of squares due to the main effect A with two degrees of freedom

$Q_B = \frac{1}{3}(x^2_{\cdot b^2} + x^2_{\cdot b} + x^2_{\cdot 1}) - 9\bar{x}^2$ sum of squares due to the main effect B with two degrees of freedom.

Hence the difference $Q - Q_A - Q_B$ with $8 - 2 - 2 = 4$ degrees of freedom gives the sum of squares due to the first-order interaction $A \times B$. The nine treatment combinations are regarded as an Abelian group G generated by a and b such that

$$a^3 = b^3 = 1.$$

Now note that the arrangement indicated here corresponds to the coset decomposition of the treatment group G :

$$G = \{b\} + a\{b\} + a^2\{b\} \quad \text{row,}$$

$$= b^2\{a\} + b\{a\} + \{a\} \quad \text{column.}$$

The corresponding subgroup gave $\{b\}$ and $\{a\}$, respectively. *The effect* $A^{\alpha}B^{\beta}$ *and the treatment* $a^{\alpha'}b^{\beta'}$ *are said to be orthogonal if* $\alpha\alpha' + \beta\beta' \equiv 0 \bmod 3$. For a given effect $A^{\alpha}B^{\beta}$ the set of all treatments that are orthogonal to $A^{\alpha}B^{\beta}$ forms a subgroup of G called *an intrablock subgroup*. In fact, it $\alpha\alpha' + \beta\beta' \equiv 0$, $\alpha\alpha' + \beta\beta'' \equiv 0$, mod 3 implies $\alpha(\alpha' + \alpha'') + \beta(\beta' + \beta'') \equiv 0 \bmod 3$. $\{b\}$ is the intrablock subgroup of G for A and $\{a\}$ is the intrablock subgroup for B. The intrablock subgroups for AB and AB^2 are a^2b, ab^2, 1 and a^2b^2, ab, 1, respectively:

AB			AB^2		
x_{a^2b}	x_{ab^2}	x_1	$x_{a^2b^2}$	x_{ab}	x_1
x_{b^2}	x_{a^2}	x_{ab}	x_b	x_{a^2}	x_{ab^2}
x_a	x_b	$x_{a^2b^2}$	x_a	x_{b^2}	x_{a^2b}

One can notice that the column (row) sums of AB coincide with the row (column) sums in AB^2. Thus

$$Q_{AB} = \frac{1}{3}(x_{a^2b} + x_{ab^2} + x_1)^2 + \frac{1}{3}(x_{b^2} + x_{a^2} + x_{ab})^2 + \frac{1}{3}(x_a + x_b + x_{a^2b^2})^2 - 9\bar{x}^2,$$

$$Q_{AB^2} = \frac{1}{3}(x_{a^2b} + x_{b^2} + x_a)^2 + \frac{1}{3}(x_{ab^2} + x_{a^2} + x_b)^2 + \frac{1}{3}(x_1 + x_{ab} + x_{a^2b^2})^2 - 9\bar{x}^2.$$

Comparing square arrangements, one can see that the four partial sums of squares Q_A, Q_B, Q_{AB}, and Q_{AB^2} are orthogonal. AB^2 and AB are sometimes called I-component and J-component of interaction repectively.

In 3^3 factorial experiments there are 27 treatment combinations. For instance, in order to have the sum of squares due to the main effect A, one has to have the intrablock subgroup of the group G of all treatment combinations. This is given by

$$b^2c^2,\ b^2c,\ b^2,\ bc^2,\ bc,\ b,\ c^2,\ c,\ 1.$$

Hence the sum of squares due to the main effect A is

$$Q_A = \frac{1}{9}(x_{b^2c^2} + x_{b^2c} + x_{b^2} + x_{bc^2} + x_{bc} + x_b + x_{c^2} + x_c + x_1)^2$$

$$+ \frac{1}{9}(x_{ab^2c^2} + x_{ab^2c} + x_{ab^2} + x_{abc^2} + x_{abc} + x_{ab} + x_{ac^2} + x_{ac} + x_a)^2$$

$$+ \frac{1}{9}(x_{a^2b^2c^2} + x_{a^2b^2c} + x_{a^2b^2} + x_{a^2bc^2} + x_{a^2b} + x_{a^2c^2} + x_{a^2c} + x_{a^2})^2$$

$$- 27\bar{x}^2$$

with two degrees of freedom.

The intrablock subgroup orthogonal to ABC^2 is given by $a^\alpha b^\beta c^\gamma$ such that

$$\alpha + \beta + 2\gamma \equiv 0 \mod 3$$

α	0	0	0	1	1	1	2	2	2
β	0	1	2	0	1	2	0	1	2
γ	0	1	2	1	2	0	2	0	1

hence the intrablock subgroup is

$$a^2b^2c,\ a^2b,\ a^2c^2,\ ab^2,\ abc^2,\ ac,\ b^2c^2,\ bc,\ 1.$$

The sum of squares due to the second order interaction ABC^2 is

$$Q_{ABC^2} = \frac{1}{9}(x_{a^2b^2c} + x_{a^2b} + x_{a^2c^2} + x_{ab^2} + x_{abc^2} + x_{ac} + x_{b^2c^2} + x_{bc} + x_1)^2$$

$$+ \frac{1}{9}(x_c + x_{b^2} + x_{bc^2} + x_{a^2} + x_{a^2b^2c^2} + x_{a^2bc} + x_{ac^2}$$

$$+ x_{ab^2c} + x_{ab})^2$$

$$+ \frac{1}{9}(x_{abc} + x_a + x_{ab^2c^2} + x_b + x_{c^2} + x_{b^2c} + x_{a^2bc^2}$$

$$+ x_{a^2c} + x_{a^2b^2})^2$$

$$- 27\bar{x}^2$$

with two degrees of freedom.

In this way one can always write down the partial sum of squares due to some interaction.

An example of the analysis of variance of an unsymmetrical factorial experiment is mentioned. Suppose a factory product was to be made experimentally at each of three temperatures for a certain stage of the process, using each of four quantities of a certain raw material and maintaining the temperature for one of two alternative periods. This is a $4 \times 3 \times 2$ factorial scheme. Such an experiment might be conducted in six randomized blocks, a block being a week in which a batch of the product was made under each of the 24 combinations of treatments. The "yield" in this case would be one measurement of strength, durability, color or other appropriate property on each of the 6×24 batches.

The analysis of variance is easy (Table 3.3). The first step is to analyze the total sum of squares of deviation with 143 degrees of freedom into components for blocks (5 degrees of freedom) treatments (23 degrees of freedom) and error (115 degrees of freedom) exactly as for an ordinary randomized block design. In accordance with the general theory, the treatment component can be subdivided into 23 separate squares for single and mutually orthogonal contrasts.

TABLE 3.3

Analysis of Variance for a 4 × 3 × 2 Factorial Experiment in Randomized Blocks

Variations	Degrees of freedom		Sum of squares	Mean sum of squares
Blocks		5		
Temperature (T)	2			
Quantities (Q)	3			
Periods (P)	1			
$T \times Q$	6			
$T \times P$	2			
$Q \times P$	3			
$T \times Q \times P$	6			
Treatment		23		
Error		115		
Total		143		

3.4 Fractional Replication of a Factorial Design

Often the chief limiting factor on planning an experiment is the total numbers of plots to be available, as this is likely to be approximately proportional to the total cost or time or effort and a second important consideration is *the desirability of keeping blocks reasonably small and homogenous*, the "reasonability" here depends on the least number of animals normally obtained in one litter, or the number of cycles of a complex sequence of operations and measurements that one man can perform in one day, or the capacity of one rack in an incubator. The possibility of conducting experiments on many factors, but without an excessive number of plots, and also of introducing *blocking* so that block size can be less than the total number of treatment combinations needs consideration. First one has to examine what will happen when only a fraction of a whole factorial combination is actually laid out and hence only the observations corresponding to this fraction are available.

In a 2^3 factorial experiment we shall consider the subgroup of treatment combinations that are orthogonal to the second-order interaction ABC (i.e., $a^\alpha b^\beta c^\gamma$) such that

$$\alpha + \beta + \gamma \equiv 0 \mod 2.$$

The solutions of this congruence are

α	0	0	1	1
β	0	1	0	1
γ	0	1	1	0

that is,

$$1,\ bc,\ ac,\ ac.$$

From (3.2.7) we see that

$$\begin{aligned}
x_1 &= \frac{1}{2}(-L_{ABC} + L_{AB} + L_{AC} + L_{BC} - L_A - L_B - L_C + 2L_0),\\
x_{bc} &= \frac{1}{2}(-L_{ABC} - L_{AB} - L_{AC} + L_{BC} - L_A + L_B + L_C + 2L_0),\\
x_{ac} &= \frac{1}{2}(-L_{ABC} - L_{AB} + L_{AC} - L_{BC} + L_A - L_B + L_C + 2L_0),\\
x_{ab} &= \frac{1}{2}(-L_{ABC} + L_{AB} - L_{AC} - L_{BC} + L_A + L_B - L_C + 2L_0).
\end{aligned} \tag{3.4.1}$$

Hence

$$- x_1 + x_{ab} + x_{ac} - x_{bc} = - 2L_{BC} + 2L_A,$$

$$- x_1 + x_{ab} - x_{ac} + x_{bc} = - 2L_{AB} + 2L_B,$$

$$- x_1 - x_{ab} + x_{ac} + x_{bc} = - 2L_{AB} + 2L_C,$$

$$x_1 + x_{ab} + x_{ac} + x_{bc} = - 2L_{ABC} + 4L_0.$$

Therefore, by taking expectations, we obtain

$$E(-x_1 + x_{ab} + x_{ac} - x_{bc}) = 2(A - BC),$$

(3.4.2) $$E(-x_1 + x_{ab} - x_{ac} + x_{bc}) = 2(B - AC),$$

$$E(-x_1 - x_{ab} + x_{ac} + x_{bc}) = 2(C - AB).$$

Thus the contrast estimating the main effect A is the same as that estimating the first-order interaction BC. In such a case A and BC are called *aliases*, and this fact is denoted by the symbol

(3.4.3) $$A \equiv BC.$$

Similarly we can see that $B \equiv AC$ and $C \equiv AB$. The equation

$$E(x_1 + x_{ab} + x_{ac} + x_{bc}) = 2(2\mu - ABC)$$

show that the general mean μ and second-order interaction ABC are alias

(3.4.4) $$ABC \equiv 1.$$

If one considers that these are equalities in the effect group $\mathcal{G}$ generated by A, B, C with $A^2 = B^2 = C^2 = 1$, all other relations can be derived from

$$(3.4.5.) \qquad ABC \equiv 1,$$

for example, $A(ABC) \equiv A^2BC \equiv BC \equiv A$. The relation $ABC \equiv 1$ is called *the defining relation of the fractional replicate*, and ABC is called *the generator* of the fractional replicate.

Although the following discussions can be extended to the case $s = p^m$ (prime power), for the sake of simplicity we confine ourselves to the case of $s = p$ at this stage. Let s be any prime number and consider an Abelian group $\mathcal{G}$ (called *the effect group*) generated by the elements $F_1, F_2, \ldots, F_n$ satisfying the relations

$$F_1^s = F_2^s = \cdots = F_n^s = I,$$

where I denotes the unit element of the group $\mathcal{G}$. If H is any element of $\mathcal{G}$, then H can be expressed uniquely in the form

$$(3.4.6) \quad H = F_1^{h_1} F_2^{h_2} \cdots F_n^{h_n} \qquad (0 \le h_i \le s - 1;\ i = 1,2,\ldots,n),$$

where $F_1, F_2, \ldots, F_n$ can be identified with the n factors of a factorial experiment in which each factor can be chosen at any one of s distinct levels. The treatment combination in which the factor F_i occurs at the level a_i is written as

$$f_1^{a_1} f_2^{a_2} \cdots f_n^{a_n} \qquad (0 \le a_i \le s - 1;\ i = 1,2,\ldots,n),$$

and the corresponding observation is denoted by

$$x_{a_1 a_2 \cdots a_n}.$$

For any given interaction H or, more precisely, for a cyclical subgroup $\{H\}$ of $\mathcal{G}$ there is a unique subgroup $g_{\{H\}}$ of the treatment group G-an Abelian group generated by $f_1, f_2, \ldots, f_n$ and satisfying $f_1^s = f_2^s = \cdots = f_n^s = I$, such that

(3.4.7)

$$f_1^{a_1} f_2^{a_2} \cdots f_n^{a_n} \in g_{\{H\}} \Leftrightarrow h_1 a_1 + h_2 a_2 + \cdots + h_n a_n \equiv 0 \quad \text{mod } s.$$

The subgroup $g_{\{H\}}$ is called *the intrablock subgroup* of G corresponding to the interaction $\{H\}$. Let

$$G = g_{\{H\}} + k_1 g_{\{H\}} + \cdots + k_{s-1} g_{\{H\}} \tag{3.4.8}$$

be a coset decomposition of G with respect to the intrablock subgroup $g_{\{H\}}$, where

$$k_i = f_1^{k_{i1}} f_2^{k_{i2}} \cdots f_n^{k_{in}} \qquad (i = 1, 2, \ldots, s - 1),$$

$$k_{i1} h_1 + k_{i2} h_2 + \cdots + k_{in} h_n \equiv i \quad \text{mod } s.$$

Let

$$X_i = \sum_{a_1 a_2 \cdots a_n} x_{a_1 a_2 \cdots a_n} \qquad (i = 1, \ldots, s - 1),$$

where the summation is taken over all $a_1, \ldots, a_n$ such that

$$f_1^{a_1} f_2^{a_2} \cdots f_n^{a_n} \in k_i g_{\{H\}}$$

or

$$a_1 h_1 + a_2 h_2 + \cdots + a_n h_n \equiv i \mod s.$$

Then one can establish a one-to-one correspondence

$$(3.4.9) \qquad \{H\} \leftrightarrow Q_H = \frac{1}{s^{n-1}} \sum_{i=0}^{s-1} X_i - \frac{1}{s^n} \left(\sum_{i=0}^{s-1} X_i \right)^2,$$

where Q_H has $(s - 1)$ degrees of freedom corresponding to the interaction H.

Suppose that *only the observations belonging to the intrablock subgroup are available*; in other words, $(1/s)$th fractional replicate of the s^n factorial experiment is available. We want to calculate the sum of squares corresponding to an interaction

$$L = F_1^{l_1} F_2^{l_2} \cdots F_n^{l_n},$$

which is assumed not to belong to $\{H\}$. Let

$$Y_i = \sum_{a_1 a_2 \cdots a_n} x_{a_1 a_2 \cdots a_n} \qquad (i = 0,1,\ldots,s - 1),$$

where the summation should be taken over all $f_1^{a_1} f_2^{a_2} \cdots f_n^{a_n}$ such that

$$(3.4.10)\qquad \begin{aligned} h_1a_1 + h_2a_2 + \cdots + h_na_n &\equiv 0 \\ l_1a_1 + l_2a_2 + \cdots + l_na_n &\equiv i \end{aligned} \qquad (\text{mod } s).$$

Since the number of treatments in $g_{\{H\}}$ is s^{n-1} and the number of treatments satisfying (3.4.10) is s^{n-2}, we obtain

$$(3.4.11)\qquad Q_L = \frac{1}{s^{n-2}} \sum_{i=0}^{s-1} Y_i - \frac{1}{s^{n-1}} \left(\sum_{i=0}^{s-1} Y_i \right)^2 \leftrightarrow \{L\}.$$

However, since (3.4.10) is equivalent to

$$(3.4.12)\qquad \begin{aligned} & h_1a_1 + h_2a_2 + \cdots + h_na_n \equiv 0 \\ & (l_1 + jh_1)a_1 + (l_2 + jh_2)a_2 \\ & \quad + \cdots + (l_n + jh_n)a_n \equiv i \end{aligned} \qquad \begin{aligned} & \text{mod } s; \\ & (i = 0,1,\ldots,s-1; \\ & j = 0,1,\ldots,s-1), \end{aligned}$$

we obtain

$$(3.4.13)\qquad Q_L = Q_{LH} = Q_{LH^2} = \cdots = Q_{LH^{s-1}} .$$

If we decompose Q_L into the sum of $s - 1$ squares of orthonormal contrasts,

$$(3.4.14)\qquad Q_L = L_L^2 + L_{L^2}^2 + \cdots + L_{L^{s-1}}^2 ,$$

then it must be

(3.4.15)

$$E(L_{L^j}) = L^j \pm L^jH \pm L^jH^2 \pm \cdots \pm L^jH^{s-1} \qquad (j = 1,2,\ldots,s-1).$$

The effects or interaction LH^h, $h = 0,1,\ldots,s-1$ are said to be *aliases* to L. The subgroup $\{H\}$ of the effect group $\mathcal{G}$ is called *the alias subgroup*.

The above arguments are applicable to the case when only the observations corresponding to a certain coset $k_j g_{\{H\}}$ are available. In fact, let

$$Y_i^{(j)} = \sum_{a_1 a_2 \cdots a_n} x_{a_1 a_2 \cdots a_n} \qquad (i = 0,1,\ldots,s-1),$$

where the summation is taken over all $a_1, a_2, \ldots, a_n$ such that

$$\text{(3.4.16)} \qquad \begin{aligned} h_1 a_1 + h_2 a_2 + \cdots + h_n a_n &\equiv j \\ l_1 a_1 + l_2 a_2 + \cdots + l_n a_n &\equiv i \end{aligned} \qquad \text{mod } s.$$

If $f_1^{a_1^\circ} f_2^{a_2^\circ} \cdots f_n^{a_n^\circ}$ is a fixed element of $k_j g_{\{H\}}$, then any element of $k_j g_{\{H\}}$ is written as

$$f_1^{a_1^\circ + c_1} f_2^{a_2^\circ + c_2} \cdots f_n^{a_n^\circ + c_n}.$$

such that

$$h_1 c_1 + h_2 c_2 + \cdots + h_n c_n \equiv 0,$$

$$\text{(3.4.17)} \qquad \begin{aligned} &l_1 c_1 + l_2 c_2 + \cdots + l_n c_n \\ &\quad \equiv i - (l_1 a_1^\circ + l_2 a_2^\circ + \cdots + l_n a_n^\circ)^2 \equiv i' \pmod{s}, \end{aligned}$$

and

$$\text{(3.4.18)} \qquad Q_L^{(j)} = \frac{1}{s^{n-2}} \sum_{i=0}^{s-1} Y_i^{(j)2} - \frac{1}{s^{n-1}} \left(\sum_{i=0}^{s-1} Y_i^{(j)} \right)^2.$$

More generally, let $H_1, H_2, \ldots, H_k$ be any k independent element of $\mathcal{G}$ and

$$H_i = F_1^{k_{i1}} F_2^{k_{i2}} \cdots F_n^{k_{in}} \qquad (i = 1,2,\ldots,k).$$

Let the subgroup of order s^k generated by $H_1, H_2, \ldots, H_k$ be $\mathcal{G}_k$. Consider the intrablock subgroup s_{n-k} of the treatment group G corresponding to $\mathcal{G}_k$. Then the elements of s_{n-k} are

$$f_1^{x_1} f_2^{x_2} \cdots f_n^{x_n}$$

such that

$$(3.4.19) \qquad \left.\begin{array}{l} h_{11}x_1 + h_{12}x_2 + \cdots + h_{1n}x_n \equiv 0 \\ h_{21}x_1 + h_{22}x_2 + \cdots + h_{2n}x_n \equiv 0 \\ \cdots\cdots\cdots\cdots\cdots\cdots\cdots\cdots \\ h_{k1}x_1 + h_{k2}x_2 + \cdots + h_{kn}x_n \equiv 0 \end{array}\right\} \bmod s.$$

Hence they form a $(1/s^k)$th fractional replicate of the totality of the possible treatment combinations. Similarly any coset of s_{n-k} in G gives a $(1/s^k)$th fractional replicate.

Let $L = F_1^{l_1} F_2^{l_2} \cdots F_n^{l_n}$ be any interaction not belonging to $\mathcal{G}_k$; then the sum of squares corresponding to the cyclic subgroup $\{L\}$ in a $(1/s^k)$th fractional replicate is given by the following : Let us suppose that a fraction corresponding to a certain coset is observed, in other words, the observations $x_{a_1 a_2 \cdots a_n}$ such that

$$(3.4.20)\qquad \left.\begin{array}{l} h_{11}a_1 + h_{12}a_2 + \cdots + h_{1n}a_n \equiv c_1 \\ h_{21}a_1 + h_{22}a_2 + \cdots + h_{2n}a_n \equiv c_2 \\ \cdots\cdots\cdots\cdots\cdots\cdots\cdots\cdots \\ h_{k1}a_1 + h_{k2}a_2 + \cdots + h_{kn}a_n \equiv c_k \end{array}\right\} \bmod s.$$

are given. Then

$$(3.4.21)\qquad Z_i = \sum_{a_1a_2\cdots a_n} x_{a_1a_2\cdots a_n} \qquad (i = 0,1,\ldots,s-1),$$

where the summation is taken over all $a_1,a_2,\ldots,a_n$ such that

$$(3.4.22)\qquad \left.\begin{array}{l} h_{11}a_1 + h_{12}a_2 + \cdots + h_{1n}a_n \equiv c_1 \\ h_{21}a_1 + h_{22}a_2 + \cdots + h_{2n}a_n \equiv c_2 \\ \cdots\cdots\cdots\cdots\cdots\cdots\cdots\cdots \\ h_{k1}a_1 + h_{k2}a_2 + \cdots + h_{kn}a_n \equiv c_k \\ l_1a_1 + l_2a_2 + \cdots + l_na_n \equiv i \end{array}\right\} \bmod s$$

and

$$(3.4.23)\qquad Q_L = \frac{1}{s^{n-k-1}} \sum_{i=0}^{s-1} Z_i^2 - \frac{1}{s^{n-k}} \left(\sum_{i=0}^{s-1} Z_i \right)^2 .$$

Hence it will be clear that

$$(3.4.24)\qquad Q_L = Q_{LH_1^{n_1}H_2^{n_2}\cdots H_k^{n_k}} \qquad (0 \le n_i \le s-1).$$

Thus the set of interactions $L\mathcal{G}_k$ is called *the alias set* of L, $\mathcal{G}_k$ is called *the alias subgroup* (also sometimes called *the fundamental subgroup*).

Since it is in general more important to estimate lower order interactions, it would be of interest to choose the generators of the fundamental subgroup $\mathcal{G}_k$ in such a way that, for a specified t, no t-factor interaction is aliased with other t-factor or lower order interaction. It is seen that for this it is necessary and sufficient that every interaction represented by an element of $\mathcal{G}_k$ should have $2t + 1$ or more factors.

3.4.1 Theorem (Fisher). A $1/p^{n-r}$ replicate of a p^n factorial design, requiring p^r plots, with the restriction that no degrees of freedom belonging to main effects shall be alias of one another, is possible for any n such that

$$n \leq \frac{p^r - 1}{p - 1}, \tag{3.4.25}$$

where p is a prime number.

This can be generalized for any positive integer L :

$$n \leq \frac{L^r - 1}{L - 1}.$$

Indeed, let us consider the factorial scheme L^n. The degrees of freedom belonging to n main effects are $n(L - 1)$. Since there is a general mean, the total number of parameters to be estimated is

$$n(L - 1) + 1.$$

Hence the number of plots L^r should be

$$n(L - 1) + 1 \leq L^r,$$

$$n \leq \frac{L^r - 1}{L - 1}.$$

The reader is referred to Plackett and Burman [1] for further details.

Weighing Designs. If the absence of all interactions is certain, the idea of the fractional replication can be carried to its logical extreme. One such example is an experiment of weighing articles on a good balance. The true weight of two articles in combination must be just the sum of their weights, and there are no interactions at all.

Given seven articles A, B, C, D, E, F, G, the obvious procedure is to make eight weights, one with an empty pan to give a zero correction and one with each article in turn. If σ is the standard deviation of the error per single weighing, the variance of the weight estimate (i.e., the difference between its observational weight and the zero correction) is $2\sigma^2$.

Now considering this to be a 2^7 factorial design, the smallest number of plots is, from Fisher's theorem (Theorem 3.4.1),

$$\frac{2^r - 1}{2 - 1} = 7 \quad \text{or} \quad 2^r = 8.$$

Thus one can select a $1/2^4$ fractional replicate of the total $2^7 = 128$. Choose the defining relations to be

$$ABC \equiv 1, \; ADE \equiv 1, \; BDF \equiv 1, \; CDG \equiv 1. \tag{3.4.26}$$

The intrablock subgroups of the treatment group orthogonal to these three factor effects ABC, ADE, BDF, CDF are given by $a^{\alpha}b^{\beta}c^{\gamma}d^{\delta}e^{\varepsilon}f^{\theta}g^{\lambda}$ such that

$$\left.\begin{array}{lr} \alpha + \beta + \gamma & \equiv 0 \\ \alpha + \delta + \varepsilon & \equiv 0 \\ \beta + \delta + \theta & \equiv 0 \\ \gamma + \delta + \lambda & \equiv 0 \end{array}\right\} \text{mod } 2.$$

Hence on summing up we obtain

$$\delta \equiv \varepsilon + \theta + \lambda \qquad \text{mod } 2.$$

Consequently

$\alpha \equiv \theta + \lambda$	$\beta \equiv \varepsilon + \lambda$	$\gamma \equiv \varepsilon + \theta$	$\delta \equiv \varepsilon + \theta + \lambda$	ε	θ	λ	
0	0	0	0	0	0	0	1
1	1	0	1	0	0	1	*abdg*
1	0	1	1	0	1	0	*acdf*
0	1	1	0	0	1	1	*bcfg*
0	1	1	1	1	0	0	*bcde*
1	0	1	0	1	0	1	*aceg*
1	1	0	0	1	1	0	*abef*
0	0	0	1	1	1	1	*defg*

Thus we get the intrablock subgroup of the treatment group G as follows :

(3.4.27) $\qquad 1, aceg, bcfg, abef, defg, acdf, bcde, abdg.$

The estimate for article A is given by

(3.4.28)
$$L_A = \frac{1}{4}(x_{aceg} + x_{abef} + x_{acdf} + x_{abdg} - x_{bcfg} - x_{defg} - x_{bcde} - x_1).$$

Hence

$$\text{Var}(L_A) = \frac{\sigma^2}{2}.$$

Similarly

(3.4.29) $$L_B = \frac{1}{4}(-x_{aceg} + x_{abef} - x_{acdf} + x_{abdg} + x_{bcfg} - x_{defg} + x_{bcde} - x_1),$$

and hence

$$\text{Var}(L_B) = \frac{\sigma^2}{2}.$$

One can see that the contrasts $L_A, L_B, L_C, \ldots, L_G$ are mutually orthogonal.

It was later pointed out by Hotelling [2] that, if for each combination all the "omitted" articles are placed in the opposite pan of the balance so that the observation relates to a difference, the variance of the estimate of the weight of A is further reduced to $\sigma^2/8$.

Now in this special case, our model becomes as follows :

(3.4.30)
$$\begin{aligned}
E(x_{aceg}) &= A - B + C - D + E - F + G - \mu \\
E(x_{abef}) &= A + B - C - D + E + F - G - \mu \\
E(x_{acdf}) &= A - B + C + D - E + F - G - \mu \\
E(x_{abdg}) &= A + B - C + D - E - F + G - \mu \\
E(x_{bcfg}) &= -A + B + C - D - E + F + G - \mu \\
E(x_{defg}) &= -A - B - C + D + E + F + G - \mu \\
E(x_{bcde}) &= -A + B + C + D + E - F - G - \mu \\
E(x_1) &= -A - B - C - D - E - F - G - \mu
\end{aligned}$$

This can be written as

$$E(\underline{x}) = H\underline{\beta}, \tag{3.4.31}$$

where

$$\underline{x}' = (x_{aceg}\ x_{abef}\ x_{acdf}\ x_{abdg}\ x_{bcfg}\ x_{defg}\ x_{bcde}\ x_1),$$

$$\underline{\beta}' = (A\ B\ C\ D\ E\ F\ G\ \mu),$$

and

$$H = \begin{Vmatrix} 1 & -1 & 1 & -1 & 1 & -1 & 1 & -1 \\ 1 & 1 & -1 & -1 & 1 & 1 & -1 & -1 \\ 1 & -1 & 1 & 1 & -1 & 1 & -1 & -1 \\ 1 & 1 & -1 & 1 & -1 & -1 & 1 & -1 \\ -1 & 1 & 1 & -1 & -1 & 1 & 1 & -1 \\ -1 & -1 & -1 & 1 & 1 & 1 & 1 & -1 \\ -1 & 1 & 1 & 1 & 1 & -1 & -1 & -1 \\ -1 & -1 & -1 & -1 & -1 & -1 & -1 & -1 \end{Vmatrix} \tag{3.4.32}$$

H being the Hadamard matrix of order 8. Hence the least-squares estimate $\underline{b}$ of $\underline{\beta}$ is obtained by

$$\underline{b} = (H'H)^{-1}H'\underline{x}. \tag{3.4.33}$$

The variance of $\underline{b}$ is given by

$$\begin{aligned} E(\underline{b} - \underline{\beta})(\underline{b} - \underline{\beta})' &= (H'H)^{-1}H'H(H'H)^{-1}\sigma^2 \\ &= (H'H)^{-1}\sigma^2 = \frac{\sigma^2}{8} I_8. \end{aligned}$$

The reader is referred to Kishen [3], Mood [4], and Banerjee [5] for further details.

3.5. Confounding

The advantage of dividing experimental units into blocks in such a way that plots in a block are inherently more alike than those in different blocks is that the precision of the analysis of experimental results then depends only on intrablock variation. Adoption of a fractional scheme means that the total number of treatments may be large and may far exceed the number of plots in the most suitable size of block. For example, a 3^3 factorial arrangement might be used in a randomized block experiment with blocks of size 27; so large a block, however, might happen to be impossible because each block would have to consist of leaves of approximately equal age on one plant or animals of one litter. Even though this size of block is not absolutely prevented by the nature of the experimental material, the commonly encountered *negative correlation between block size and homogeneity of plots within a block* may go far to eliminate all advantages from blocking. No general rule can be given for the suitable size of block: blocks of size 27 may be very satisfactory when they are 27 parallel bacterial cultures, but hopelessly heterogeneous when they are 27 successive weeks of factory operation in which alternative combinations of raw materials for a certain process are to be tried. In a study of response to alternative methods of instructing children, blocks of 2 of identical age and parentage -- that is, twins -- might be assembled in order to eliminate family heterogeneity, and in such a case block size 2 is the maximum. In a study of plant nutrition in a greenhouse, with a single pot as the plot, block size might rest upon the convenience of the experimenter, and heterogeneity would probably increase only slowly with block size; but arrangements of bench space would impose some constraints. In an experiment involving tests on human blood samples, with each block consisting of samples drawn from one subject on one occasion, heterogeneity within the block might be almost independent of block size, but the block size would depend upon the persuasiveness of the experimenter.

Confounding has been devised in order to permit gaining the advantages of small blocks without seriously limiting the use of the factorial scheme of treatments. The underlying idea is that *certain interactions of little importance should be chosen for sacrifice -- confounded with blocks*. Instead of the imprecision consequent on the use of large and heterogeneous blocks spread over all treatment contrasts, it is concentrated on those confounded interactions in such a way that the remaining contrasts have the full precision of smaller blocks. If it is desirable, the confounded interactions can be different in different blocks, so that the loss of information would be partial for a large number of interactions instead of total for a few. In combination with fractional replication, confounding enables factorial design to be exploited to the full, with great flexibility and adaptability to special circumstances. - Quoted from D. J. Finney: *An Introduction to The Theory of Experimental Design*, 1963.

Suppose that a 2^3 design is wanted, but four plots per block can be available. The experiment might be arranged so as to have equal numbers of blocks of two types :

(i) 1, ab, ac, bc,

(ii) a, b, c, abc.

The contrasts for the six elements A, B, C, AB, AC, BC of the effect group are seen to be orthogonal to the block differences :

$$L_A = \frac{1}{2}(x_{abc} + x_{ab} + x_{ac} - x_{bc} + x_a - x_b - x_c - x_1),$$

(3.5.1)

$$L_{AB} = \frac{1}{2}(x_{abc} + x_{ab} - x_{ac} - x_{bc} - x_a - x_b + x_c + x_1).$$

The three-factor interaction ABC corresponds to the difference of blocks of type (ii) and type (i). The interaction ABC is said to be *confounded with blocks*. In other words, the contrast

(3.5.2) $$L_{ABC} = \frac{1}{2}(x_{abc} - x_{ab} - x_{ac} - x_{bc} + x_a + x_b + x_c + x_1)$$

estimating the three-factor interaction *ABC* belongs to the degrees of freedom of block effects. Unless one has some information on the block effects, little information on this interaction can be obtained, and it has been sacrificed in order to be able to investigate the remaining six degrees of freedom with precision appropriate to blocks of size 4. However, if the total number of plots is $8r$, so that the experiment has r blocks of each of the two types, information on the confounded interaction can be obtained, that is, since we have r blocks of each type, the block totals are denoted by

$$B_{11}, B_{12}, \ldots, B_{1r} \quad \text{for type (i)}$$

and

$$B_{21}, B_{22}, \ldots, B_{2r} \quad \text{for type (ii).}$$

If r is sufficiently large, one can proceed as follows : group together pairs of blocks as complete replicates, *by choice of pairs to be inherently similar*, and choose one block at random from each "superblock" to be of type (i), leaving the other as type (ii). The sum of squares of block totals

$$Q = \frac{1}{4}\sum_{\alpha=1}^{r} B_{1\alpha}^2 + \frac{1}{4}\sum_{\alpha=1}^{r} B_{2\alpha}^2 - \frac{1}{8r} B_{..}^2 \qquad \text{(d.f. } 2r - 1\text{)},$$

where

$$B_{1.} = \sum_{\alpha=1}^{r} B_{1\alpha} \ ; \quad B_{2.} = \sum_{\alpha=1}^{r} B_{2\alpha} \ ; \quad B_{..} = B_{1.} + B_{2.}$$

is partitioned as follows : Let $\left(B_{1\alpha}, B_{2\alpha}\right)$, $\alpha = 1,\ldots,r$ be superblocks. The sum of squares due to superblocks (between superblocks) is given by

(3.5.3)
$$Q_1 = \frac{1}{8} \sum_{\alpha=1}^{r} B_{\cdot\alpha}^2 - \frac{1}{8r} B_{\cdot\cdot}^2 \qquad \text{(d.f. } r - 1\text{)},$$

where $B_{\cdot\alpha} = B_{1\alpha} + B_{2\alpha}$, and the sum of squares due to ABC is given by

(3.5.4)
$$Q_2 = \frac{1}{4r}\left(B_{1\cdot}^2 + B_{2\cdot}^2\right) - \frac{1}{8r} B_{\cdot\cdot}^2 \qquad \text{(d.f. 1)}.$$

Hence the sum of squares due to interblock error is given by

(3.5.5)
$$Q - Q_1 - Q_2 = \frac{1}{4} \sum_{\alpha=1}^{r} B_{1\alpha}^2 + \frac{1}{4} \sum_{\alpha=1}^{r} B_{2\alpha}^2 - \frac{1}{8} \sum_{\alpha=1}^{r} B_{\cdot\alpha}^2$$
$$- \frac{1}{4r}\left(B_{1\cdot}^2 + B_{2\cdot}^2\right) + \frac{1}{8r} B_{\cdot\cdot}^2 \qquad \text{(d.f. } r - 1\text{)}.$$

The analysis of variance for a 2^3 factorial experiment confounding the three-factor interaction is shown in Table 3.4.

From the point of view of constructing designs, confounding can be regarded as a special type of fractional replication in which blocks play the part of an additional factor. For example, the 2^3 design can be formed by introducing a quasifactor X to represent the difference between the two types of block. Consider the half-replicate of 2^4 with alias subgroup 1, $ABCX$. The resultant combinations of factors orthogonal to the generator are

$$1,\ ab,\ ac,\ ax,\ bc,\ bx,\ cx,\ abcx.$$

If x is interpreted as "put in blocks of type (ii)" and the absence of x as "put in blocks of type (i)," then the confounding scheme for 2^3 is obtained, and $ABC \equiv X$; that is, the three-factor interaction ABC is the alias of the block effect. Also

$$AB \equiv CX, \quad AC \equiv BX, \quad BC \equiv AX.$$

TABLE 3.4

Analysis of Variance for a 2^3 Factorial Experiment Confounding the Three-Factor Interaction

Variation	Degrees of Freedom	Sum of Squares	Mean Sum of Squares
Between superblocks	$r - 1$		
ABC	1		
Interblock error	$r - 1$		
Blocks	$2r - 1$		
A	1		
B	1		
C	1		
AB	1		
AC	1		
BC	1		
Error	$6(r - 1)$		
Total	$8r - 1$		

Suppose now that two quasifactors X and Y are introduced, and the quarter-replicate of 2^5 with aliases determined from

$$ABX \equiv ACY \equiv BCXY \equiv 1$$

is taken. The treatments orthogonal to the alias subgroup are $a^{\alpha}b^{\beta}c^{\gamma}x^{\delta}y^{\varepsilon}$ such that

$$\alpha + \beta + \delta \equiv 0 \quad \text{and} \quad \alpha + \gamma + \varepsilon \equiv 0 \quad \text{mod } 2.$$

Hence

$$1,\ axy,\ bx,\ aby,\ cy,\ acx,\ bcxy,\ abc.$$

The absence of both x and y, the presence of x alone, y alone, and xy may be taken to correspond to four types of block, giving

$$\begin{aligned} &\text{(i)} && 1,\ abc; \\ x \to\ &\text{(ii)} && b,\ ac; \\ y \to\ &\text{(iii)} && c,\ ab; \\ xy \to\ &\text{(iv)} && a,\ bc. \end{aligned}$$

This is a 2^3 design in blocks of size 2, which can be executed in multiples of four blocks. It confounds AB, AC, BC simultaneously. As a matter of fact,

$$A \equiv BX \equiv CY \equiv ABCXY,$$

$$B \equiv AX \equiv ABCY \equiv CXY,$$

$$C \equiv ABCX \equiv AY \equiv BXY.$$

Any main effect has as aliases the interactions of each of the others with block differences. If such interactions can be reasonably ignored, this confounding scheme enables one to estimate each of the main effects A, B, and C to be as well as the three-factor interaction ABC, although AB, AC, BC have been sacrificed.

From the point of view of the analysis of variance, the total seven degrees of freedom are divided into three degrees of freedom corresponding to the main effect, one degree of freedom corresponding to the three-factor interaction ABC, and three degrees of freedom corresponding to three two-factor interactions AB, AC, BC.

In order to confound in blocks of size 2^p, an additional $(n - p)$ quasifactors X, Y,... should be taken, and the main effects and interactions corresponding to $(2^{n-p} - 1)$ degrees of freedom between blocks are confounded. Any $1/2^{n-p}$ replicate of a design for 2^{2n-p} then gives a possible scheme of confounding, *provided that no elements of the alias subgroup consist of quasifactors alone.* For example, for confounding 2^5 in blocks of size 8, quasifactors X and Y must be introduced and a quarter-replicate of 2^7 formed.

Finally we consider the confounding of 3^n design. For example, to confound the 27 treatments of a 3^3 experiment in blocks of size 9, a confounding subgroup of order 9 must be chosen (see Table 3.5). The three-factor interactions are $AB^{\beta'}C^{\gamma'}$, where β', γ' = 1, 2. If ABC^2 is chosen, the intrablock subgroup of treatments consists of all $a^\alpha b^\beta c^\gamma$ such that

$$\alpha + \beta + 2\gamma \equiv 0 \quad \text{mod } 3.$$

Hence one obtains

$$1,\ ab^2,\ a^2b,\ ac,\ a^2c^2,\ a^2b^2c,\ b^2c^2,\ bc,\ abc^2.$$

Multiplication of these nine elements by any elements $a^\alpha b^\beta c^\gamma$ satisfying

$$\alpha + \beta + 2\gamma \equiv 1 \quad \text{mod } 3,$$

$$\alpha + \beta + 2\gamma \equiv 2 \quad \text{mod } 3$$

gives the other two blocks.

The partition of degrees of freedom is shown in Table 3.6.

The sum of squares corresponding to the main effect A is obtained in the following manner : The treatment orthogonal to A is the set of $a^\alpha b^\beta c^\gamma$ such that $\alpha \equiv 0$ mod 3. Hence

$$1,\ b,\ b^2,\ c,\ c^2,\ bc,\ bc^2,\ b^2c,\ b^2c^2,$$

Let

TABLE 3.5

System of Blocks for Confounding ABC^2 in a 3^3 Experiment

I	II	III
1	a	a^2
ab^2	a^2b^2	b^2
a^2b	b	ab
ac	a^2c^2	c
a^2c^2	c	ac^2
bc	abc	a^2bc
b^2c^2	ab^2c^2	$a^2b^2c^2$
abc^2	a^2bc^2	bc^2
a^2b^2c	b^2c	ab^2c

TABLE 3.6

Partition of Degrees of Freedom in a 3^3 Experiment

Sources of Variation	Degrees of Freedom
Blocks ($+ABC^2$)	2
A	2
B	2
C	2
$A \times B$	4
$A \times C$	4
$B \times C$	4
ABC, AB^2C, AB^2C^2	6
Total	26 = 27 - 1

$$X_1 \equiv x_1 + x_b + x_{b^2} + x_c + x_{c^2} + x_{bc} + x_{bc^2} + x_{b^2c} + x_{b^2c^2},$$

(3.5.6) $$X_a \equiv x_a + x_{ab} + x_{ab^2} + x_{ac} + x_{ac^2} + x_{abc} + x_{abc^2} + x_{ab^2c} + x_{ab^2c^2},$$

$$X_{a^2} \equiv x_{a^2} + x_{a^2b} + x_{a^2b^2} + x_{a^2c} + x_{a^2c^2} + x_{a^2bc} + x_{a^2bc^2} + x_{a^2b^2c} + x_{a^2b^2c^2}.$$

Then the sum of squares due to A is given by

(3.5.7) $$Q_A = \frac{1}{9}(X_1^2 + X_a^2 + X_{a^2}^2) - 27\bar{x}^2.$$

One can check that the treatments in X_1, X_a, X_{a^2} are evenly distributed in three blocks:

BLOCKS

	I	II	III
X	$1, bc, b^2c$	b, c^2, b^2c^2	b^2, c, bc^2
X_a	ab^2, ac, abc^2	a, abc, ab^2c^2	ab, ac^2, ab^2c
X_{a^2}	a^2b, a^2c^2, a^2b^2c	a^2b^2, a^2c, a^2bc^2	$a^2, a^2bc, a^2b^2c^2$

Hence any contrast depending on X_1, X_a, X_{a^2} is orthogonal to any contrast depending on block totals, that is,the two degrees of freedom carried by the main effect A are orthogonal to the two degrees of freedom carried by blocks, (pp.17-19).

REFERENCES

1. R. L. Plackett and J. P. Burman, *Biometrika, 33,* 305 (1946).

2. H. Hotelling, *Ann. Math. Statist., 15,* 297 (1944).

3. K. Kishen, *Ann. Math. Statist., 17,* 432 (1946).

4. A. M. Mood, *Ann. Math. Statist., 17,* 432 (1946).

5. K. S. Banerjee, *Advances in the Theory and Construction of Weighing Designs,* presented at the International Symposium of Design of Experiments and Linear Models, Colorado State University, Fort Collins, Colorado, March 19-25, 1973.

Chapter 4

THEORY OF BLOCK DESIGNS -- INTRABLOCK ANALYSIS

4.1. Linear Estimation

We begin this section with the following lemma:

Lemma 4.1.1. For any given rectangular matrix $A(m \times n)$ and a vector $\underline{y}(n \times 1)$, a solution $\underline{x}(m \times 1)$ of the equation

$$AA'\underline{x} = A\underline{y} \tag{4.1.1}$$

always exists. The solution $\underline{x}$ may not be unique, but $A'\underline{x}$ is unique.

Proof. First we remark that

$$r(AA') = r(A).$$

Since $r(AA':A\underline{y}) = r(A)$, it follows that

$$r(AA' : A\underline{y}) = r(AA').$$

This is the necessary and sufficient condition for the existence of a solution of the linear equation (4.1.1).

Let $\underline{x}_1$ and $\underline{x}_2$ be two solutions of (4.1.1). Then

$$AA'\underline{x}_1 = A\underline{y} = AA'\underline{x}_2. \tag{4.1.2}$$

Hence we get

(4.1.3) $$AA'(\underline{x}_1 - \underline{x}_2) = 0.$$

If we put

(4.1.4) $$\underline{z} = A'(\underline{x}_1 - \underline{x}_2),$$

then it follows from (4.1.3) that

$$(\underline{x}_1' - \underline{x}_2')AA'(\underline{x}_1 - \underline{x}_2) = 0$$

or

$$\underline{z}'\underline{z} = 0.$$

Thus we get $\underline{z} = 0$, that is,

$$A'\underline{x}_1 = A'\underline{x}_2.$$

Lemma 4.1.2. Given a vector $\underline{c}(n \times 1)$ and a matrix $A(m \times n)$, we can find a unique decomposition of $\underline{c}$,

(4.1.5) $$\underline{c} = \underline{b} + \underline{e},$$

such that

(4.1.6) $$A\underline{e} = 0 \quad \text{and} \quad \underline{b} = A'\underline{l},$$

where $\underline{b}(n \times 1)$, $\underline{e}(n \times 1)$, and $l(m \times 1)$.

Proof. Let

(4.1.7) $$\underline{c}' = \underline{e}' + \underline{l}'A.$$

Then

(4.1.8) $$\underline{c}'A' = \underline{e}'A' + \underline{l}'AA'.$$

So, if $A\underline{e} = 0$, then

(4.1.9) $$\underline{c}'A' = \underline{l}'AA' \quad \text{or} \quad AA'\underline{l} = A\underline{c}.$$

Now $A'\underline{l}$ is determined uniquely from (4.1.9) (Lemma 4.1.1). Put

$$\underline{b} = A'\underline{l}. \tag{4.1.10}$$

and put

$$\underline{c} - \underline{b} \equiv \underline{e}.$$

Then

$$A\underline{e} = A\underline{c} - A\underline{b} = A\underline{c} - AA'\underline{l} = 0,$$

$$\underline{b}'\underline{e} = \underline{l}'A\underline{e} = 0;$$

that is, $\underline{e}$ is orthogonal to the vector space generated by the row vector of A and also to $\underline{b}$. Thus the decomposition

$$\underline{c} = \underline{e} + \underline{b}$$

is unique.

Let $\underline{y}(n \times 1)$ be a random vector such that

$$E(\underline{y}) = A'\underline{p}, \tag{4.1.11}$$

where

$$A = \begin{Vmatrix} a_{11} & a_{12} & \cdots & a_{1n} \\ a_{21} & a_{22} & \cdots & a_{2n} \\ \cdots & \cdots & \cdots & \cdots \\ a_{m1} & a_{m2} & \cdots & a_{mn} \end{Vmatrix}$$

is a known constant matrix and $\underline{p}' = (p_1 \cdots p_m)$ is unknown and to be estimated.

<u>Definition 4.1.1</u> We say that a linear parametric function $\underline{l}'\underline{p}$ is *estimable* or *linearly estimable* if there exsits a linear function $\underline{c}'\underline{y}$ such that

(4.1.12) $$E(\underline{c}'\underline{y}) = \underline{l}'\underline{p}$$

identically in $\underline{p}$. Hence, if $\underline{l}'\underline{p}$ is estimable, that is,

$$\underline{c}'A'\underline{p} = \underline{l}'\underline{p}$$

identically with respect to $\underline{p}$, then of necessity

(4.1.13) $$\underline{l} = A\underline{c}.$$

Thus *the necessary and sufficient condition for* $\underline{l}'\underline{p}$ *to be estimable is*

(4.1.14) $$r(A) = r(A:\underline{l}).$$

Let us further assume that $y_1, y_2, \ldots, y_n$ are mutually independent random variables with common variance σ^2, (unknown).

Definition 4.1.2 A linear function $\underline{c}'\underline{y}$ is said to *belong to error* if

(4.1.15) $$E(\underline{c}'\underline{y}) = 0 \quad \text{(independently of } \underline{p}.)$$

The necessary and sufficient condition for $\underline{c}'\underline{y}$ *to belong to error is*

(4.1.16) $$A\underline{c} = 0;$$

that is, the row vectors of A are orthogonal to $\underline{c}$; in other words, the vector $\underline{c}$ lies in the vector space E, which is orthogonal to the vector space generated by the row vectors of A.

Definition 4.1.3 The vector space generated by the row vectors of A is called *the estimation space*. The set of linear forms of $A\underline{y}$ is called *the estimation set*. The vector space E is called *the error space* and the set of linear forms of the type $\underline{c}'\underline{y}$, $\underline{c} \in E$, is called *the error set*.

There will be no confusion in denoting the estimation space by A. Since A and E are orthogonal to each other, hence $A \cap E = 0$, and consequently $A \cup E$ is the whole space of vectors of n-dimensions.

Theorem 4.1.1 If $\underline{l}'\underline{p}$ is estimable, then there exists a unique linear form $\underline{d}'\underline{y}$ belonging to the estimation set such that $E(\underline{d}'\underline{y}) = \underline{l}'\underline{p}$ and $\underline{d}'\underline{y}$ is the best linear unbiased estimator of $\underline{l}'\underline{p}$.

Proof. Since $\underline{l}'\underline{p}$ is estimable, there exists a linear form $\underline{c}'\underline{y}$ such that

$$E(\underline{c}'\underline{y}) = \underline{l}'\underline{p}.$$

Decompose $\underline{c}'$ into $\underline{c}' = \underline{d}' + \underline{e}'$, where $\underline{d}$ belongs to A and $\underline{e} \in E$, (i.e., $A\underline{e} = 0$). Then

$$E(\underline{c}'\underline{y}) = E(\underline{d}'\underline{y}) + E(\underline{e}'\underline{y}) = E(\underline{d}'\underline{y}) = \underline{l}'\underline{p}.$$

Suppose there is a second vector $\underline{d}^*$ in A such that

$$E(\underline{d}^{*\prime}\underline{y}) = \underline{l}'\underline{p}.$$

Then we have

$$E[(\underline{d} - \underline{d}^*)'\underline{y}] = 0.$$

Thus $\underline{d} - \underline{d}^*$ must belong to the error space E. But $\underline{d} - \underline{d}^*$ belongs to the estimation space. Hence

$$\underline{d} = \underline{d}^*.$$

Let $\underline{c}'\underline{y}$ be any estimator of $\underline{l}'\underline{p}$, that is, $E(\underline{c}'\underline{y}) = \underline{l}'\underline{p}$. Then

$$\begin{aligned} Var(\underline{c}'\underline{y}) = Var(\underline{d}'\underline{y} + \underline{e}'\underline{y}) &= Var(\underline{d}'\underline{y}) + Var(\underline{e}'\underline{y}) \\ &\geqq Var(\underline{d}'\underline{y}). \end{aligned}$$

The equality is attained if and only if $\underline{c} = \underline{d}$, that is,

$$Var(\underline{e}'\underline{y}) = \underline{e}'\underline{e}\cdot\sigma^2 = 0 \nRightarrow \underline{e} = 0.$$

Suppose $\underline{l}'\underline{p}$ is estimable. Then the best linear unbiased estimator is given by $\underline{d}'\underline{y}$, where $\underline{d}' = \underline{q}'A$. On the other hand, using (4.1.13), we obtain

$$A\underline{d} = \underline{l},$$

and hence

$$AA'\underline{q} = \underline{l}. \tag{4.1.17}$$

Since A and $\underline{l}$ are given, this last equation yields the desired solution and (4.1.17) has a unique solution $A'\underline{q} = \underline{d}$.

Theorem 4.1.2. If $\hat{\underline{p}}$ is a solution of $AA'\underline{p} = A\underline{y}$ and $\underline{l}'\underline{p}$ is estimable, then $\underline{l}'\hat{\underline{p}}$ is the best linear unbiased estimator.

Proof. Since $\underline{l}'\underline{p}$ is estimable, there exists a $\underline{q}$ such that

$$AA'\underline{q} = \underline{l},$$

and hence

$$\underline{l}'\hat{\underline{p}} = \underline{q}'AA'\hat{\underline{p}} = \underline{q}'A\underline{y},$$

that is, $\underline{l}'\hat{\underline{p}}$ belongs to the estimation set. Thus by Theorem 4.1.1, so long as

$$E(\underline{l}'\hat{\underline{p}}) = \underline{l}'\underline{p},$$

$\underline{l}'\underline{p}$ is the best linear unbiased estimator. Now

$$E(\underline{l}'\hat{\underline{p}}) = \underline{q}'AE(\underline{y}) = \underline{q}AA'\underline{p} = \underline{l}'\underline{p}.$$

Remark : Theorem 4.1.2 asserts that $\underline{l}'\hat{\underline{p}}$ is the best linear unbiased estimator of $\underline{l}'\underline{p}$ if $\hat{\underline{p}}$ is the least-squares estimator of $\underline{p}$ (Gauss-Markov theorem).

The variance of the best linear unbiased estimator $\underline{l}'\hat{\underline{p}}$ is

(4.1.18) $$Var(\underline{l}'\hat{\underline{p}}) = \underline{q}'AA'\underline{q}\cdot\sigma^2 = \underline{q}'\underline{l}\cdot\sigma^2.$$

4.2. General Block Designs

There are given b blocks consisting of more or less homogeneous experimental units or plots. Suppose the jth block consist of k_j plots. Given v treatments whose effects have the magnitudes $\tau_1, \tau_2, \ldots, \tau_v$, our objective is to estimate or test some hypotheses concerning τ_i.

The design is said to be *a complete block design* if v is the number of plots in every block, and the design is said to be an *incomplete block design* if v is greater than the number of plots in a block.

There are $n = \sum_{j=1}^{b} k_j$ experimental units and therefore n observations $y_1 \cdots y_n$. The usual model is

$$y_u = \gamma + \tau_i + \beta_j + e_{ij},$$

if the uth plot belonging to the jth block has been applied the ith treatment. τ_i is called *the treatment effect*, β_j is called *the block effect*, and e_{ij} is called *the residual (or error) variate*.

In *intrablock analysis* we consider the block effect β_j as a constant, whereas in *interblock analysis* we consider the block effect β_j as a random variable. Here in Chapter IV we shall be concerned with intrablock analysis.

Let the observation vector be $\underline{y}$ whose uth component is y_u and let the incidence matrices for treatments and blocks be

(4.2.1) $$\Phi = \left\| \underline{\zeta}_1 \underline{\zeta}_2 \cdots \underline{\zeta}_v \right\| \quad \text{and} \quad \Psi = \left\| \underline{\eta}_1 \underline{\eta}_2 \cdots \underline{\eta}_b \right\|$$

respectively. Then our linear model is

(4.2.2) $$\underline{y} = \gamma \underline{j}_n + \Phi \underline{\tau} + \Psi \underline{\beta} + \underline{e}.$$

(4.2.3)

Grand total $G = \underline{j}_n' \underline{y}$,

Treatment totals $$\underline{T} = \begin{bmatrix} T_1 \\ T_2 \\ \vdots \\ T_v \end{bmatrix} = \Phi' \underline{y},$$

Block totals $$\underline{B} = \begin{bmatrix} B_1 \\ B_2 \\ \vdots \\ B_b \end{bmatrix} = \Psi' \underline{y}.$$

The noraml equation of the least-squares estimation is

(4.2.4) $$\left\| \begin{matrix} \underline{j}_n' \\ \Phi' \\ \Psi' \end{matrix} \right\| \cdot \left\| \underline{j}_n \Phi \Psi \right\| \cdot \begin{bmatrix} g \\ \underline{t} \\ \underline{b} \end{bmatrix} = \left\| \begin{matrix} \underline{j}_n' \\ \Phi' \\ \Psi' \end{matrix} \right\| \underline{y},$$

that is

(4.2.5) $$\left\| \begin{matrix} n & \underline{r}' & \underline{k}' \\ \underline{r} & D_r & N \\ \underline{k} & N' & D_k \end{matrix} \right\| \cdot \begin{bmatrix} g \\ \underline{t} \\ \underline{b} \end{bmatrix} = \begin{bmatrix} G \\ \underline{T} \\ \underline{B} \end{bmatrix}.$$

In detail

$$ng + \underline{r}'\underline{t} + \underline{k}'\underline{b} = G,$$

(4.2.6) $$\underline{r}g + D_r\underline{t} + N\underline{b} = \underline{T},$$

$$\underline{k}g + N'\underline{t} + D_k\underline{b} = \underline{B}.$$

There are $v + b + 1$ linear equations in the whole, but they are linearly dependent. Indeed

$$ng + \underline{r}'\underline{t} + \underline{k}'\underline{b} = \underline{j}_v'(\underline{r}g + D_r\underline{t} + N\underline{b}) = \underline{j}_b'(\underline{k}g + N'\underline{t} + D_k\underline{b}).$$

Thus we consider only

(4.2.7a) $$\underline{r}g + D_r\underline{t} + N\underline{b} = \underline{T},$$

(4.2.7b) $$\underline{k}g + N'\underline{t} + D_k\underline{b} = \underline{B}.$$

If we multiply ND_k^{-1} with the expression (4.2.7b) from the left, we have

(4.2.8) $$\underline{r}g + ND_k^{-1}N'\underline{t} + N\underline{b} = ND_k^{-1}\underline{B}.$$

Hence by subtracting (4.2.8) from (4.2.7a), we get

(4.2.9) $$(D_r - ND_k^{-1}N')\underline{t} = \underline{T} - ND_k^{-1}\underline{B} \equiv \underline{Q},$$

where

(4.2.10) $$Q_\alpha = T_\alpha - n_{\alpha 1}\frac{B_1}{k_1} - n_{\alpha 2} \times \frac{B_2}{k_2} - \cdots - n_{\alpha b}\frac{B_k}{k_b} \qquad (\alpha = 1,2,\ldots,v),$$

is called *the adjusted treatment total*, and obviously

(4.2.11) $$\sum_{\alpha=1}^{v} Q_\alpha = 0.$$

Let

(4.2.12) $$D_r - ND_k^{-1}N' \equiv C = \|c_{ij}\|.$$

Then

(4.2.13)
$$c_{ii} = r_i - \frac{n_{i1}^2}{k_1} - \cdots - \frac{n_{ib}^2}{k_b},$$
$$c_{ij} = -\frac{n_{i1}n_{j1}}{k_1} - \cdots - \frac{n_{ib}n_{jb}}{k_b} \qquad (i \neq j).$$

<u>Assumption.</u> If, in particular, the block sizes are constant and are equal to k, and $n_{ij} = 0$ or 1, then

(4.2.14)
$$c_{ii} = r_i(1 - \tfrac{1}{k}); \qquad c_{ij} = -\frac{\lambda_{ij}}{k},$$

where λ_{ij} is the number of times the ith and jth treatments occur together in the same block.

In this case *the adjusted normal equation* becomes

(4.2.15)
$$C\underline{t} = \underline{Q}$$

or

(4.2.16)
$$\left\|\begin{matrix} r_1(1-\frac{1}{k}) & -\frac{\lambda_{12}}{k} & \cdots & -\frac{\lambda_{1v}}{k} \\ -\frac{\lambda_{12}}{k} & r_2(1-\frac{1}{k}) & \cdots & -\frac{\lambda_{2v}}{k} \\ \vdots & \vdots & \ddots & \vdots \\ -\frac{\lambda_{1v}}{k} & -\frac{\lambda_{2v}}{k} & \cdots & r_v(1-\frac{1}{k}) \end{matrix}\right\| \cdot \left|\begin{matrix} t_1 \\ t_2 \\ \vdots \\ t_v \end{matrix}\right| = \left|\begin{matrix} Q_1 \\ Q_2 \\ \vdots \\ Q_v \end{matrix}\right| .$$

Under our assumption, it follows that

(4.2.17)
$$NN' = \left\|\begin{matrix} r_1 & \lambda_{12} & \cdots & \lambda_{1v} \\ \lambda_{12} & r_2 & \cdots & \lambda_{2v} \\ \vdots & \vdots & \ddots & \vdots \\ \lambda_{1v} & \lambda_{2v} & \cdots & r_v \end{matrix}\right\| .$$

The row sums and column sums of C *are zero.* In fact,

$$C\underline{j}_v = D_r\underline{j}_v - ND_k^{-1}N\,\underline{j}_v = \underline{r} - ND_k^{-1}\underline{k} = \underline{r} - N\underline{j}_b = \underline{r} - \underline{r} = 0.$$

Hence

(4.2.18) $$r(C) \leqq v - 1.$$

Thus *in general all treatment effects cannot be estimated.*

If $\underline{l}'\underline{\tau}$ is estimable, then

(4.2.19*) $$\underline{l}' = \underline{p}'C$$

because

$$\underline{d}'E(y) = \gamma\underline{d}'\underline{j}_n + \underline{d}'\Phi\underline{\tau} + \underline{d}'\Psi\underline{\beta} = \underline{l}'\underline{\tau}$$

whence

$$\left.\begin{array}{l}\underline{d}'\underline{j}_n = 0,\\ \underline{d}'\Psi = 0;\end{array}\right. \rightarrow \underline{d} = \left(\Phi - \frac{1}{k}\Psi N'\right)\underline{\Phi},$$

then the relation $\underline{l}' = \underline{d}'\Phi$, yields

$$\underline{l}' = \underline{p}'C.$$

and hence

(4.2.19) $$\underline{l}'\underline{j}_v = \underline{p}'C\underline{j}_v = 0 \quad \text{or} \quad \sum_{\alpha=1}^{v} l_\alpha = 0.$$

This shows that if a linear function $\underline{l}'\underline{\tau}$ is estimable, then this must be a *treatment contrast.*

Definition 4.2.1. A treatment and a block are said to be *associated* if the treatment occurs in the block.

Two treatments (or two blocks) or a block and a treatment are said to be *connected* if one can find a chain between the two treatments (or two blocks) or the block and the treatment consisting alternatively of a treatment and a block such that any two consecutive members of the chain are associated.

Thus if the treatments i_0 and i_n are connected, then we should be able to find a chain

$$i_0 j_1 i_1 j_2 i_2 \cdots i_{n-1} j_n i_n$$

such that the block j_ρ is associated with the treatments $i_{\rho-1}$ and i_ρ.

Definition 4.2.2. A design is said to be *connected* if all the treatments and blocks are connected with each other.

We can similarly define a *connected portion* of a design.

Thus a design will either be connected or will break up into different connected portions, where any two members belonging to different portions are unconnected.

Theorem 4.2.1. For a *connected design* every treatment contrast is estimable.

Proof. Consider a treatment contrast

$$l_1\tau_1 + l_2\tau_2 + \cdots + l_v\tau_v, \qquad \sum_{\alpha=1}^{v} l_\alpha = 0.$$

Then

$$l_1\tau_1 + \cdots + l_v\tau_v = l_1(\tau_1 - \tau_v) + l_2(\tau_2 - \tau_v) + \cdots + l_{v-1}(\tau_{v-1} - \tau_v).$$

Hence it will be sufficient to show that we can estimate the difference of any two treatment effects, say τ_{i_0} and τ_{i_n}.

Now by the definition of connectedness, there exists a chain

$$i_0 j_1 i_1 j_2 i_2 \cdots i_{n-1} j_n i_n$$

such that treatments $i_{\rho-1}$ and i_ρ occur in the block j_ρ. The difference of the observations corresponding to $i_{\rho-1}$ and i_ρ from the block j_ρ is an estimate of $\tau_{i_{\rho-1}} - \tau_{i_\rho}$. Putting $\rho = 1,2,\ldots,n$ and adding up, we get an estimate $\tau_{i_0} - \tau_{i_n}$. QED.

For any contrast $l_1\tau_1 + \cdots + l_v\tau_v$, its best linear unbiased estimator must be of the form

(4.2.20) $$p_1Q_1 + \cdots + p_vQ_v$$

because $C\underline{t} = \underline{Q}$ has a solution

(4.2.21) $$\underline{\hat{t}} = C^*\underline{Q}.$$

We shall show that C^* turns out to be a *conditional inverse* of C, that is,

(4.2.22) $$CC^*C = C.$$

Since

$$\underline{Q} = \Phi'(I - \frac{1}{k}B)\underline{y}$$

and

$$r\left(\Phi'(I - \frac{1}{k}B)\right) = r\left(\Phi'(I - \frac{1}{k}B)\Phi\right) = r(C),$$

if one can show that $r(C) = v - 1$ (Theorem 4.2.2), then

$$\underline{j}_v'\Phi'(I - \frac{1}{k}B) = 0$$

or

$$\underline{j}_v'\underline{Q} = 0$$

is the only linear relation between $Q_1,\ldots,Q_v$.

From the relation

$$CC^*\underline{C} = \underline{Q}$$

or

$$(CC^* - I)\underline{Q} = 0$$

it follows that each row of $CC^* - I$ is a multiple of $\underline{j}_v'$, and hence

$$CC^* - I = D_{\underline{a}}J_v, \tag{4.2.23}$$

where

$$D_{\underline{a}} = \begin{Vmatrix} a_1 & & & 0 \\ & a_2 & & \\ & & \ddots & \\ 0 & & & a_v \end{Vmatrix}, \qquad J_v = \begin{Vmatrix} 1 & \cdots & 1 \\ & \cdots & \\ 1 & \cdots & 1 \end{Vmatrix},$$

and consequently

$$CC^*C = C.$$

<u>Theorem 4.2.2.</u> For a connected design

$$r(C) = v - 1 \tag{4.2.24}$$

<u>Proof.</u> By (4.2.19*), any linear function $\underline{l}'\underline{\tau}$ that is estimable must be a linear combination of $\underline{c}_1'\underline{\tau},\ldots,\underline{c}_v'\underline{\tau}$, where

$$C = \left\| \begin{array}{c} \underline{c}_1' \\ \underline{c}_2' \\ \vdots \\ \underline{c}_v' \end{array} \right\| .$$

Since we know that the dimension of the set of estimable linear functions of $\underline{\tau}$ is $v - 1$, the number of linearly independent linear forms among $\underline{c}_1'\underline{\tau},\ldots,\underline{c}_v'\underline{\tau}$ is exactly $v - 1$; that is, $r(C) = v - 1$.

Remark : For a general design that breaks up into g connected portions we can show that

(4.2.25) $$r(C) = v - g.$$

The parameters with which we are concerned here are

$$r;\ \tau_1 \cdots \tau_v;\ \beta_1 \cdots \beta_b.$$

They are $v + b + 1$ in number. Mathematically the roles of $\underline{\tau}$ and $\underline{\beta}$ are symmetrical to each other.

Theorem 4.2.3. For a connected design the number of degrees of freedom belonging to estimates is $v + b - 1$.

Proof. The best linear estimates of the parameters are linear combinations of

$$G;\ T_1, T_2, \ldots, T_v;\ B_1, B_2, \ldots, B_b,$$

and we have the relations

$$G = T_1 + T_2 + \cdots + T_v = B_1 + B_2 + \cdots + B_b.$$

Hence not more than $b + v - 1$ of the best linear estimates are linearly independent. Thus the dimension of the estimation space or the number of degrees of freedom belonging to the best linear estimates $\leqq b + v - 1$.

Then consider the set of linear functions

$$Q_1, Q_2, \ldots, Q_v : B_1, B_2, \ldots, B_b,$$

where we have seen that the functions Q are linear combinations of functions T and B. For a connected design the number of degrees of freedom belonging to the functions Q is $v - 1$.

Since $cov(B_i, B_j) = 0$ for $i \neq j$, B_i is orthogonal to B_j if $i \neq j$. Orthogonality implies linear independence in the normal theory. Hence the number of degrees of freedom belonging to the functions B is b. Since $cov(B_i, Q_\alpha) = 0$, the number of degrees of freedom belonging to the linear set generated by $Q_1, \ldots, Q_v : B_1, \ldots, B_b$ is $b + v - 1$. Therefore the number of degrees of freedom belonging to the best linear estimates $\geqq b + v - 1$.

Hence the number of degrees of freedom belonging to the best linear estimate is exactly equal to $v + b - 1$. Q.E.D.

The estimation space is generated by the column vectors of

$$A' \equiv \| j_n \Phi \Psi \|$$

The number of degrees of freedom belonging to the error = $n - (b + v - 1) = n - b - v + 1$.

4.3. Partition of Degrees of Freedom and Sum of Squares

Connected Design. First we construct the correction term

$$\frac{G^2}{n}. \tag{4.3.1}$$

Then we have

$$(4.3.2) \qquad S_b = \frac{B_1^2}{k_1} + \cdots + \frac{B_b^2}{k_b} - \frac{G^2}{n} = \underline{y}'\left(\Psi D_{\underline{k}}^{-1}\Psi' - \frac{1}{n}\underline{1}\,\underline{1}'\right)\underline{y}.$$

The coefficient matrix

$$\left(\Psi D_{\underline{k}}^{-1}\Psi' - \frac{1}{n}\underline{1}\,\underline{1}'\right)$$

is idempotent, and

$$r\left(\Psi D_{\underline{k}}^{-1}\Psi' - \frac{1}{n}\underline{1}\,\underline{1}'\right) = tr\left(\Psi D_{\underline{k}}^{-1}\Psi' - \frac{1}{n}\underline{1}\,\underline{1}'\right) = b - 1$$

and also

$$\left(\Psi D_{\underline{k}}^{-1}\Psi' - \frac{1}{n}\underline{1}\,\underline{1}'\right)\underline{1}\,\underline{1}' = \Psi D_{\underline{k}}^{-1}\,\underline{k}\,\underline{1}' - \underline{1}\,\underline{1}' = \Psi\underline{j}_b\underline{1}' - \underline{1}\,\underline{1}' = 0;$$

that is, any component of the vector

$$\left(\Psi D_{\underline{k}}^{-1}\Psi' - \frac{1}{n}\underline{1}\,\underline{1}'\right)\underline{y}$$

is a contrast.

Next we consider *the sum of squares due to treatment adjusted.* Now

$$\underline{Q} = \left(\Phi' - ND_{\underline{k}}^{-1}\Psi'\right)\underline{y} \equiv K\underline{y}$$

$$K\underline{1} = \Phi'\left(I - \Psi D_{\underline{k}}^{-1}\Psi'\right)\underline{1} = \underline{r} - ND_{\underline{k}}^{-1}\,\underline{k} = \underline{r} - N\underline{j}_b$$

$$(4.3.3) \qquad = \underline{r} - \underline{r} = 0$$

and

$$K\left(\Psi D_{\underline{k}}^{-1}\Psi' - \frac{1}{n}\underline{1}\,\underline{1}'\right) = \Phi'\left(\Psi D_{\underline{k}}^{-1}\Psi' - \frac{1}{n}\underline{1}\,\underline{1}'\right) - ND_{\underline{k}}^{-1}\Psi'\left(\Psi D_{\underline{k}}^{-1}\Psi' - \frac{1}{n}\underline{1}\,\underline{1}'\right)$$

$$= ND_{\underline{k}}^{-1}\Psi' - \frac{1}{n}\underline{r}\,\underline{1}' - ND_{\underline{k}}^{-1}\Psi' + \frac{1}{n}ND_{\underline{k}}^{-1}\,\underline{k}\,\underline{1}' = 0;$$

That is, all the functions of Q are contrasts and orthogonal to the contrasts in S_b^2.

$$Var(Q) = K\ Var(\underline{y}K') = KK'\sigma^2 = C\sigma^2.$$

Thus we take

$$(4.3.4) \qquad S_t^2 = \underline{Q}'C^*\underline{Q} = \hat{\underline{t}}'\underline{Q} = \underline{Q}'\hat{\underline{t}},$$

where $\underline{t}$ is a solution of the adjusted normal equation

$$C\hat{\underline{t}} = \underline{Q}.$$

Since we can write

$$(4.3.5) \qquad S_t^2 = \underline{y}'K'C^*K\underline{y}$$

and

$$(4.3.6) \qquad K'C^*KK'C^*K = K'C^*CC^*K = K'C^*K$$

because we can select the conditional inverse C^* in such a way that

$$(4.3.7) \qquad CC^*C = C \quad \text{and} \quad C^*CC^* = C^*$$

and

$$(4.3.8) \qquad r(K'C^*K) = tr(CC^*) = v - 1.$$

We can easily confirm that all the components of $K\underline{y}$ are contrasts and orthogonal to those in S_b^2.

The analysis of variance is shown in Table 4.1.

In the analysis of variance the sum of squares due to error, S_e^2, should be obtained by subtraction.

Remark 1. For a connected design, $r(C) = v - 1$ (theorem 4.1.2). Hence the homogeneous linear equation

TABLE 4.1

Analysis of Variance

Source of variation	Degrees of Freedom	Sum of Squares
Treatment adjusted	$v - 1$	$S_t^2 = \underline{t}'\underline{Q}$
Block unadjusted	$b - 1$	$S_b^2 = \sum_{j=1}^{b} \frac{B_{.j}}{k_{.j}} - \frac{G^2}{n}$
Error	$n - b - v + 1$	S_e^2 (by subtraction)
Total	$n - 1$	$\sum y_i^2 - \frac{G^2}{n}$

(4.3.9) $$C\underline{t} = 0$$

has a nontrivial solution. There is only one linearly independent solution, and every other solution is a multiple of this. As a matter of fact, one such solution is $\underline{j}_v$. If $\underline{t}_0$ is a particular solution of

(4.3.10) $$C\underline{t} = \underline{Q},$$

then the general solution of (4.3.10) must be the form

(4.3.11) $$\underline{t} = \underline{t}_0 + \theta\underline{j}_v.$$

where θ is arbitrary. The best linear estimate of $\underline{l}'\underline{\tau}$ is

(4.3.12) $$l_1t_1 + \cdots + l_vt_v = \underline{l}\,\underline{t}_0 + \theta(l_1 + \cdots + l_v) = \underline{l}'\underline{t}_0;$$

that is, the best linear estimate remains unaltered.

Remark 2. The effect of taking the restriction

(4.3.13) $$\tau_1 + \cdots + \tau_v = 0$$

is simply to measure the treatment effects from their mean.

Remark 3. $C\underline{t} = \underline{Q}$. If C is symmetric, then

$$C\underline{t}\underline{t}'C = \underline{Q}\underline{Q}'.$$

Thus

$$C\ Var(\underline{t})C = Var(\underline{Q})$$

or

(4.3.14) $$C\ Var(\underline{t})C = \sigma^2 C.$$

Thus we can put

$$Var(\underline{t}) = \sigma^2 C^*,$$

where C^* is a conditional inverse of C satisfying

(4.3.16) $$CC^*C = C.$$

If $\underline{l}'\underline{\tau}$ is estimable, then

$$\underline{l}' = \underline{b}'C,$$

and hence

(4.3.17) $$\underline{l}'\underline{t} = \underline{b}'C\underline{t} = \underline{b}'CC^*\underline{Q},$$

(4.3.18) $$Var(\underline{l}'\underline{t}) = \underline{b}'CC^*\ Var(Q)C^{*\prime}C'\underline{b} = \underline{b}'CC^*CC^{*\prime}C'\underline{b}\sigma^2$$
$$= \underline{b}'CC^{*\prime}C'\underline{b}\sigma^2 = \underline{l}'C^*\ \underline{l}\cdot\sigma^2,$$

Lemma 4.3.1. If C is symmetric and C^* is a conditional inverse of C, then $C^{*\prime}$ is also a conditional inverse of C.

Hence

(4.3.19) $$Var(\underline{l}'\underline{t}) = \underline{l}'C^*\underline{l}\cdot\sigma^2.$$

Let

$$(4.3.20)\qquad \begin{aligned} t_1 &= c^*_{11}Q_1 + c^*_{12}Q_2 + \cdots + c^*_{1v}Q_v \\ &\cdots\cdots\cdots\cdots\cdots\cdots, \\ t_v &= c^*_{v1}Q_1 + c_{v2}Q_2 + \cdots + c^*_{vv}Q_v. \end{aligned}$$

A more general solution is given by

$$(4.3.21)\qquad \underline{t} = C^*\underline{Q} + \theta\underline{j}'\underline{Q} = (C^* + \theta\underline{j}')\underline{Q}.$$

An estimate of a contrast $\underline{l}'\underline{\tau}$ is the same for all solutions

$$(4.3.22)\qquad \underline{l}'\underline{t} = \underline{l}'C^*\underline{Q} + \underline{l}'a\underline{j}'\underline{Q} = \underline{l}'C^*\underline{Q}.$$

<u>Theorem 4.3.1</u>

$$(4.3.23)\qquad Var(\underline{l}'\underline{t}) = \underline{l}'C^*\underline{l}\,\sigma^2 = \sigma^2 \sum_{\alpha,\beta=1}^{v} l_\alpha l_\beta c^*_{\alpha\beta}$$

Hence

$$(4.3.24)\qquad Var(t_\alpha - t_\beta) = \sigma^2(c^*_{\alpha\alpha} + c^*_{\beta\beta} - c^*_{\alpha\beta} - c^*_{\beta\alpha}).$$

<u>Corollary 4.3.1.</u> To test the hypothesis $H : \tau_1 - \tau_2 = 0$, use the following :

$$(4.3.25)\qquad t = \frac{t_1 - t_2}{s\sqrt{c^*_{11} + c^*_{22} - c^*_{12} - c^*_{21}}}$$

referring to the t-distribution with $n - b - v + 1$ degrees of freedom where

$$(4.3.26)\qquad s^2 = \frac{1}{n - b - v + 1} S_e^2.$$

Unconnected Design. If a design is not connected, it will break up into g connected portions. The contrasts between the treatment effects in the ρth portion will be estimable. Let the number of treatments in the ρth portion be v_ρ. Then we get $v_\rho - 1$ estimable functions of the treatment effects. On the whole, the number of estimable linear functions of the treatment effects is

$$\sum_{\rho=1}^{g} (v_\rho - 1) = v - g. \tag{4.3.27}$$

Thus there are at least $v - g$ linearly independent linear functions that are estimable.

The sum of the functions Q from the ρth portion is zero. Hence there are g linearly independent linear relations satisfied by the functions of Q. Therefore not more than $v - g$ of $Q_1,\ldots,Q_v$ are linearly independent. Thus it follows that

$$r(C) = v - g. \tag{4.3.28}$$

One can summarize the result as follows :

	Degrees of Freedom
$Q_1,\ldots,Q_v$	$v - g$
$B_1,\ldots,B_b$	b
	$v + b - g$

The number of degrees of freedom belonging to the estimable set is $v + b - g$, and hence the number of degrees of freedom belonging to the error is $n - v - b + g$.

4.4. Balanced Incomplete Block Design

Definition 4.4.1. An arrangement of v treatments in b blocks of k experimental units each is called *a balanced incomplete block design* (BIBD) if it satisfies the following three conditions :

(1) Each block contains k different treatments,

(2) Each treatment occurs in r blocks, and

(3) Any two treatments occur together in λ blocks.

The parameters v, b, r, k and λ are given constants for a BIBD.

An example of a BIBD : $v = b = 7$, $r = k = 3$, $\lambda = 1$.

Block	Treatment
I	1, 2, 4
II	2, 3, 5
III	3, 4, 6
IV	4, 5, 7
V	5, 6, 1
VI	6, 7, 2
VII	7, 1, 3

There are $bk = vr$ experimental units in the whole. Let θ be one of the treatments; it occurs in r blocks. There are $r(k - 1)$ other experimental units in these blocks.

$$r \left\{ \begin{array}{llll} \theta & * & * \ldots & * \\ \theta & * & * \ldots & * \\ & & & \\ \theta & \underbrace{* \quad\; * \ldots \;\, *}_{k-1} & & \end{array} \right.$$

Each of the $(v - 1)$ treatments other than θ will occur λ times in these experimental units. Hence we get

$$r(k - 1) = \lambda(v - 1). \tag{4.4.1}$$

It is obvious that for a randomized block design,

$$k = v, \qquad \lambda = r = b.$$

Conversely, *if* λ = r *for a BIBD, then it must be a randomized block design.* Indeed, from $\lambda(v - 1) = r(k - 1)$ and $\lambda = r$, it follows that $k = v$. Hence from $bk = vr$ and $k = v$, we have $b = r$.

Theorem 4.4.1. (Fisher's inequality) :

$$b \geqq v \quad \text{or} \quad r \geqq k.$$

Proof. Suppose we have a BIBD and $\lambda \neq r$, that is, the design is not a randomized block design. Let

$$N = \begin{Vmatrix} n_{11} & \cdots & n_{1b} \\ \cdot & \cdots & \cdot \\ n_{v1} & \cdots & n_{vb} \end{Vmatrix}$$

be the incidence matrix of the design. If $b < v$, then we can add $v - b$ null columns to N to get a square matrix

$$(4.4.2) \qquad N_1 = \begin{Vmatrix} n_{11} & \cdots & n_{1b} & 0 & \cdots & 0 \\ \vdots & & \vdots & \vdots & & \vdots \\ n_{v1} & \cdots & n_{vb} & 0 & \cdots & 0 \end{Vmatrix},$$

and obviously

$$N_1N_1' = NN' = \begin{Vmatrix} r & \lambda & \cdots & \lambda \\ \lambda & r & \cdots & \lambda \\ \vdots & \vdots & \ddots & \vdots \\ \lambda & \lambda & \cdots & r \end{Vmatrix}.$$

Hence

$$(4.4.3) \qquad det(N_1N_1') = det(NN') = rk(r - \lambda)^{v-1} \neq 0.$$

Since $det(N_1) = 0$, this is a contradiction.

Alternative Proof.

$$\text{Rank } N = \text{Rank } NN' = v. \tag{4.4.4}$$

On the other hand, since

$$\text{Rank } N \leqq b, \tag{4.4.5}$$

if $b < v$, then

$$\text{Rank } N < v,$$

which contradicts (4.4.4). Q.E.D.

Thus, for example, $v = 16$, $b = 8$, $r = 3$, $k = 6$, $\lambda = 1$ is impossible even though these satisfy $bk = vr$ and (4.4.1), except in the critical case when the design is a randomized block design.

Definition 4.4.2. A BIBD is said to be symmetrical if $b = v$ and therefore $r = k$.

Theorem 4.4.2. For a symmetrical BIBD, if v is even, the $r - \lambda$ must be a perfect square.

Proof. Let N be the incidence matrix. Then

$$det(NN') = rk(r - \lambda)^{v-1}.$$

Since N is a square matrix and $r = k$ for a symmetrical BIBD,

$$(r - \lambda)^{v-1} = \left(\frac{det\ N}{r}\right)^2. \tag{4.4.6}$$

Therefore $r - \lambda$ must be a perfect square. Q.E.D.

Example 1 : $v = b = 22$, $r = k = 7$, $\lambda = 2$ is impossible.

Example 2 : $v = b = 46$, $r = k = 10$, $\lambda = 2$ is impossible.

Theorem 4.4.3. (Fisher). Any two blocks of a symmetrical BIBD have exactly λ treatments in common.

Proof. For any BIBD with the incidence matrix N, we get

$$N'N = \begin{Vmatrix} k & \alpha_{12} & \cdots & \alpha_{1b} \\ \alpha_{21} & k & \cdots & \alpha_{2b} \\ \vdots & \vdots & \ddots & \vdots \\ \alpha_{b1} & \alpha_{b2} & \cdots & k \end{Vmatrix}, \tag{4.4.7}$$

where α_{ij} is the number of treatments common to the ith and jth blocks. Thus we have to show that for a *symmetrical BIBD*

$$\alpha_{ij} = \lambda \quad \text{for} \quad \text{any } i \neq j. \tag{4.4.8}$$

Now for a symmetrical BIBD, $b = v$, $r = k$. Let G be the $v \times v$-matrix with all ites elements being unity. We get

$$NG = \begin{Vmatrix} r & \cdots & r \\ \cdot & \cdot & \cdot \\ r & \cdots & r \end{Vmatrix} = rG. \tag{4.4.9}$$

Similarly $GN = kG$. Therefore in our case, where $r = k$,

$$NG = GN. \tag{4.4.10}$$

We knew that

$$NN' = (r - \lambda)I + \lambda G,$$

and hence

(4.4.11) $NN'N = (r - \lambda)N + \lambda GN = N\{(r - \lambda)I + \lambda G\} = N \cdot NN'$.

If $r \neq \lambda$, then N is nonsingular, and hence from (4.4.11) it follows that

(4.4.12) $N'N = NN' = (r - \lambda)I + \lambda G$.

This means that $\alpha_{ij} = \lambda$ for all $i \neq j$. Q.E.D.

Given a symmetrical BIBD with the parameters $v = b$, $r = k$, λ, we can derive a new BIBD with the parameters

(4.4.13) $v^* = v - k,\ b^* = b - 1,\ r^* = r,\ k^* = k - \lambda,\ \lambda^* = \lambda$

by deleting one block and all the treatments contained in it.

Example : $v = b = 16$, $r = k = 6$, $\lambda = 2$.

Blocks

	1	2	3	4	5	6	7	8	9	10	11	12	13	14	15	16
Treatments (1,2,...,16)	~~2~~	1	1	1	6	~~5~~	~~5~~	~~5~~	10	~~9~~	~~9~~	~~9~~	14	~~13~~	~~13~~	~~13~~
	~~3~~	~~3~~	~~2~~	~~2~~	7	7	6	6	11	11	10	10	15	15	14	14
	~~4~~	~~4~~	~~4~~	~~3~~	8	8	8	7	12	12	12	11	16	16	16	15
	~~5~~	6	7	8	1	~~2~~	~~3~~	~~4~~	1	~~2~~	~~3~~	~~4~~	1	~~2~~	~~3~~	~~4~~
	~~9~~	10	11	12	~~9~~	10	11	12	~~5~~	6	7	8	~~5~~	6	7	8
	~~13~~	14	15	16	~~13~~	14	15	16	~~13~~	14	15	15	~~9~~	10	11	12

$v^* = 10,\quad b^* = 15,\quad r^* = 6,\quad k^* = 4,\quad \lambda^* = 2.$

Again, if we delete one block from a symmetrical BIBD and retain from the other blocks those treatments that are contained in the first (deleted) block, we get a new BIBD with the parameters

(4.4.14) $\quad v' = k, \quad b' = b - 1, \quad r' = r - 1, \quad k' = \lambda, \quad \lambda' = \lambda - 1.$

Example : $v = b = 19, \quad r = k = 9, \quad \lambda = 4;$
$v' = 9, \quad b' = 18, \quad r' = 8, \quad k' = 4, \quad \lambda' = 3.$

The normal equation for treatments in a BIBD is

(4.4.15)
$$C\underline{t} = \underline{Q},$$

where

(4.4.16)
$$c_{\alpha\alpha} = (1 - \frac{1}{k}), \qquad c_{\alpha\beta} = -\frac{\lambda}{k} \quad \text{for } \alpha \neq \beta.$$

Or in detail

(4.4.17)
$$\begin{array}{l} r(1 - \frac{1}{k})t_1 - \frac{\lambda}{k} t_2 - \cdots - \frac{\lambda}{k} t_v = Q_1, \\ -\frac{\lambda}{k} t_1 + r(1 - \frac{1}{k})t_2 - \cdots - \frac{\lambda}{k} t_v = Q_2, \\ \cdots\cdots\cdots\cdots\cdots\cdots\cdots\cdots, \\ -\frac{\lambda}{k} t_1 - \frac{\lambda}{k} t_2 - \cdots + r(1 - \frac{1}{k})t_v = Q_v. \end{array}$$

Now we shall add another restriction :

(4.4.18)
$$t_1 + t_2 + \cdots + t_v = 0.$$

Then the αth equation of (4.4.17) reduced to

$$\left\{r(1 - \frac{1}{k}) + \frac{\lambda}{k}\right\} t_\alpha = Q_\alpha$$

or

(4.4.19)
$$\{r(k - 1) + \lambda\}t_\alpha = kQ_\alpha.$$

Using the relation $r(k - 1) = \lambda(v - 1)$, we get

(4.4.20)
$$t_\alpha = \frac{k}{v\lambda} Q_\alpha = \frac{kr}{v\lambda} \frac{Q_\alpha}{r}.$$

Therefore the calculations of the functions Q are only necessary in the case of practical application, and these should be done as follows :

(4.4.21) $$Q_\alpha = T_\alpha - \sum \text{(average of block totals in which the } \alpha\text{th treatment occurs)}$$

By (4.3.24) we get

(4.4.22) $$Var(t_\alpha - t_\beta) = (c^*_{\alpha\alpha} + c^*_{\beta\beta} - c^*_{\alpha\beta} - c^*_{\beta\alpha})\sigma^2$$

$$= \frac{2k}{v\lambda}\sigma^2 = \frac{kr}{v\lambda}\frac{2}{r}\sigma^2 = \frac{1}{E}\frac{2}{r}\sigma^2,$$

where E, called the *efficiency factor* of the design, was defined as

(4.4.23) $$E = \frac{v\lambda}{kr}.$$

On the other hand, from the relation

$$\lambda(v - 1) = r(k - 1)$$

it follows that

$$\frac{\lambda}{r} = \frac{k - 1}{v - 1}.$$

Hence we get

(4.4.24) $$E = \frac{v(k - 1)}{k(v - 1)} = \frac{1 - (1/k)}{1 - (1/v)}.$$

Since, for a proper BIBD, $k < v$,

$$\frac{1}{k} > \frac{1}{v},$$

whence we have

(4.4.25) $$E = \frac{1 - (1/k)}{1 - (1/v)} < 1.$$

If it were a randomized block design, then $E = 1$ and

$$Var(t_\alpha - t_\beta) = \frac{2}{r}\sigma^2. \tag{4.4.26}$$

So, using a balanced incomplete block design, the variance of the difference $t_\alpha - t_\beta$ will become larger by $(1/E)$.

In fact, for a randomized block design (RBI), we have

$$t_\alpha = \frac{Q_\alpha}{r} = \frac{1}{r}\left(T_\alpha - \frac{B_1 + \cdots + B_b}{k}\right) = \frac{T_\alpha}{r} - \frac{G}{n} \tag{4.4.27}$$

$$= \frac{1}{r}\left[T_\alpha - \frac{1}{k}(n_{\alpha 1}B_1 + \cdots + n_{\alpha b}B_b)\right] \qquad (b = r),$$

and for a BIBD,

$$C^* = \begin{Vmatrix} \frac{1}{Er} & 0 & \cdots & 0 \\ 0 & \frac{1}{Er} & \cdots & 0 \\ \vdots & \vdots & \ddots & \vdots \\ 0 & 0 & \cdots & \frac{1}{Er} \end{Vmatrix}. \tag{4.4.28}$$

Thus we get

$$Var(t_\alpha \quad t_\beta) - \frac{2}{Er}\sigma^2 \quad \text{for BIBD.}$$

$$Var(r_\alpha - t_\beta) = \frac{2}{r}\sigma^2 \quad \text{for RBD.}$$

The analysis of variance for a BIBD is shown in Table 4.2.

For testing the null hypothesis

$$H_0 : \tau_1 = \cdots = \tau_v = 0,$$

TABLE 4.2.

Analysis of Variance of a BIBD

Source of Variation	Degrees of Freedom	Sum of Squares	Mean Sum of Squares
Treatment adjusted	$v - 1$	$\sum_{\alpha=1}^{v} t_\alpha Q_\alpha = S_t^2$	$s_t^2 = \frac{S_t^2}{v - 1}$
Blocks unadjusted	$b - 1$	$\sum_{j=1}^{b} \frac{B_j^2}{k} - \frac{G^2}{n} = S_b^2$	
Error	$n - v - b + 1$	S_e^2 (by subtraction)	$s_e^2 = \frac{S_e^2}{n - v - b + 1}$
Total	$n - 1$	$\sum y_i^2 - \frac{G^2}{n}$	

the statistic

$$F = \frac{s_t^2}{s_e^2} \tag{4.4.29}$$

should be referred to

$$F_{(n - v - b + 1)}^{(v - 1)}$$

and

$$t = \frac{t_2 - t_1}{\sqrt{\text{Estimated } Var(t_1 - t_2)}} = \frac{(1/Er)(Q_2 - Q_1)}{\sqrt{2s_e^2/Er}} \tag{4.4.30}$$

$$= \frac{Q_2 - Q_1}{s_e \sqrt{2Er}}$$

should be referred to $t_{n-v-b+1}$.

Suppose we have one control and v treatments. We would like to estimate the differences between the control and other treatments. For that purpose, we can use a design like this : Take a BIBD with v treatments, r replications and other parameters b, k, and λ, and then add the control to each block; for example, $v = b = 16$, $r = k = 6$, $\lambda = 2$, $v' = 16 + 1 = 17$, $b' = 16$, $r_0 = 16$ every other $r_i = 6$, $k' = k + 1 = 7$, $\lambda_i = 6$, $\lambda_{ij} = 2$,

$$(4.4.31)\qquad \begin{aligned} c_{00} &= b\left(1 - \frac{1}{k+1}\right), & c_{0i} &= -\frac{r}{k+1}, \\ c_{ii} &= r\left(1 - \frac{1}{k+1}\right), & c_{ij} &= -\frac{\lambda}{k+1}. \end{aligned}$$

Hence the normal equations become

$$(4.4.32)\qquad \begin{array}{l} b(1 - \frac{1}{k+1})t_0 - \frac{r}{k+1}t_1 - \cdots - \frac{r}{k+1}t_v = Q_0 \\ -\frac{r}{k+1}t_0 + r(1 - \frac{1}{k+1})t_1 - \cdots - \frac{\lambda}{k+1}t_v = Q_1 \\ -\frac{r}{k+1}t_0 - \frac{\lambda}{k+1}t_1 - \cdots - \frac{\lambda}{k+1}t_v = Q_2 \\ \cdots\cdots\cdots\cdots\cdots\cdots\cdots, \\ -\frac{r}{k+1}t_0 - \frac{\lambda}{k+1}t_1 - \cdots + r(1 - \frac{1}{k+1})t_v = Q_v. \end{array}$$

Case I. Suppose we take the following constraint :

$$(4.4.33)\qquad t_1 + t_2 + \cdots + t_v = 0.$$

From the first equation of (4.4.32), we get

$$bkt_0 - r(t_1 + \cdots + t_v) = (k + 1)Q_0,$$

or

$$(4.4.34)\qquad t_0 = \frac{k+1}{bk} Q_0,$$

From the $(\alpha + 1)$st equation of (4.4.32), we get

$$-rt_0 + (kr + \lambda)t_\alpha - \lambda(t_1 + \cdots + t_v) = (k + 1)Q_\alpha,$$

or

$$t_\alpha = \frac{k + 1}{kr + \lambda} Q_\alpha + \frac{r(k + 1)}{bk(kr + \lambda)} Q_0, \tag{4.4.35}$$

where

$$Q_0 = T_0 - \frac{G}{k + 1}. \tag{4.4.36}$$

Hence

$$\begin{aligned} t_\alpha - t_0 &= \frac{k + 1}{kr + \lambda} Q_\alpha + \frac{r(k + 1)}{bk(kr + \lambda)} Q_0 - \frac{k + 1}{bk} Q_0 \\ &= \frac{k + 1}{kr + \lambda} Q_\alpha - \frac{\lambda(k + 1)}{r(kr + \lambda)} Q_0. \end{aligned} \tag{4.4.37}$$

Case II. We could have taken the constraint in another way, for example,

$$t_0 = 0. \tag{4.4.38}$$

Then the above equations become

$$t_1 + \cdots + t_v = -\frac{k + 1}{r} Q_0,$$

$$(kr + \lambda)t_\alpha = (k + 1)Q_\alpha - \frac{\lambda(k + 1)}{r} Q_0,$$

and hence

$$t_\alpha = \frac{k + 1}{kr + \lambda} Q_\alpha - \frac{\lambda(k + 1)}{r(kr + \lambda)} Q_0. \tag{4.4.39}$$

Thus we can show that the estimates of a contrast are the same by both methods.

As for the variance of contrast, method I gives us

$$C^* = \left\| \begin{array}{ccccc} \frac{k+1}{bk} & 0 & 0 & \cdots & 0 \\ \frac{r(k+1)}{bk(kr+\lambda)} & \frac{k+1}{kr+\lambda} & 0 & \cdots & 0 \\ \frac{r(k+1)}{bk(kr+\lambda)} & 0 & \frac{k+1}{kr+\lambda} & \cdots & 0 \\ \cdots & \cdots & \cdots & \cdots & \cdots \\ \frac{r(k+1)}{bk(kr+\lambda)} & 0 & 0 & \cdots & \frac{k+1}{kr+\lambda} \end{array} \right\|,$$

and hence we get

$$(4.4.40) \qquad Var(t_0 - t_\alpha) = \left[\frac{k+1}{bk} + \frac{k+1}{kr+\lambda} - \frac{r(k+1)}{bk(kr+\lambda)}\right]\sigma^2,$$

$$(4.4.41) \qquad Var(t_1 - t_2) = 2\,\frac{k+1}{kr+\lambda}\,\sigma^2.$$

Method II gives us

$$(4.4.42) \qquad C^* = \left\| \begin{array}{ccccc} 0 & 0 & 0 & \cdots & 0 \\ -\frac{\lambda(k+1)}{r(kr+\lambda)} & \frac{k+1}{kr+\lambda} & 0 & \cdots & 0 \\ -\frac{\lambda(k+1)}{r(kr+\lambda)} & 0 & \frac{k+1}{kr+\lambda} & \cdots & 0 \\ \cdots & \cdots & \cdots & \cdots & \cdots \\ -\frac{\lambda(k+1)}{r(kt+\lambda)} & 0 & 0 & \cdots & \frac{k+1}{kr+\lambda} \end{array} \right\|$$

and

$$Var(t_0 - t_1) = \left[\frac{k+1}{kr+\lambda} + \frac{\lambda(k+1)}{r(kr+\lambda)}\right]\sigma^2.$$

This can be shown to be the same as (4.4.40).

4.5. Randomization of a Balanced Incomplete Block Design

In this case the sum of squares due to treatment is

$$(4.5.1)\qquad S_e^2 = \frac{k}{v\lambda}\,\underline{Q}'\underline{Q} = \frac{k}{vr\lambda}\,\underline{x}'\left(T - \frac{1}{k}BT\right)\left(T - \frac{1}{k}TB\right)\underline{x},$$

and the sum of squares due to error is

$$(4.5.2)\qquad S_e^2 = \underline{x}'\left[I - \frac{1}{k}B - \frac{k}{vr\lambda}\left(T - \frac{1}{k}BT\right)\left(T - \frac{1}{k}TB\right)\right]\underline{x}.$$

Under the null hypothesis H, we have

$$(4.5.3)\qquad S_t^2 = \frac{k}{vr\lambda}\,\underline{e}'\left(T - \frac{1}{k}BT\right)\left(T - \frac{1}{k}TB\right)\underline{e} + \frac{2k}{vr\lambda}\,\underline{e}'\left(T - \frac{1}{k}BT\right) \times \left(T - \frac{1}{k}TB\right)\underline{\pi} + \frac{k}{vr\lambda}\,\underline{\pi}'\left(T - \frac{1}{k}BT\right)\left(T - \frac{1}{k}TB\right)\underline{\pi}$$

and

$$(4.5.4)\qquad S_e^2 = \underline{e}'\left[I - \frac{1}{k}B - \frac{k}{vr\lambda}\left(T - \frac{1}{k}BT\right)\left(T - \frac{1}{k}TB\right)\right]\underline{e} + 2\underline{e}'\left[I - \frac{1}{k}B - \frac{k}{vr\lambda}\left(T - \frac{1}{k}BT\right)\left(T - \frac{1}{k}TB\right)\right]\underline{\pi} + \underline{\pi}'\left[I - \frac{1}{k}B - \frac{k}{vr\lambda}\left(T - \frac{1}{k}BT\right)\left(T - \frac{1}{k}TB\right)\right]\underline{\pi}.$$

The distribution of $\chi_1^2 = S_t^2/\sigma^2$ before randomization is

$$(4.5.5)\qquad \exp\left(-\frac{\lambda_1}{2}\right)\sum_{\mu=0}^{\infty}\frac{\left(\frac{\lambda_1}{2}\right)^{\mu}}{\mu!}\,\frac{\left(\frac{\chi_1^2}{2}\right)^{[(v-1)/2]+\mu-1}}{\Gamma\left(\frac{v-1}{2}+\mu\right)}\exp\left(-\frac{\chi_1^2}{2}\right)d\left(\frac{\chi_1^2}{2}\right),$$

where

(4.5.6) $$\lambda_1 = \frac{1}{\sigma^2} \frac{k}{vr\lambda} \underline{\pi}'(T - \frac{1}{k}BT)(T - \frac{1}{k}TB)\underline{\pi} = \frac{k}{\sigma^2 v\lambda} \underline{\pi}'T\underline{\pi}.$$

Likewise the distribution of $\chi_2^2 = S_e^2/\sigma^2$ before randomization is given by

(4.5.7) $$exp\left(-\frac{\lambda_2}{2}\right)\sum_{\nu}^{\infty} \frac{\left(\frac{\lambda_2}{2}\right)^{\nu}}{\nu!} \frac{\left(\frac{\chi_2^2}{2}\right)^{[(n-v-b+1)/2]+\nu-1}}{\Gamma\left(\frac{n-v-b+1}{2} + \nu\right)} exp\left(-\frac{\chi_2^2}{2}\right) d\left(\frac{\chi_2^2}{2}\right),$$

where

(4.5.8) $$\lambda_2 = \frac{1}{\sigma^2} \underline{\pi}'\left[I - \frac{1}{k}B - \frac{k}{vr\lambda}(T - \frac{1}{k}BT)(T - \frac{1}{k}TB)\right]\underline{\pi} = \frac{\underline{\pi}'\underline{\pi}}{\sigma^2} - \lambda_1.$$

The null distribution of the statistic

$$F = \frac{n - v - b + 1}{v - 1} \frac{S_t^2}{S_e^2}$$

before the randomization is given by

(4.5.9) $$\frac{\Gamma\left(\frac{n-b}{2}\right)}{\Gamma\left(\frac{v-1}{2}\right)\Gamma\left(\frac{n-v-b+1}{2}\right)} \left(\frac{v - 1}{n - v - b + 1}F\right)^{[(v-1)/2]-1}$$

$$\times \left(1 + \frac{v - 1}{n - v - b + 1}F\right)^{-[(n-b)/2]} d\left(\frac{v - 1}{n - v - b + 1}F\right)$$

$$\times\ exp\left(-\frac{\Delta}{2\sigma^2}\right) \sum_{l=0}^{\infty} \frac{\left(\frac{\Delta}{2\sigma^2}\right)^{l}}{l!} \left(1 + \frac{v - 1}{n - v - b + 1}F\right)^{-l}$$

$$\times \sum_{\mu+\nu=l} \frac{l!}{\mu!\nu!} \theta^{\mu} (1 - \theta)^{\nu} \left(\frac{v - 1}{n - v - b + 1}F\right)^{\mu}$$

$$\times \frac{\Gamma[(v-1)/2]\ \Gamma[(n-v-b+1)/2]\ \Gamma[(n-b)/2+\mu+\nu]}{\Gamma[(n-b)/2]\ \Gamma[(v-1)/2+\mu]\ \Gamma[(n-v-b+1)/2+\nu]}$$

where

$$\theta = \frac{k}{v\lambda}\,\Delta^{-1}\,\underline{\pi}'T\underline{\pi}. \tag{4.5.10}$$

We now calculate the mean and the variance of θ with respect to the permutation distribution due to the randomization. Let

$$\Phi = \begin{Vmatrix} \Phi_1 \\ \Phi_2 \\ \vdots \\ \Phi_b \end{Vmatrix}$$

$$\Phi_p = \begin{Vmatrix} \zeta_{1,(p-1)k+1} & \zeta_{2,(p-1)k+1} & \cdots & \zeta_{v,(p-1)k+1} \\ \vdots & \vdots & & \vdots \\ \zeta_{1,(p-1)k+k} & \zeta_{2,(p-1)k+k} & \cdots & \zeta_{v,(p-1)k+k} \end{Vmatrix} \tag{4.5.11}$$

and let

$$T_{pq} = \left\| t_{ij}^{pq} \right\| = \Phi_p \Phi_q'. \tag{4.5.12}$$

Then

$$E(\theta) = \frac{k}{v\lambda}\,\Delta^{-1}(k!)^{-b}\underline{\pi}' \sum_{\underline{\sigma}\in\mathbf{G}_k} U_{\underline{\sigma}} T U_{\underline{\sigma}}'\underline{\pi}\ . \tag{4.5.13}$$

Since

$$S_{\sigma_p} T_{pq} S_{\sigma_q} = I_k \quad \text{if} \quad p = q,$$

and

$$\sum_{\underline{\sigma}} S_{\sigma_p} T_{pq} S_{\sigma_p}' = (k!)^{b-2}\left[(k-1)!\right]^2 G_k T_{pq} G_k \quad \text{if} \quad p \neq q,$$

it follows that

$$\underline{\pi}'\sum_{\underline{\sigma}} U_{\underline{\sigma}} T U_{\underline{\sigma}}'\underline{\pi} = (k!)^b\,\underline{\pi}'\underline{\pi}. \tag{4.5.14}$$

Thus we have

$$E(\theta) = \frac{k}{v\lambda}\,\Delta^{-1}\underline{\pi}'\underline{\pi} = \frac{k}{v\lambda}. \tag{4.5.15}$$

Next we have to calculate the variance of θ. Now we have the formal expansion as follows :

$$
\begin{aligned}
(\underline{\pi}'U_{\sigma}TU'_{\sigma}\underline{\pi})^2 &= \Big(\Delta + \sum_{p\neq q} \underline{\pi}^{(p)'}S_{\sigma_p}T_{pq}S'_{\sigma_q}\underline{\pi}^{(q)}\Big)^2 \\
&= \Delta^2 + 2\Delta \sum_{p\neq q} \underline{\pi}^{(p)'}S_{\sigma_p}T_{pq}S'_{\sigma_q}\underline{\pi}^{(q)} \\
&\quad + \Big(\sum_{p\neq q} \underline{\pi}^{(p)'}S_{\sigma_p}T_{pq}S'_{\sigma_q}\underline{\pi}^{(q)}\Big)^2 \\
&= \Delta^2 + 2\Delta \sum_{p\neq q} \underline{\pi}^{(p)'}S_{\sigma_p}T_{pq}S'_{\sigma_q}\underline{\pi}^{(q)} \\
&\quad + \sum_{p\neq q} \Big(\underline{\pi}^{(p)'}S_{\sigma_p}T_{pq}S'_{\sigma_q}\underline{\pi}^{(q)}\underline{\pi}^{(p)'}S_{\sigma_p}T_{pq}S'_{\sigma_q}\underline{\pi}^{(q)} \\
&\qquad + \underline{\pi}^{(p)'}S_{\sigma_p}T_{pq}S'_{\sigma_q}\underline{\pi}^{(q)}\underline{\pi}^{(q)'}S_{\sigma_q}T_{qp}S'_{\sigma_p}\underline{\pi}^{(p)}\Big) \\
&\quad + \sum_{p\neq q\neq r} \Big(\underline{\pi}^{(p)'}S_{\sigma_p}T_{pq}S'_{\sigma_q}\underline{\pi}^{(q)}\underline{\pi}^{(p)'}S_{\sigma_p}T_{pr}S'_{\sigma_r}\underline{\pi}^{(r)} \\
&\qquad + \underline{\pi}^{(p)'}S_{\sigma_p}T_{pq}S'_{\sigma_q}\underline{\pi}^{(q)}\underline{\pi}^{(q)'}S_{\sigma_q}T_{qr}S'_{\sigma_r}\underline{\pi}^{(r)}\Big) \\
&\quad + \sum_{p\neq q\neq r} \Big(\underline{\pi}^{(p)'}S_{\sigma_p}T_{pq}S'_{\sigma_q}\underline{\pi}^{(q)}\underline{\pi}^{(r)'}S_{\sigma_r}T_{rp}S'_{\sigma_p}\underline{\pi}^{(p)} \\
&\qquad + \underline{\pi}^{(p)'}S_{\sigma_p}T_{pq}S'_{\sigma_q}\underline{\pi}^{(q)}\underline{\pi}^{(r)'}S_{\sigma_r}T_{rq}S'_{\sigma_q}\underline{\pi}^{(q)}\Big)
\end{aligned}
\tag{4.5.16}
$$

$$+ \sum_{p \neq q \neq r \neq s} \underline{\pi}^{(p)'} S_{\sigma_p} T_{pq} S_{\sigma_q} \underline{\pi}^{(q)} \underline{\pi}^{(r)'} S_{\sigma_r} T_{rs} S'_{\sigma_s} \underline{\pi}^{(s)},$$

and the terms being linear with respect to some S_{σ_p}, vanish in the average. It is easy to see that

$$(4.5.17) \quad E\left(\underline{\pi}' U_{\underline{\sigma}} T U'_{\underline{\sigma}} \underline{\pi}\right)^2 = \Delta^2 + \sum_{p \neq q} \left[E\left(\underline{\pi}^{(p)'} S_{\sigma_p} T_{pq} S'_{\sigma_q} \underline{\pi}^{(q)}\right)^2 \right.$$

$$\left. + E\left(\underline{\pi}^{(p)'} S_{\sigma_p} T_{pq} S_{\sigma_q} \underline{\pi}^{(q)} \underline{\pi}^{(q)'} S_{\sigma_q} T_{qp} S_{\sigma_p} \underline{\pi}^{(p)}\right) \right]$$

$$= \Delta^2 + 2 \sum_{p \neq q} E\left(\underline{\pi}^{(p)'} S_\sigma T_{pq} S'_\tau \underline{\pi}^{(q)}\right)^2.$$

Since

$$\underline{\pi}^{(p)'} S_\sigma T_{pq} S_\tau \underline{\pi}^{(q)} = \sum_{i=1}^{k} t^{pq}_{ii} \pi^{(p)}_{\sigma(i)} \pi^{(q)}_{\tau(i)} + \sum_{i \neq j} t^{pq}_{ij} \pi^{(p)}_{\sigma(i)} \pi^{(q)}_{\tau(j)},$$

it follows that

$$(4.5.18) \quad \left(\underline{\pi}^{(p)'} S_\sigma T_{pq} S'_\tau \underline{\pi}^{(q)}\right)^2 = \sum_{i=1}^{k} t^{pq}_{ii} \pi^{(p)2}_{\sigma(i)} \pi^{(q)2}_{\tau(i)}$$

$$+ \sum_{i \neq j} t^{pq}_{ii} t^{pq}_{jj} \pi^{(p)}_{\sigma(i)} \pi^{(p)}_{\tau(j)} \pi^{(q)}_{\tau(i)} \pi^{(q)}_{\tau(j)}$$

$$+ 2 \sum_{l \neq i \neq j} t^{pq}_{ll} t^{pq}_{ij} \pi^{(p)}_{\sigma(l)} \pi^{(p)}_{\sigma(i)} \pi^{(q)}_{\tau(l)} \pi^{(q)}_{\tau(j)}$$

$$+ 2 \sum_{i \neq j} t^{pq}_{ii} t^{pq}_{ij} \pi^{(p)2}_{\sigma(i)} \pi^{(q)}_{\tau(i)} \pi^{(q)}_{\tau(j)}$$

$$+ 2 \sum_{i \neq j} t^{pq}_{jj} t^{pq}_{ij} \pi^{(p)}_{(i)} \pi^{(p)}_{\tau(j)} \pi^{(q)2}_{\tau(j)}$$

$$+ 2 \sum_{i \neq j} t^{pq}_{jj} t^{pq}_{ij} \pi^{(p)}_{\sigma(i)} \pi^{(p)}_{\sigma(j)} \pi^{(q)2}_{\tau(j)}$$

$$+ 2 \sum_{i \neq j} t_{ij}^{pq} \pi_{\sigma(i)}^{(p)2} \pi_{\tau(j)}^{(q)2}$$

$$+ \sum_{i \neq j} t_{ij}^{pq} t_{ji}^{pq} \pi_{\sigma(i)}^{(p)} \pi_{\tau(j)}^{(p)} \pi_{\tau(i)}^{(q)} \pi_{\sigma(j)}^{(q)}$$

$$+ \sum_{i \neq j \neq l} \left[t_{ij}^{pq} t_{il}^{pq} \pi_{\sigma(i)}^{(p)2} \pi_{\tau(j)}^{(q)} \pi_{\tau(l)}^{(q)} \right.$$

$$\left. + t_{ij}^{pq} t_{li}^{pq} \pi_{\sigma(i)}^{(p)} \pi_{\sigma(l)}^{(p)} \pi_{\tau(j)}^{(q)} \pi_{\tau(i)}^{(q)} \right]$$

$$+ \sum_{i \neq j \neq l} \left[t_{ij}^{pq} t_{jl}^{pq} \pi_{\sigma(i)}^{(p)} \pi_{\tau(j)}^{(p)} \pi_{\tau(j)}^{(q)} \pi_{\tau(l)}^{(q)} \right.$$

$$\left. + t_{ij}^{pq} t_{lj}^{pq} \pi_{\sigma(i)}^{(p)} \pi_{\sigma(i)}^{(p)} \pi_{\tau(j)}^{(q)2} \right]$$

$$+ \sum_{i \neq j \neq l \neq m} t_{ij}^{pq} t_{lm}^{pq} \pi_{\sigma(i)}^{(p)} \pi_{\sigma(l)}^{(p)} \pi_{\tau(j)}^{(q)} \pi_{\tau(m)}^{(q)}.$$

Now we have

$$E\left[\pi_{\sigma(i)}^{(p)2} \pi_{\tau(i)}^{(q)2}\right] = \frac{1}{k^2} \Delta_p \Delta_q,$$

$$E\left[\pi_{\sigma(i)}^{(p)} \pi_{\sigma(j)}^{(p)} \pi_{\tau(i)}^{(q)} \pi_{\tau(j)}^{(q)}\right] = \frac{1}{k^2(k-1)^2} \Delta_p \Delta_q,$$

$$E\left[\pi_{\sigma(l)}^{(p)} \pi_{\sigma(i)}^{(p)} \pi_{\tau(l)}^{(q)} \pi_{\tau(j)}^{(q)}\right] = \frac{1}{k^2(k-1)^2} \Delta_p \Delta_q,$$

$$E\left[\pi_{\sigma(i)}^{(p)2} \pi_{\tau(i)}^{(q)} \pi_{\tau(j)}^{(q)}\right] = E\left[\pi_{\sigma(i)}^{(p)} \pi_{\sigma(j)}^{(q)} \pi_{\tau(j)}^{(q)2}\right] = \frac{-1}{k^2(k-1)} \Delta_p \Delta_q,$$

$$E\left[\pi_{\sigma(i)}^{(p)2} \pi_{\tau(j)}^{(q)2}\right] = \frac{1}{k^2} \Delta_p \Delta_q,$$

$$E\left[\pi^{(p)}_{\sigma(i)}\pi^{(p)}_{\sigma(j)}\pi^{(q)}_{\sigma(i)}\pi^{(q)}_{\sigma(j)}\right] = \frac{1}{k^2(k-1)^2}\,\Delta_p\Delta_q,$$

$$E\left[\pi^{(p)2}_{\sigma(i)}\pi^{(q)}_{\tau(j)}\pi^{(q)}_{\tau(l)}\right] = E\left[\pi^{(p)}_{\sigma(i)}\pi^{(p)}_{\sigma(l)}\pi^{(q)2}_{\tau(j)}\right] = \frac{-1}{k^2(k-1)}\,\Delta_p\Delta_q,$$

$$E\left[\pi^{(p)}_{\sigma(i)}\pi^{(p)}_{\sigma(l)}\pi^{(q)}_{\tau(j)}\pi^{(q)}_{\tau(i)}\right] = E\left[\pi^{(p)}_{\sigma(i)}\pi^{(p)}_{\sigma(j)}\pi^{(q)}_{\tau(j)}\pi^{(q)}_{\tau(l)}\right]$$

$$= \frac{1}{k^2(k-1)^2}\,\Delta_p\Delta_q,$$

(4.5.19) $$E\left[\pi^{(p)}_{\sigma(i)}\pi^{(p)}_{\sigma(l)}\pi^{(q)}_{\tau(j)}\pi^{(q)}_{\tau(m)}\right] = \frac{1}{k^2(k-1)^2}\,\Delta_p\Delta_q.$$

Thus we get

(4.5.20) $$E(\underline{\pi}^{(p)'}S_{\sigma_p}T_{pq}S'_{\sigma_q}\underline{\pi}^{(q)}) = V_{pq}\Delta_p\Delta_q ,$$

where

(4.5.21) $$V_{pq} = \frac{1}{k^2}\sum_{i,j=1}^{k} t^{pq}_{ij} + \frac{1}{k^2(k-1)^2}\left[\sum_{i\neq j} t^{pq}_{ii}t^{pq}_{jj} + \right.$$

$$+ 2\sum_{l\neq i\neq j} t^{pq}_{ll}t^{pq}_{ij} + \sum_{i\neq j} t^{pq}_{ij}t^{pq}_{ji} + \sum_{i\neq j\neq l} t^{pq}_{ij}t^{pq}_{jl}$$

$$\left. + \sum_{i\neq j\neq l\neq m} t^{pq}_{ij}t^{pq}_{lm}\right] - \frac{1}{k^2(k-1)}\left[2\sum_{i\neq j} t^{pq}_{ii}t^{pq}_{ij}\right.$$

$$\left. + 2\sum_{i\neq j} t^{pq}_{jj}t^{pq}_{ij} + \sum_{i\neq j\neq l}\left(t^{pq}_{ij}t^{pq}_{il} + t^{pq}_{ij}t^{pq}_{lj}\right)\right].$$

Since

$$t^{pq}_{ii}t^{pq}_{ij} = t^{pq}_{jj}t^{pq}_{ij} = 0 \quad \text{if} \quad i \neq j,$$

(4.5.22)

$$t^{pq}_{ij}t^{pq}_{il} = t^{pq}_{ij}t^{pq}_{lj} = 0 \quad \text{if} \quad i \neq l,$$

we finally get

(4.5.23) $$V_{pq} = \frac{\lambda_{pq}}{k^2} \quad \frac{\lambda^2_{pq} - \lambda_{pq}}{k^2(k-1)^2}, \quad \lambda_{pq} = \sum_{i,j=1}^{k} t^{pq}_{ij}.$$

Consequently

$$V(\theta) = 2\frac{k}{v^2\lambda^2}\frac{1}{k-1}W, \tag{4.5.24}$$

with

$$W = \frac{1}{\Delta^2}\left[\frac{1}{k^2(k-1)}\Delta_2 + \frac{k-2}{k(k-1)}\Delta_1 - \sum_{p=1}^{b}\Delta_p^2\right], \tag{4.5.25}$$

$$\Delta_1 = \sum_{p,q}\lambda_{pq}\Delta_p\Delta_q, \quad \Delta_2 = \sum_{p,q}\lambda_{pq}^2\Delta_p\Delta_q \tag{4.5.26}$$

For a randomized block design where $v = k$, $r = \lambda = b$, it is easy to see that $\lambda_{pq} = k$.

Hence

$$W = \frac{1}{\Delta^2}\left\{\Delta^2 - \sum_{p=1}^{b}\Delta_p^2\right\} = \frac{b-1}{b}\left(1 - \frac{V}{b}\right),$$

and therefore

$$V(\theta) = 2\frac{b-1}{b^3(k-1)}\left(1 - \frac{V}{b}\right).$$

This again confirms the result obtained on page 102.

In a similar argument as in a randomized block design we can approximate the permutation distribution of θ due to randomization by a β-distribution:

$$\frac{\Gamma[(\mu+\nu)/2]}{\Gamma(\mu/2)\Gamma(\nu/2)}\theta^{(\mu/2)-1}(1-\theta)^{(\nu/2)-1}\,d\theta, \tag{4.5.27}$$

where

$$\mu = \phi(v-1) \quad \text{and} \quad \nu = \phi(n - v - b + 1),$$

with

$$\phi = \frac{v\lambda - k}{vr}W^{-1} - \frac{2k}{vr(k-1)}. \tag{4.5.28}$$

If, in particular, the variance of plot effects within each block is constant, that is,

$$\Delta_p = \Delta_0, \qquad (p = 1,\ldots,b),$$

then it turns out that

$$W = \frac{1}{b^2}\left[\frac{1}{k^2(k-1)}\sum_{p,q}\lambda_{pq}^2 + \frac{k-2}{k(k-1)}\sum_{p,q}\lambda_{pq} - b\right]. \tag{4.5.29}$$

Since

$$\sum_{p,q}\lambda_{pq}^2 = tr(N'N)^2 = tr(NN')^2 = vr^2 + v(v-1)\lambda^2,$$

$$\sum_{p,q}\lambda_{pq} = \underline{1}'N'N\underline{1} = r^2v, \tag{4.5.30}$$

it follows that

$$W = \frac{r + \lambda(k-1) + rk(k-2) - k(k-1)}{bk(k-1)} \tag{4.5.31}$$

Therefore

$$\begin{aligned}\phi &= \frac{v-k}{vr}W^{-1} - \frac{2k}{vr(k-1)} \\ &= \frac{(v\lambda - k)(k-1)}{r + \lambda(k-1) + rk(k-2) - k(k-1)} - \frac{2k}{vr(k-1)} \\ &= 1 - \frac{2k}{vr(k-1)} = 1 - \frac{2}{b(k-1)}\ .\end{aligned} \tag{4.5.32}$$

In other words, if the variance within blocks approaches uniformity as $b \to \infty$, then $\phi \sim 1$. In such a circumstance we may take

$$\frac{\Gamma[(n-k)/2]}{\Gamma[(v-1)/2]\Gamma[(n-v-b+1)/2]}\theta^{[(v-1)/2]-1} \times (1-\theta)^{[(n-v-b+1)/2]-1}\, d\theta \tag{4.5.33}$$

as an approximation to the permutation distribution of θ.

Thus averaging (4.5.9) with respect to (4.5.33) we obtain an absolute null distribution of F as

$$\text{(4.5.34)} \qquad \frac{\Gamma[(n-b)/2]}{\Gamma[(v-1)/2]\Gamma[(n-v-b+1)/2]} \left(\frac{v-1}{n-v-b+1}F\right)^{[(v-1)/2]-1}$$

$$\times \left(1 + \frac{v-1}{n-v-b+1}F\right)^{-(n-b)/2} d\left(\frac{v-1}{n-v-b+1}F\right),$$

which is the usual central F-distribution.

So far we have treated the randomization procedure for a complete block (Section 2.3), Latin-square (Section 2.5) and a balanced incomplete block design in the present section. However our method is heuristic. A mathematically rigorous treatment of the randomization procedure will be given in Chapter 6.

4.6. The Lattice Design

We have seen already that for a BIBD, $bk = vr$, $\lambda(v - 1) = r(k - 1)$, $b > v$ and hence $r > k$. Therefore we can deduce an inequality between v and r:

$$\lambda(v - 1) \leqq r(r - 1),$$

$$v - 1 \leqq \frac{r(r-1)}{\lambda} \leqq r(r - 1),$$

or

$$\text{(4.6.1)} \qquad v \leqq r^2 - r + 1.$$

This means that a large number of treatments cannot be managed with a few replications in BIBD (e.g., $r = 4$, then $v = 13$).

r	3	4	5	6	7	8	9
max v	7	13	21	31	43	57	73

To remedy this point, we consider the lattice design, in which $v = k^2$. Arrange $v = k$ treatments in a $k \times k$ square as follows (e.g., $k = 5$):

1	2	3	4	5
6	7	8	9	10
11	12	13	14	15
16	17	18	19	20
21	22	23	24	25

take k blocks, each block containing k treatments in the same row, and then take k blocks, each block containing k treatments in the same column. Next consider $k - 1$ mutually orthogonal Latin squares of order k and corresponding to each square, take k blocks, each block containing k treatments corresponding to the same letter. Thus we have a BIBD such that

$$(4.6.2) \quad v = k^2, \quad b = k(k + 1), \quad r = k + 1, \quad k = k, \quad \lambda = 1.$$

This is called *a balanced lattice design.*

Suppose we have $k + 1$ other treatments $A_1, A_2, \ldots, A_{k+1}$ and add A_i to each block of the ith replication in the balanced lattice design. Then we get a design such that

$$(4.6.3) \quad \begin{aligned} &v^* = k^2 + k + 1, \quad b^* = k(k + 1), \quad k^* = k + 1, \\ &r^* = k + 1 \text{ or } k, \ \lambda^* = 1. \end{aligned}$$

Consider a lattice design with m orthogonal replications.

$$(4.6.4) \quad v = k^2, \quad b = mk, \quad r = m, \quad k = k, \quad \lambda_{ij} = 1 \text{ or } 0 \quad (m \leq k + 1),$$

that is, rows, columns, and $(m - 2)$ mutually orthogonal Latin squares. This is not a BIBD, but a partially balanced incomplete block design (PBIBD).

The analysis of this design is as follows :

$S_r(t_i)$ = the sum of treatment effects for those treatments that occur in the same row as the ith treatment.

$S_c(t_i)$ = the sum of treatment effects for those treatments that occur in the same column as the ith treatment.

$S_j(t_i)$ = the sum of treatment effects for those treatments that correspond to the same letter as the ith treatment in the jth Latin square.

The sums $S_r(Q_i)$, $S_c(Q_i)$ and $S_j(Q_i)$ are defined in a similar manner. The normal equation is

$$C\underline{t} = \underline{Q}, \tag{4.6.5}$$

where

$$c_{ii} = r_i(1 - \frac{1}{k}) = m(1 - \frac{1}{k}), \tag{4.6.6}$$

and for $i \neq j$

$$c_{ij} = \begin{cases} -\dfrac{\lambda_{ij}}{k} = -\dfrac{1}{k} & \text{if the treatments } i \text{ and } j \text{ occur together in the same row or column, or correspond to the same letter in a Latin square;} \\ 0 & \text{otherwise.} \end{cases} \tag{4.6.7}$$

The ith adjusted normal equation may be written as

$$\left(m(1 - \frac{1}{k}) + \frac{m}{k}\right)t_i - \frac{1}{k}\Big(S_r(t_i) + S_c(t_i) + S_1(t_i) + \cdots + S_{m-2}(t_i)\Big) = Q_i$$

or

$$mt_i - \frac{1}{k}\left[S_r(t_i) + S_c(t_i) + S\ (t_i) + \cdots + S_{m-2}(t_i)\right] = Q_i. \tag{4.6.8}$$

Here we add a restriction :

$$t_1 + \cdots + t_v = 0. \tag{4.6.9}$$

Summing up (4.6.8) for all treatments that are in the same row as i, we get

$$mS_r(t_i) - S_r(t_i) = S_r(Q_i),$$

and therefore

$$S_r(t_i) = \frac{1}{m - 1} S_r(Q_i). \tag{4.6.10}$$

Similarly we get

$$\begin{aligned} S_c(t_i) &= \frac{1}{m - 1} S_c(Q_i), \\ S_j(t_i) &= \frac{1}{m - 1} S_j(Q_i). \end{aligned} \tag{4.6.11}$$

Hence from (4.6.8) it follows that

$$t_i = \frac{1}{m} Q_i + \frac{1}{m(m - 1)k}\left\{S_r(Q_i) + S_c(Q_i) + S_1(Q_i) + \cdots + S_{m-2}(Q_i)\right\}. \tag{4.6.12}$$

We now calculate the variance of $t_i - t_j$:

(i) $\lambda_{ij} = 1$:

$$c^*_{ii} = \frac{1}{m} + \frac{1}{(m - 1)k}, \quad c^*_{ij} = \frac{1}{m(m - 1)k},$$

$$Var(t_i - t_j) = \left[\frac{2}{m} + \frac{2}{(m - 1)k} - \frac{2}{m(m - 1)k}\right]\sigma^2$$

$$= \frac{2\sigma^2}{m}\left(1 + \frac{1}{k}\right). \tag{4.6.13}$$

(ii) $\lambda_{ij} = 0$:

$$c^*_{ii} = \frac{1}{m} + \frac{1}{(m-1)k}, \qquad c^*_{ij} = 0,$$

$$Var(t_i - t_j) = \left(\frac{2}{m} + \frac{2}{(m-1)k}\right)\sigma^2$$

$$= 1 + \frac{1}{k[1 - (1/m)]} \tag{4.6.14}$$

Suppose that m orthogonal replications are used n times each. Then

$$v = k, \qquad b = mnk, \qquad r = mn, \qquad k = k, \tag{4.6.15}$$
$$\lambda_{ij} = n \quad \text{or} \quad 0.$$

Hence the numbers c_{ii} and c_{ij} should be multiplied by n. In the new adjusted normal equations each coefficient on the left is multiplied by n. Thus the new solutions can be obtained from the old by replacing Q_i by Q_i/n. Therefore

$$t_i = \frac{1}{mn} Q_i + \frac{1}{mn(m-1)k} \tag{4.6.16}$$
$$\times \left[S_r(Q_i) + S_c(Q_i) + S\ (Q_i) + \cdots + S_{m-2}(Q_i)\right]$$

and

$$Var(t_i - t_j) = \begin{cases} \frac{1}{mn}(1 + \frac{1}{k}) & \text{if} \quad \lambda_{ij} = n, \\ \frac{1}{mn}\left[1 + \frac{m}{(m-1)k}\right] & \text{if} \quad \lambda_{ij} = 0. \end{cases} \tag{4.6.17}$$

If we had used mn orthogonal replications, then

$$Var(t_i - t_j) = \begin{cases} \frac{2\sigma^2}{mn}(1 + \frac{1}{k}) & \text{if } \lambda_{ij} = 1, \\ \frac{2\sigma^2}{mn}\left[1 + \frac{mn}{(mn-1)k}\right] & \text{if } \lambda_{ij} = 0. \end{cases} \tag{4.6.18}$$

Obviously

$$1 + \frac{mn}{(mn - 1)k} \leqq 1 + \frac{m}{(m - 1)k} \quad \text{for} \quad m \geqq 1, \tag{4.6.19}$$

as was to be expected.

4.7. Partially Balanced Incomplete Block Design

Definition 4.7.1. Given v treatments $1,2,\ldots,v$, a relation satisfying the following conditions is said to be *an association* with m associate classes :

1. Any two treatments are either first or second or .... mth associates.
2. Any treatment has n_i ith associates, $i = 1,2,\ldots,m$.
3. If two treatments are ith associates, then the number of treatments that are jth associates of the first and jth associates of the first and kth associates of the second is p^i_{jk} $(i,j,k = 1,2,\ldots,m)$.

We get a *partially balanced incomplete block design* if the following conditions exist :

1. Each block contains k different treatments.
2. Each treatment occurs in r blocks.
3. Any two treatments that are ith associates occur together in λ_i blocks, $i = 1,2,\ldots,m$.

In a degenerate case when $m = 1$, this reduces to a BIBD. The cases when $m = 2$ are important for practical applications.

Example of association scheme. There are $v = mn$ treatments, and let us write them down in the following scheme :

Group 1	1	2	3 $\cdots$ n
Group 2	$n + 1$	$n + 2$	$n + 3 \cdots 2n$
$\cdots$	$\cdots$	$\cdots$	$\cdots$
Group m	$(m - 1)n + 1$	$(m - 1)n + 2$	$(m - 1)n + 3 \cdots mn$

Now define : treatments belonging to the same group are the first associates, and treatments belonging to different groups are the second associates. In this case, number of associate classes is 2, and

$$n_1 = n - 1, \quad n_2 = n(m - 1),$$

$$p_{11}^1 = n - 2, \quad p_{12}^1 = 0, \quad p_{11}^2 = 0, \quad p_{12}^2 = n - 1,$$

$$p_{21}^1 = 0, \quad p_{22}^1 = n(m - 1), \quad p_{21}^2 = n - 1, \quad p_{22}^2 = n(m - 2).$$

This association is called *a group-divisible association.*

The parameters of the association are

$$v; \quad n_i (i = 1,2,\ldots,m); \quad p_{jk}^i \ (i,j,k = 1,2,\ldots,m),$$

and the additional design parameters are

$$b; \ r; \ k; \ \lambda_i (i = 1,2,\ldots,m).$$

It should be

$$n_1 + n_2 + \cdots + n_m = v - 1 \tag{4.7.1}$$

and

$$p_{jk}^i = p_{kj}^i \tag{4.7.2}$$

(symmetry). Furthermore,

(4.7.3) $$\sum_{k=1}^{m} p_{jk}^{i} = \begin{cases} n_j & \text{if } i \neq j, \\ n_i - 1 & \text{if } i = j, \end{cases}$$

and

(4.7.4) $$n_i p_{jk}^{i} = n_j p_{ik}^{j} = n_k p_{ij}^{k}.$$

In fact, for a fixed treatment θ there are n_i $\theta^{(i)}$'s which are ith associates and n_j $\theta^{(j)}$'s which are jth associates and there are $n_i p_{jk}^{i}$ lines connecting $\theta^{(i)}$ and $\theta^{(j)}$. On the other hand, there are $n_j p_{ik}^{j}$ lines connecting $\theta^{(i)}$ and $\theta^{(j)}$. Hence

$$n_i p_{jk}^{i} = n_j p_{ik}^{j}.$$

Finally

(4.7.5) $$n_1\lambda_1 + n_2\lambda_2 + \cdots + n_m\lambda_m = r(k - 1).$$

A treatment may be regarded as the oth associate of its own. Thus we establish the following conventions :

(4.7.6) $$n_0 = 1, \quad p_{jk}^{o} = n_j\delta_{jk}, \quad p_{ok}^{i} = \delta_{ik}.$$

Under these conventions, we have

(4.7.7) $$\sum_{i=0}^{m} n_i = v,$$

(4.7.8) $$\sum_{k=0}^{m} p_{jk}^{i} = n_j,$$

and if we put $r \equiv \lambda_0$, then

(4.7.9) $$\sum_{i=0}^{m} n_i\lambda_i = rk.$$

Let A_0 be the identity matrix of order v. Also let A_i be a symmetrical matrix such that the element in the αth row and the βth column is unity if the treatments α and β are ith associates and zero otherwise :

$$(4.7.10) \qquad A_i = \left\| a_{\alpha i}^{\beta} \right\| = \left\| \begin{matrix} a_{1i}^1 & a_{1i}^2 & \cdots & a_{1i}^v \\ a_{2i}^1 & a_{2i}^2 & \cdots & a_{2i}^v \\ \cdots & \cdots & \cdots & \cdots \\ a_{vi}^1 & a_{vi}^2 & \cdots & a_{vi}^v \end{matrix} \right\|,$$

where

$$a_{\alpha i}^{\beta} = \begin{cases} 1 \text{ if treatments } \alpha \text{ and } \beta \text{ are } i\text{th associates,} \\ 0 \text{ otherwise.} \end{cases}$$

The matrices $A_1, A_2, \ldots, A_m$ are called *association matrices*. They are all symmetric having 1 or 0 elements, and

$$(4.7.11) \qquad A_0 + A_1 + A_2 + \cdots + A_m = G.$$

Hence we can see that

$$c_0 A_0 + c_1 A_1 + \cdots + c_m A_m = 0$$

if and only if $c_0 = c_1 = \cdots = c_m = 0$ (linear independence).

Further we have

$$(4.7.12) \qquad A_j A_k = p_{jk}^0 A_0 + p_{jk}^1 A_1 + \cdots + p_{jk}^m A_m = \sum_{i=0}^{m} p_{jk}^i A_i.$$

Thus the linear closure of $A_0, A_1, \ldots, A_m$ is a linear associative algebra, it is called *the association algebra* and is denoted by $\mathfrak{A}$.

Let

$$(4.7.13) \qquad P_k = \left\| \begin{matrix} p_{0k}^0 & p_{0k}^1 & \cdots & p_{0k}^m \\ p_{1k}^0 & p_{1k}^1 & \cdots & p_{1k}^m \\ \cdots & \cdots & \cdots & \cdots \\ p_{mk}^0 & p_{mk}^1 & \cdots & p_{mk}^m \end{matrix} \right\| \qquad (k = 0,1,\ldots,m).$$

Since

$$A_jA_k = \sum_{i=0}^{m} p_{jk}^i A_i \qquad (j = 0,1,\ldots,m),$$

or

$$(4.7.14) \qquad \begin{bmatrix} A_0 \\ A_1 \\ \vdots \\ A_m \end{bmatrix} A_k = P_k \begin{bmatrix} A_0 \\ A_1 \\ \vdots \\ A_m \end{bmatrix},$$

it follows that

$$(4.7.15) \qquad \begin{bmatrix} A_0 \\ A_1 \\ \vdots \\ A_m \end{bmatrix} A_iA_k = P_i \begin{bmatrix} A_0 \\ A_1 \\ \vdots \\ A_m \end{bmatrix} A_k = P_iP_k \begin{bmatrix} A_0 \\ A_1 \\ \vdots \\ A_m \end{bmatrix},$$

which means that the mapping $A_k \to P_k$ is a regular representation the association algebra $\mathfrak{A}$. Hence

$$(4.7.16) \qquad P_jP_k = \sum_{i=0}^{m} p_{jk}^i P_i.$$

Let $A = c_0A_0 + c_1A_1 + \cdots + c_mA_m$ be any element of the association algebra and let $f(\lambda)$ be a polynomial. Then we can express $f(A)$ in the form

(4.7.17) $$f(A) = l_0A_0 + l_1A_1 + \cdots + l_mA_m.$$

If $P = c_0P_0 + c_1P_1 + \cdots + c_mP_m$ is the representation of A, then

(4.7.18) $$f(P) = l_0P_0 + l_1P_1 + \cdots + l_mP_m.$$

Now let $f(\lambda)$ be the minimum function of A. Then $f(\lambda)$ is the monic polynomial of the least degree for which $f(A) = 0$. Since $f(A)$ can be expressed in the form of (4.7.17), $f(A) = 0$ means that

(4.7.19) $$l_0 = l_1 = \cdots = l_m = 0.$$

Thus we get

(4.7.20) $$f(P) = 0.$$

Therefore the minimum function $g(\lambda)$ of P divides $f(\lambda)$.

In a similar manner we can show that $g(\lambda)$ is divisible by $f(\lambda)$, and therefore

(4.7.21) $$f(\lambda) \equiv g(\lambda).$$

Since A is symmetric, all the characteristic roots are real and its minimum function is simply the product of linear factors corresponding to the distinct characteristic roots. Hence the distinct characteristic roots of A and those of P must be the same.

Let $\theta_0, \theta_1, \ldots, \theta_m$ be the characteristic roots of P. Then A has the same characteristic roots with the respective multiplicities $\alpha_0, \alpha_1, \ldots, \alpha_m$, and

$$
\begin{aligned}
& \alpha_0 + \alpha_1 + \cdots + \alpha_m = v, \\
tr(A) &= \alpha_0\theta_0 + \alpha_1\theta_1 + \cdots + \alpha_m\theta_m = vc_0, \\
tr(A^2) &= \alpha_0\theta_0^2 + \alpha_1\theta_1^2 + \cdots + \alpha_m\theta_m^2 = vu_0, \\
& \cdots\cdots\cdots\cdots\cdots, \\
tr(A^m) &= \alpha_0\theta_0^m + \alpha_1\theta_1^m + \cdots + \alpha_m\theta_m^m = vw_0.
\end{aligned} \tag{4.7.22}
$$

Thus we can calculate the multiplicities α_s' in terms of p_{jk}^i.

The adjusted normal equation is

$$C\underline{t} = \underline{Q}, \tag{4.7.23}$$

where

$$c_{\alpha\alpha} = (1 - \frac{1}{k}), \qquad c_{\alpha\beta} = -\frac{\lambda_{\alpha\beta}}{k}.$$

Hence

$$C = r(1 - \frac{1}{k})A_0 - \frac{\lambda_1}{k}A_1 - \cdots - \frac{\lambda_m}{k}A_m. \tag{4.7.24}$$

Since

$$NN' = rA_0 + \lambda_1 A_1 + \cdots + \lambda_m A_m,$$

we get

$$kC = krA_0 - NN' = krI_v - NN'. \tag{4.7.25}$$

Theorem 4.7.1. If e is a characteristic root of C, then $k(r - e)$ is a characteristic root of NN', and the converse is also true.

Proof. If e is a characteristic root of C, then

$$|C - I_v e| = 0$$

or

$$|kC - ke\, I_v| = 0,$$

or by (4.7.25),

$$\left|NN' - k(r - e)I_v\right| = 0.$$

Q.E.D.

The converse is also true.

If the design is connected, then 0 is a characteristic root of C with multiplicity 1. Hence kr is a characteristic root of NN' with multiplicity 1.

Let

$$P = rP_0 + \lambda_1 P_1 + \cdots + \lambda_m P_m. \tag{4.7.26}$$

Then kr is a characteristic root of P. Let us define

$$p_{ij} = rp^j_{i0} + \lambda_1 p^j_{i1} + \cdots + \lambda_m p^j_{im} \tag{4.7.27}$$

and

$$P = \|p_{ij}\|,$$

$$\sum_{i=0}^{m} p_{ij} = rn_0 + \lambda_1 n_1 + \cdots + \lambda_m n_m = rk. \tag{4.7.28}$$

Consider

$$P - \theta I_{m+1} = \begin{Vmatrix} p_{00} - \theta & p_{01} & \cdots & p_{0m} \\ p_{10} & p_{11} - \theta & \cdots & p_{1m} \\ \cdots & \cdots & \cdots & \cdots \\ p_{m0} & p_{m1} & \cdots & p_{mm} - \theta \end{Vmatrix}. \tag{4.7.29}$$

Then

$$\left|P - \theta I_{m+1}\right| = (rk - \theta) \begin{vmatrix} p^*_{11} - \theta & p^*_{12} & \cdots & p^*_{1m} \\ p^*_{21} & p^*_{22} - \theta & \cdots & p^*_{2m} \\ \cdots & \cdots & \cdots & \cdots \\ p^*_{m1} & p^*_{m2} & \cdots & p^*_{mm} - \theta \end{vmatrix}, \tag{4.7.30}$$

where

$$p^*_{ij} = p_{ij} - p_{i0}$$

$$= r(p^j_{i0} - p^o_{i0}) + \lambda_1(p^j_{i1} - p^o_{i1})$$

$$+ \cdots + \lambda_m(p^j_{im} - p^o_{im})$$

(4.7.31) $$= r\delta_{ij} + \lambda_1 p^j_{i1} + \cdots + \lambda_m p^j_{im} - n_i\lambda_i.$$

We already knew that

(4.7.32) $$b \geqq \text{rank } N = \text{rank } NN'.$$

Theorem 4.7.2. A necessary condition for $b < v$ is that NN' be singular or

(4.7.33) $$|p^*_{ij}| = 0.$$

Let $\theta_0 \equiv kr$, $\theta_1\ \theta_2, \ldots, \theta_m$, be the characteristic roots if P. Then these are also the characteristic roots of NN' with the respective multiplicities $\alpha_0 = 1$, $\alpha_1, \alpha_2, \ldots, \alpha_m$ (assuming that the design is connected). Thus

(4.7.34) $$|NN' - \theta I_v| = (rk - \theta)(\theta_1 - \theta)^{\alpha_1}(\theta_2 - \theta)^{\alpha_2} \cdots (\theta_m - \theta)^{\alpha_m}.$$

If NN' is singular, it must have at least one zero root, say $\theta_i = 0$. Thus

$$\text{rank } NN' \geqq v - \alpha_i.$$

Consequently

(4.7.35) $$b \geqq \text{rank } NN' \geqq v - \alpha_i.$$

This is a result obtained by Fisher.

4.8. Partially Balanced Incomplete Block Design with Two Associate Classes

Let θ be a characteristic root of P^*:

(4.8.1)

$$|P^* - \theta I_2|$$

$$= \begin{vmatrix} r + \lambda_1 p^1_{11} + \lambda_2 p^1_{12} - n_1\lambda_1 - \theta & \lambda_1 p^2_{11} + \lambda_2 p^2_{12} - n_1\lambda_1 \\ \lambda_1 p^1_{21} + \lambda_2 p^1_{22} - n_2\lambda_2 & r + \lambda_1 p^2_{21} + \lambda_1 {}^{2}_{22} - n_2\lambda_2 - \theta \end{vmatrix} = 0.$$

Adding the second row to the first, we get

$$\begin{vmatrix} r - \lambda_1 - \theta & r - \lambda_2 - \theta \\ \lambda_1 p^1_{21} + \lambda_2 p^1_{22} - n_2\lambda_2 & r + \lambda_1 p^2_{21} + \lambda_2 p^2_{22} - n_2\lambda_2 - \theta \end{vmatrix} = 0.$$

Subtracting the first column from the second, we get

(4.8.2)

$$\begin{vmatrix} r - \lambda_1 - \theta & \lambda_1 - \lambda_2 \\ (\lambda_1 - \lambda_2)p^1_{12} & r - \lambda_2 + (\lambda_1 - \lambda_2)(p^2_{12} - p^1_{12}) - \theta \end{vmatrix} = 0.$$

Hence one characteristic root of NN' is rk, and the other two are the roots of the following quadratic equation

$$(r - \theta)^2 + [(\lambda_1 - \lambda_2)(p^2_{12} - p^1_{12}) - (\lambda_1 + \lambda_2)](r - \theta)$$

(4.8.3)

$$+ [(\lambda_1 - \lambda_2)(\lambda_2 p^1_{12} - \lambda_2 p^2_{12} - \lambda_1 p^2_{12}) + \lambda_1\lambda_2] = 0.$$

If we put

(4.8.4) $\beta = p^1_{12} + p^2_{12}, \quad \gamma = p^2_{12} - p^1_{12}, \quad \Delta = \gamma^2 + 2\beta + 1,$

then

$$(4.8.5)\qquad \begin{aligned} r - \theta_1 &= \tfrac{1}{2}[(\lambda_1 - \lambda_2)(-\gamma - \sqrt{\Delta}) + \lambda_1 + \lambda_2],\\ r - \theta_2 &= \tfrac{1}{2}[(\lambda_1 - \lambda_2)(-\gamma + \sqrt{\Delta}) + \lambda_1 + \lambda_2]. \end{aligned}$$

Hence

$$(4.8.6)\qquad \theta_1 - \theta_2 = (\lambda_1 - \lambda_2)\sqrt{\Delta},$$

$$|NN' - I_v\theta| = (rk - \theta)(\theta_1 - \theta)^{\alpha_1}(\theta_2 - \theta)^{\alpha_2},$$

$$(4.8.7)\qquad \begin{aligned} 1 + \alpha_1 + \alpha_2 &= v,\\ rk + \alpha_1\theta_1 + \alpha_2\theta_2 &= vr, \end{aligned}$$

$$(4.8.8)\qquad \alpha_1(\theta_1 - \theta_2) = vr - kr - (v - 1)\Big\{r - \tfrac{1}{2}[(\lambda_1 - \lambda_2) \times (-\gamma + \sqrt{\Delta}) + \lambda_1 + \lambda_2]\Big\}.$$

Thus we get

$$(4.8.9)\qquad \begin{aligned} \alpha_1 &= \frac{n_1 + n_2}{2} - \frac{(n_1 - n_2) + \gamma(n_1 + n_2)}{2\sqrt{\Delta}},\\ \alpha_2 &= \frac{n_1 + n_2}{2} + \frac{(n_1 - n_2) + \gamma(n_1 + n_2)}{2\sqrt{\Delta}}, \end{aligned}$$

where α_1 *and* α_2 *must be positive integers.*

Now we shall investigate the class of all PBIBDs with two associate classes, when

(4.8.10)
(i) $r < k$.
(ii) $\lambda_1 = 1, \quad \lambda_2 = 0$.

If $r < k$, then $b < v$, and hence NN' must be singular. Therefore P^* must have a zero characteristic root.

$$NN' = rA_0 + \lambda_1 A_1 + \lambda_2 A_2,$$

$$P = rP_0 + \lambda_1 P_1 + \lambda_2 P_2,$$

$$|P - \theta I_3| = |(rk - \theta)\ P^* - \theta I_2|,$$

$$|P^* - \theta I_2| = (r - \theta)^2 + [(\lambda_1 - \lambda_2)(p_{12}^2 - p_{12}^1) - (\lambda_1 + \lambda_2)] \times (r - \theta) + [(\lambda_1 - \lambda_2)(\lambda_2 p_{12}^1 - \lambda_1 p_{12}^2) + \lambda_1\lambda_2].$$

Thus we have

$$r^2 + [(\lambda_1 - \lambda_2)(p_{12}^2 - p_{12}^1) - (\lambda_1 + \lambda_2)]r + [(\lambda_1 - \lambda_2)(\lambda_2 p_{12}^1 - \lambda_1 p_{12}^2) + \lambda_1\lambda_2] = 0. \tag{4.8.11}$$

Since $\lambda_1 = 1$ and $\lambda_2 = 0$, this relation reduces to

$$r^2 + (p_{12}^2 - p_{12}^1 - 1)r - p_{12}^2 = 0$$

or

$$rp_{12}^1 - (r - 1)p_{12}^2 = r(r - 1). \tag{4.8.12}$$

Since

$$n_1 p_{12}^1 + n_2 p_{12}^2 = n_1(n_2 - p_{22}^1) + n_2 p_{12}^2 = n_1 n_2 - n_1 p_{22}^1 + n_2 p_{12}^2 = n_1 n_2, \tag{4.8.13}$$

from (4.8.12) and (4.8.13), we find that

$$p_{12}^1 = \frac{n_2(r - 1)(r + n_1)}{n_1(r - 1) + n_2 r}, \tag{4.8.14a}$$

$$p_{12}^2 = \frac{rn_1(n_2 - r + 1)}{n_1(r - 1) + n_2 r}, \tag{4.8.14b}$$

$$v = n_2 + 1 + r(k - 1), \tag{4.8.15}$$

and

(4.8.16) $$b = \frac{vr}{k} = \frac{r^2(k - 1) + n_2 r}{k} = r^2 + \frac{r(n_2 - r + 1)}{k}.$$

Since b and r are positive integers, we should have

(4.8.17) $$r(n_2 - r + 1) = sk,$$

where s should be an integer. Then it follows that

(4.8.18a) $$n_2 = r - 1 + \frac{sk}{r},$$

(4.8.18b) $$b = r^2 + s,$$

(4.8.18c) $$v = \frac{k(r^2 + s)}{r},$$

and

(4.8.19a) $$p^1_{11} = k - r - 1 + \frac{r(r - 1)^2(k - 1)}{s + r(r - 1)},$$

(4.8.19b) $$p^2_{12} = r(k - 1) - \frac{r^2(r - 1)(k - 1)}{s + r(r - 1)}.$$

From (4.8.19), $r(r - 1)^2(k - 1)$ and $r^2(r - 1)(k - 1)$ must both be multiples of $s + r(r - 1)$. Hence their difference

$$r(r - 1)(k - 1)$$

must be a multiple of $s + r(r - 1)$. Thus we introduce a parameter t by

(4.8.20) $$t = \frac{r(r - 1)(k - 1)}{s + r(r - 1)}.$$

Then

(4.8.21) $$s = \frac{r(r - 1)(k - t - 1)}{t}.$$

Now we can express all the parameters in terms of r, k, and t as follows :

$$v = \frac{k[(r - 1)(k - 1) + t]}{t}, \quad b = \frac{r[(r - 1)(k - 1) + t]}{t},$$

$$r = r, \quad k = k, \quad \lambda_1 = 1, \quad \lambda_2 = 0,$$

$$n_1 = r(k - 1), \quad n_2 = \frac{(r - 1)(k - 1)(k - t)}{t},$$

$$P^1 = \begin{Vmatrix} (t - 1)(r - 1) + k - 2 & (r - 1)(k - t) \\ (r - 1)(k - t) & \frac{(r - 1)(k - 1)(k - t - 1)}{t} \end{Vmatrix},$$

(4.8.22)

$$P^2 = \begin{Vmatrix} rt & r(k - t - 1) \\ r(k - t - 1) & [(r - 1)(k - 1)(k - 2t) + t(rt - k)]\frac{1}{t} \end{Vmatrix}.$$

We shall show that

$$1 \leqq t \leqq r. \tag{4.8.23}$$

Proof. Let θ and ϕ be two treatments that are first associates. Since $\lambda_1 = 1$, they occur together in exactly one block $B(\theta\ \phi)$. The other $k - 2$ treatments occurring in this block must be first associates of both θ and ϕ. Hence

$$p_{11}^1 \geqq k - 2,$$

that is,

$$(t - 1)(r - 1) \geqq 0.$$

Since $r > 1$, we have

$$t \geqq 1. \tag{4.8.24}$$

Since θ and ϕ have once occurred together, they cannot occur together in any other blocks. Let the $r - 1$ other blocks in which θ appears be

$$B_1(\theta),\ B_2(\theta),\ldots,B_{r-1}(\theta)$$

and let

$$B_1(\phi),\ B_2(\phi),\ldots,B_{r-1}(\phi)$$

be the other $r - 1$ blocks in which ϕ appears. Any treatment that is a first associate of θ and ϕ and does not occur in $B(\theta,\phi)$ must appear in one and only one of the blocks $B_i(\theta)$ and similarly in one of the blocks $B_j(\theta)$. On the other hand, any two blocks cannot have more than one treatment in common. Any one of the blocks $B_i(\theta)$ cannot have more than $r - 1$ treatments that are first associates of both θ and ϕ, since no two of them can occur together in the blocks $B_j(\phi)$.

Hence

$$p_{11}^1 \leqq k - 2 + (r - 1)^2,$$

(4.8.25)

$$(t - 1)(r - 1) \leqq (r - 1)^2,$$

that is,

$$t \leqq r,$$

$$\alpha_1 = \frac{n_1 + n_2}{2} - \frac{(n_1 - n_2) + \gamma(n_1 + n_2)}{2\sqrt{\Delta}},$$

$$\gamma = p_{12}^2 - p_{12}^1,\quad \beta = p_{12}^2 + p_{12}^1,\quad \Delta = \gamma^2 + 2\beta + 1,$$

and in our case α_1 turns out to be

(4.8.26)
$$\alpha_1 = \frac{r(r - 1)k(k - 1)}{t(k + r - t - 1)} .$$

Suppose $r = 3$, then

$$\alpha_1 = \frac{6k(k - 1)}{t(k - t + 2)}$$

must be a positive integer.

$$t = 1, \quad \frac{6k(k - 1)}{k + 1} \quad (k = 2,3,5);$$

$$t = 2, \quad 3(k - 1);$$

$$t = 3, \quad 2k.$$

Special Case : $r = 2, \quad k = k, \quad \lambda_1 = 1, \quad \lambda_2 = 0$

Series I, $t = 1$:

$$v = k^2, \quad b = 2k,$$

$$n_1 = 2(k - 1), \quad n_2 = (k - 1)^2,$$

$$P^1 = \begin{Vmatrix} k - 2 & k - 1 \\ k - 1 & (k - 1)(k - 2) \end{Vmatrix},$$

$$P^2 = \begin{Vmatrix} 2 & 2k - 4 \\ 2k - 4 & (k - 1)(k - 2) + (2 - k) \end{Vmatrix}.$$

Series II, $t = 2$:

$$v = \frac{k(k + 1)}{2}, \quad b = k + 1,$$

$$n_1 = 2(k - 1), \quad n_2 = \frac{(k - 1)(k - 2)}{2},$$

$$P^1 = \begin{Vmatrix} k - 1 & k - 2 \\ k - 2 & \frac{(k - 2)(k - 3)}{2} \end{Vmatrix},$$

$$P^2 = \begin{Vmatrix} 4 & 2k - 6 \\ 2k - 6 & \dfrac{(k - 1)(k - 4) + 2(4 - k)}{2} \end{Vmatrix}.$$

Remark. Series I is the lattice design with two replications. The particular case when $k = 4$ in series II is a *singly linked design* proposed by Youden [1].

×	1	2	3	4
1	×	5	6	7
2	5	×	8	9
3	6	8	×	10
4	7	9	10	×

Two treatments in the same row are first associates. If we take rows as blocks, then

$b = 5, \quad v = 10, \quad r = 2, \quad k = 4,$

$\lambda_1 = 1, \quad \lambda_2 = 0.$

4.9. Analysis of Variance of Partially Balanced Incomplete Block Design

The adjusted normal equation is

(4.9.1) $$C\underline{t} = \underline{Q},$$

where

$$C = Dr - ND_k^{-1} N'.$$

Thus

(4.9.2) $$C = (r - \frac{r}{k})A_0 - \frac{\lambda_1}{k}A - \cdots - \frac{\lambda_m}{k}A_m.$$

The adjusted normal equation may be written in the form

(4.9.3) $$(a_0A_0 + a_1A_1 + \cdots + a_mA_m)\underline{t} = \underline{Q},$$

where

$$a_0 = r(1 - \frac{1}{k}), \quad a_i = -\frac{\lambda_i}{k}.$$

A restriction

(4.9.4) $$t_1 + t_2 + \cdots + t_v = 0$$

may be written in the form

(4.9.5) $$(A_0 + A_1 + \cdots + A_m)\underline{t} = 0.$$

Multiplying A_u with both sides of (4.9.3) from the left, we get

(4.9.6) $$\sum_{s=0}^{m} \sum_{\nu=0}^{m} a_\nu p_{u\nu}^s A_s \underline{t} = A_u \underline{Q}.$$

Since

$$\sum_{\nu=0}^{m} a_\nu p_{u\nu}^o = a_u n_u,$$

we multiply (4.9.5) by $a_u n_u$ and subtract it from (4.9.6), getting

(4.9.7) $$(a_{u1}A_1 + a_{u2}A_2 + \cdots + a_{um}A_m)\underline{t} = A_u\underline{Q},$$

where

$$a_{uu} = -\frac{1}{k}\left[\lambda_1 p_{u1}^u + \lambda_2 p_{u2}^u + \cdots + \lambda_m p_{um}^u - \lambda_u n_u - r(k - 1)\right],$$

(4.9.8)

$$a_{ui} = -\frac{1}{k}\left[\lambda_1 p_{u1}^i + \lambda_2 p_{u2}^i + \cdots + \lambda_m p_{um}^i - \lambda_u n_u\right] \qquad (u \neq i).$$

Let $S_u(t_\alpha)$ be the sum of the treatment-effects of the uth associates of the αth treatment and let $S_u(Q_\alpha)$ be the sum of the adjusted yield totals of the uth associates of the αth treatment. Then it is easily seen that

$$(4.9.9) \qquad A_u \underline{t} = \begin{bmatrix} S_u(t_1) \\ S_u(t_2) \\ \vdots \\ S_u(t_v) \end{bmatrix} \quad \text{and} \quad A_u \underline{Q} = \begin{bmatrix} S_u(Q_1) \\ S_u(Q_2) \\ \vdots \\ S_u(Q_v) \end{bmatrix}.$$

The αth equation in (4.9.7) may be written as

$$(4.9.10) \qquad a_{u1}A\ (t_\alpha) + a_{u2}S_2(t_\alpha) + \cdots + a_{um}S_m(t_\alpha) = S_u(Q_\alpha)$$

$$(u = 1,2,\ldots,m).$$

Solving (4.9.10) with respect to $S_u(t_\alpha)$, α being fixed, we get

$$(4.9.11) \qquad S_u(t_\alpha) = a^*_{u1}S_1(Q_\alpha) + a^*_{u2}S_2(Q_\alpha) + \cdots + a^*_{um}S_m(Q_\alpha)$$

$$(u = 1,2,\ldots,m).$$

On the other hand, from

$$\left(r(1 - \frac{1}{k})A_0 - \frac{\lambda_1}{k}A_1 - \cdots - \frac{\lambda_m}{k}A_m\right)\underline{t} = \underline{Q}$$

we have

$$(4.9.12) \qquad r(1 - \frac{1}{k})t_\alpha - \frac{\lambda_1}{k}S_1(t_\alpha) - \cdots - \frac{\lambda_m}{k}S_m(t_\alpha) = Q_\alpha$$

$$(\alpha = 1,2,\ldots,v).$$

Therefore it follows that

$$(4.9.13) \qquad r(1 - \frac{1}{k})t_\alpha = Q_\alpha + \frac{\lambda_1}{k}S_1(t_\alpha) + \cdots + \frac{\lambda_m}{k}S_m(t_\alpha)$$

$$= Q_\alpha + \frac{1}{k}(\lambda_1,\ldots,\lambda_m)A^* \begin{bmatrix} S_1(Q_\alpha) \\ S_2(Q_\alpha) \\ \vdots \\ S_m(Q_\alpha) \end{bmatrix}.$$

Thus

$$(4.9.14)\qquad \sum_{\alpha=1}^{v} t_\alpha Q_\alpha = \frac{k}{r(k-1)} \sum_{\alpha=1}^{v} Q_\alpha^2 + \frac{k}{r(k-1)}$$

$$\times\ (\lambda_1,\ldots,\lambda_m)A^* \begin{bmatrix} \sum_{\alpha=1}^{v} Q_\alpha S_1(Q_\alpha) \\ \vdots \\ \sum_{\alpha=1}^{v} Q_\alpha S_m(Q_\alpha) \end{bmatrix}.$$

In the actual calculation we must first write down $\|a_{ui}\|$ and then convert this matrix, getting $\|a^*_{ui}\|$.

We already knew that if

$$t_\alpha = c^*_{\alpha 1}Q_1 + c^*_{\alpha 2}Q_2 + \cdots + c^*_{\alpha v}Q_v,$$

then

$$Var(t_\alpha - t_\beta) = (c^*_{\alpha\alpha} + c^*_{\beta\beta} - c^*_{\alpha\beta} - c^*_{\beta\alpha})\sigma^2,$$

and, from (4.9.13),

$$(4.9.15)\qquad c^*_{\alpha\alpha} = c^*_{\beta\beta} = \frac{k}{r(k-1)},$$

and if treatments α and β are the uth associates, then

$$(4.9.16)\qquad c^*_{\alpha\beta} = c^*_{\beta\alpha} = \frac{\lambda_1 a^*_{1u} + \lambda_2 a^*_{2u} + \cdots + \lambda_m a^*_{mu}}{r(k-1)}.$$

The Case of Two Associate Classes. We shall carry out the calculation a little further for the case of two associate classes.

In this special case we have

(4.9.17)
$$\begin{aligned}
a_{11} &= r(1 - \tfrac{1}{k}) + \frac{n_1\lambda_1}{k} - \frac{p^1_{11}\lambda_1}{k} - \frac{p^1_{12}\lambda_2}{k},\\
a_{12} &= \frac{n_1\lambda_1}{k} - \frac{p^2_{11}\lambda_1}{k} - \frac{p^2_{12}\lambda_2}{k},\\
a_{21} &= \frac{n_2\lambda_2}{k} - \frac{p^1_{21}\lambda_1}{k} - \frac{p^1_{22}\lambda_2}{k},\\
a_{22} &= r(1 - \tfrac{1}{k}) + \frac{n_2\lambda_2}{k} - \frac{p^2_{21}\lambda_1}{k} - \frac{p^2_{22}\lambda_2}{k}.
\end{aligned}$$

Now using the relations

(4.9.18)
$$\begin{aligned}
a_{11} &= r(1 - \tfrac{1}{k}) + \frac{\lambda_1}{k} + \frac{p^1_{12}}{k}(\lambda_1 - \lambda_2),\\
a_{12} &= \frac{p^2_{12}}{k}(\lambda_1 - \lambda_2),\\
a_{21} &= \frac{p^1_{21}}{k}(\lambda_1 - \lambda_2),\\
a_{22} &= r(1 - \tfrac{1}{k}) + \frac{\lambda_2}{k} + \frac{p^2_{21}}{k}(\lambda_2 - \lambda_1).
\end{aligned}$$

We shall put

$$r(k - 1) \equiv a, \quad p^1_{12} \equiv f, \quad p^2_{12} \equiv g,$$

and

$$\Delta = \begin{vmatrix} a_{11} & a_{12} \\ a_{21} & a_{22} \end{vmatrix},$$

Then we get

$$k^2\Delta = \begin{vmatrix} a + \lambda_1 + f(\lambda_1 - \lambda_2) & g(\lambda_1 - \lambda_2) \\ -f(\lambda_1 - \lambda_2) & a + \lambda_2 - g(\lambda_1 - \lambda_2) \end{vmatrix}$$

$$= \begin{vmatrix} a + \lambda_1 & a + \lambda_2 \\ -f(\lambda_1 - \lambda_2) & a + \lambda_2 - g(\lambda_1 - \lambda_2) \end{vmatrix}$$

(4.9.19) $$= (a + \lambda_1)(a + \lambda_2) + (\lambda_1 - \lambda_2)\left[f(a + \lambda_2) - g(a + \lambda_1)\right].$$

Since

$$a^*_{11} = \frac{a_{22}}{\Delta}, \quad a^*_{12} = -\frac{a_{12}}{\Delta}, \quad a^*_{21} = -\frac{a_{21}}{\Delta}, \quad a^*_{22} = \frac{a_{11}}{\Delta},$$

we have

(4.9.20) $$r\left(1 - \frac{1}{k}\right)t_\alpha = Q_\alpha + \frac{1}{k}(\lambda_1 a^*_{11} + \lambda_2 a^*_{21})S_1(Q_\alpha) + \frac{1}{k}(\lambda_1 a^*_{12} + \lambda_2 a^*_{22})S_2(Q_\alpha) \qquad (\alpha = 1,2,\ldots,v)$$

or

(4.9.21) $$r(k - 1)t_\alpha = kQ_\alpha + c_1S_1(Q_\alpha) + c_2S_2(Q_\alpha),$$

where

(4.9.22) $$(c_1,c_2) = (\lambda_1,\lambda_2)\begin{Vmatrix} a^*_{11} & a^*_{12} \\ a^*_{21} & a^*_{22} \end{Vmatrix}$$

and

(4.9.23) $$\Delta c_1 = \lambda_1 a_{22} - \lambda_2 a_{21}.$$

Hence

$$k\Delta c_1 = \lambda_1(a + \lambda_2) + (\lambda_1 - \lambda_2)(f\lambda_2 - g\lambda_1),$$

$$k\Delta c_2 = \lambda_2(a + \lambda_1) + (\lambda_1 - \lambda_2)(f\lambda_2 - g\lambda_1),$$

(4.9.24)

$$Q_\alpha + S_1(Q_\alpha) + S_2(Q_\alpha) = 0.$$

Thus finally,

$$r(k - 1)t_\alpha = (k - c_2)Q_\alpha + (c_1 - c_2)S_1(Q_\alpha). \tag{4.9.25}$$

From (4.9.21) one can see that

$$c^*_{\alpha\alpha} = \frac{k}{r(k - 1)}, \tag{4.9.26}$$

and if α and β are the uth associates, then

$$c^*_{\alpha\beta} = \frac{c_u}{r(k - 1)}, \tag{4.9.27}$$

$$Var(t_\alpha - t_\beta) = (c^*_{\alpha\alpha} + c^*_{\beta\beta} - c^*_{\alpha\beta} - c^*_{\beta\alpha})\sigma^2 = \frac{2\sigma^2}{r}\,\frac{k - c_u}{k - 1}. \tag{4.9.28}$$

<u>Calculations of $S_1(Q_\alpha)$ and $S_2(Q_\alpha)$</u>

1. Group divisible : $v = mn$

				Row Sum
Q_1	Q_2	$\cdots$	Q_n	G_1
Q_{n+1}	Q_{n+2}	$\cdots$	Q_{2n}	G_2
$\cdots$	$\cdots$	$\cdots$	$\cdots$	$\cdots$
$Q_{(m-1)n+1}$	$Q_{(m-1)n+2}$		Q_{mn}	G_m

Then

$$S_1(Q_\alpha) = G_\beta - Q_\alpha \quad \text{if} \quad \alpha = (\beta - 1)n + i,$$

$$t_\alpha = \frac{k - c_1}{r(k - 1)}Q_\alpha + \frac{c_1 - c_2}{r(k - 1)}G_\beta.$$

2. Triangular design : Let $R(\alpha)$ and $C(\alpha)$ be the row sum and the column sum respectively, in which α occurs. Then

$$S_1(Q_\alpha) = R(\alpha) + C(\alpha) - 2Q_\alpha.$$

Hence

$$t_\alpha = \frac{k - 2c_1 + c_2}{r(k - 1)}Q_\alpha + \frac{c_1 - c_2}{r(k - 1)}[R(\alpha) + C(\alpha)].$$

3. Lattice design with three replications :

$$S_1(Q_\alpha) = R(\alpha) + C(\alpha) + L(\alpha) - 2Q_\alpha.$$

Example of the Application of a Group Divisible Design

We explain the examples of group divisible designs given by Clyde Young Kramer and Ralph Allan Bradley [2] and [3].

Let V_{ij} denote the jth treatment in the ith group. We assume the following linear model :

$$x_{ijs} = \mu + \tau_{ij} + \beta_s + e_{ijs},$$

where μ stands for the grand mean, τ_{ij} is the effect of V_{ij}, and β_s is the block effect subject to the restrictions

$$\sum_{i=1}^{m} \sum_{j=1}^{n} \tau_{ij} = 0, \qquad \sum_{s=1}^{b} \beta_s = 0,$$

and e_{ijs} is the residual variate.

We shall be concerned with a group divisible PBIBD with the parameters

$$v = mn, \quad b, \quad r, \quad k, \quad \lambda_1, \quad \lambda_2,$$

$$P^1 \equiv \|p^1_{jk}\| = \begin{Vmatrix} n - 2 & 0 \\ 0 & n(m-1) \end{Vmatrix},$$

$$P^2 \equiv \|p^2_{jk}\| = \begin{Vmatrix} 0 & n-1 \\ n-1 & n(m-2) \end{Vmatrix},$$

$$n_1 = n - 1, \quad n_2 = n(m - 1),$$

$$(n - 1)\lambda_1 + n(m - 1)\lambda_2 = r(k - 1),$$

or

$$\lambda_1 + rk - r = n\lambda_1 + n(m - 1)\lambda_2,$$

and

$$k^2\Delta = v\lambda_2(\lambda_1 + rk - r),$$

$$c_1 = \frac{v\lambda_1\lambda_2}{k\Delta} = \frac{kv\lambda_1\lambda_2}{v\lambda_2(\lambda_1 + rk - r)} = \frac{k\lambda_1}{\lambda_1 + rk - r},$$

$$c_2 = \frac{v\lambda_1\lambda_2 - (rk - r)(\lambda_1 - \lambda_2)}{k\Delta} = \frac{k\{v\lambda_1\lambda_2 - (rk - r)(\lambda_1 - \lambda_2)\}}{v\lambda_2(\lambda_1 + rk - r)}$$

From the general theory we get

$$t_{ij} = \frac{k - c_1}{r(k - 1)}Q_{ij} + \frac{c_1 - c_2}{r(k - 1)} \sum_{p=1}^{n} Q_{ip},$$

or

$$v\lambda_2(\lambda_1 + rk - r)t_{ij} = kv\lambda_2Q_{ij} + k(\lambda_1 - \lambda_2) \sum_{p=1}^{n} Q_{ip}.$$

Summing up the preceding equation with respect to j, we have

$$v\lambda_2(\lambda_1 + rk - r) \sum_{j=1}^{n} t_{ij} = [kv\lambda_2 + nk(\lambda_1 - \lambda_2)] \sum_{p=1}^{n} Q_{ip},$$

or

$$\sum_{p=1}^{n} Q_{ip} = \frac{v\lambda_2}{k} \sum_{j=1}^{n} t_{ij}.$$

Thus it follows that

$$kv\lambda_2 Q_{ij} = v\lambda_2(\lambda_1 + rk - r)t_{ij} - v\lambda_2(\lambda_1 - \lambda_2) \sum_{i=1}^{n} t_{ij},$$

or

$$Q_{ij} = \frac{\lambda_1 + rk - r}{k} t_{ij} - \frac{\lambda_1 - \lambda_2}{k} \sum_{j=1}^{n} t_{ij}.$$

The adjusted sum of squares due to treatment is

$$S_t^2 = \sum_{i=1}^{m} \sum_{j=1}^{n} t_{ij} Q_{ij} = \frac{\lambda_1 + rk - r}{k} \sum_{i=1}^{m} \sum_{j=1}^{n} t_{ij}^2 - \frac{\lambda_1 - \lambda_2}{k} \times \sum_{i=1}^{m} \left(\sum_{j=1}^{n} t_{ij} \right)^2 .$$

Now let

$$T_{ij} = \underline{\zeta}_{ij}' \underline{x}, \qquad B_{ij\cdot} = \sum_{l=1}^{b} n_{ijl} \underline{n}_l' \underline{x},$$

then

$$t_{ij} = \frac{k}{\lambda_1 + rk - r} T_{ij} - \frac{1}{\lambda_1 + rk - r} B_{ij\cdot} + \frac{k(\lambda_1 - \lambda_2)}{v\lambda_2(\lambda_1 + rk - r)} \sum_{p=1}^{n} T_{ip} - \frac{\lambda_1 - \lambda_2}{v\lambda_2(\lambda_1 + rk - r)} \times \sum_{p=1}^{n} B_{ip} .$$

The sum of squares due to block is

$$S_b^2 = \frac{1}{k} \sum_{s=1}^{b} B_s^2 - \frac{G^2}{rv},$$

and

$$\text{total sum of squares} = \sum_{ijs}\sum\sum x_{ijs}^2 - \frac{G^2}{rv}.$$

Thus the sum of squares due to error is

$$S_e^2 = S^2 - S_t^2 - S_b^2.$$

We have seen that

$$V(t_\alpha - t_\beta) = \frac{2\sigma^2}{r}\frac{k - c_u}{k - 1}$$

if the two treatments α and β are uth associates. Hence

$$\begin{cases} Var(t_{ij} - t_{ij'}) = \dfrac{2k\sigma^2}{\lambda_1 + rk - r}, & E_1 = \dfrac{\lambda_1 + rk - r}{rk}; \\ Var(t_{ij} - t_{i'j'}) = \dfrac{2k\sigma^2(\lambda_1 + \lambda_2 v - \lambda_2)}{v\lambda(\lambda + rk - r)}, & E_2 = \dfrac{v\lambda_2(\lambda_1 + rk - r)}{rk(\lambda_1 + \lambda_2 v - \lambda_2)}. \end{cases}$$

Here we shall assume that

$$\tau_{ij} = \alpha_j + \gamma_j + \delta_{ij},$$

$$\sum_i \alpha_i = \sum_j \gamma_j = \sum_i \delta_{ij} = \sum_j \delta_{ij} = 0.$$

Corresponding to these equations, we have

$$t_{ij} = a_i + c_j + d_{ij}$$

by putting

$$a_i = \frac{1}{n}\sum_{j=1}^{n} t_{ij} = t_{i\cdot},$$

$$c_j = \frac{1}{m}\sum_{i=1}^{m} t_{ij} = t_{\cdot j},$$

$$d_{ij} = t_{ij} - a_i - c_j,$$

$$S_t^2 = \frac{\lambda_1 + rk - r}{k}\sum_i\sum_j (a_i + c_j + d_{ij})^2 - \frac{n^2(\lambda_1 - \lambda_2)}{k}\sum_{i=1}^{m} a_i^2$$

$$= \frac{n\lambda_2 v}{k} \sum_{i=1}^{m} a_{i}^2 + \frac{m(\lambda_1 + rk - r)}{k} \sum_{j=1}^{n} c_{j}^2 + \frac{\lambda_1 + rk - r}{k} \sum_i \sum_j d_{ij}^2.$$

Thus we can test the hypotheses

$$H_0(A) : \alpha_1 = \cdots = \alpha_m,$$
$$H_0(C) : \gamma_1 = \cdots = \gamma_n,$$
$$H_0(AC) : \delta_{11} = \cdots = \delta_{mn}.$$

The analysis of variance for the two-factor factorial is given in Table 4.3.

TABLE 4.3

Analysis of Variance for the Two-Factor Factorial

Source	Degrees of Freedom	Sum of Squares
A-factor (adjusted)	$m - 1$	$\frac{n\lambda_2 v}{k} \sum_{i=1}^{m} t_{i\cdot}^2$
C-factor (adjusted)	$n - 1$	$\frac{m(\lambda_1 + rk - r)}{k} \sum_{j=1}^{m} t_{\cdot j}^2$
AC-interaction (adjusted)	$(m - 1)(n - 1)$	$\frac{\lambda_1 + rk - r}{k} \sum_{i=1}^{m} \sum_{j=1}^{n} (t_{ij} - t_{i\cdot} - t_{\cdot j})^2$
Blocks (unadjusted)	$b - 1$	$\frac{1}{k} \sum_s B_s^2 - \frac{G^2}{rv}$
Error	$vr - v - b + 1$	S_e^2 (by subtraction)
Total	$vr - 1$	$\sum_s\sum_i\sum_j x_{ijs}^2 - \frac{G^2}{rv}$

Numerical Example

$v = 8,\quad r = 3,\quad k = 4,\quad b = 6,$

$m = 4,\quad n = 2,\quad \lambda_1 = 3,\quad \lambda_2 = 1$

V_{11}	V_{12}	V_{21}	V_{22}	B_1
29	38	40	33	140
V_{31}	V_{32}	V_{41}	V_{42}	B_2
28	37	24	24	113
V_{11}	V_{12}	V_{31}	V_{32}	B_3
20	37	26	33	116
V_{21}	V_{22}	V_{41}	V_{42}	B_4
23	22	24	15	84
V_{11}	V_{12}	V_{41}	V_{42}	B_5
27	41	36	28	132
V_{21}	V_{22}	V_{31}	V_{32}	B_6
37	26	29	37	129
				G
				714

4.10. Association Algebra and Relationship Algebra of a Partially Balanced Incomplete Block Design

We have seen already that the mapping

$$(\alpha) : A_i \to P_i,$$

where

$$P_i = \left\| p_{\alpha i}^{\beta} \right\|, \tag{4.10.1}$$

gives the regular representation of the association algebra $\mathfrak{a}$.

Since $\mathfrak{a}$ is the so-called abstract counterpart of the matrix algebra $\mathfrak{A}$ generated by symmetric association matrices A_i, it is *completely reducible* in the field of all rational numbers, and hence in any field of characteristic zero. On the other hand, Shur's lemma shows us that any irreducible representation of a commutative algebra in an algebraically closed field must be linear. Hence any irreducible representation of a commutative matrix algebra in a field containing all the characteristic roots of the matrices must be linear. From the general theory of algebra we know that any representation of a completely reducible algebra decomposes into irreducible constituents, each of which is equivalent to one of the irreducible constituents of the regular representation ($\mathfrak{a}$).

The regular representation ($\mathfrak{a}$) must decompose into inequivalent $m + 1$ linear representations in the field of all complex numbers. But these linear constituents are the characteristic roots of the symmetric matrices A_i, and hence they are all real. Thus, after all, the regular representation ($\mathfrak{a}$) decomposes into $m + 1$ inequivalent linear representations in the field of all real numbers. Even more, if the characteristic roots of all association matrices are all rational, then ($\mathfrak{a}$) decomposes into $m + 1$ linear inequivalent representations in the field of all rational numbers.

On account of the fact that

(4.10.2) $$A_i G_v = G_v A_i = n_i G_v$$

we can choose a nonsingular matrix C of order $m + 1$ in the field of all real numbers and of the form

$$(4.10.3) \qquad C = \begin{Vmatrix} 1 & 1 & \cdots & 1 \\ c_{10} & c_{11} & \cdots & c_{1m} \\ \vdots & \vdots & & \vdots \\ c_{m0} & c_{m1} & \cdots & c_{mm} \end{Vmatrix}$$

in such a way that simultaneously

$$C\begin{bmatrix} A_0 \\ A_1 \\ \vdots \\ A_m \end{bmatrix} A_u = CP_uC^{-1}C\begin{bmatrix} A_0 \\ A_1 \\ \vdots \\ A_m \end{bmatrix}$$

and

$$(4.10.4) \qquad CP_uC^{-1} = \begin{Vmatrix} z_{0u} & & & & 0 \\ & z_{1u} & & & \\ & & z_{2u} & & \\ & & & \ddots & \\ 0 & & & & z_{mu} \end{Vmatrix} \qquad (u = 0,1,\ldots,m),$$

where

$$(4.10.5) \qquad z_{0u} = \boldsymbol{n}_u.$$

Of course,

$$z_{00} = z_{10} = \cdots = z_{m0} = 1.$$

Let

$$A_u^* = \sum_{t=0}^{m} c_{ut}A_t.$$

Then

$$A_i A_u^* = \sum_{t=0}^{m} c_{ut} A_i A_t = \sum_{t=0}^{m} c_{ut} \sum_{k=0}^{m} p_{it}^k A_k$$

$$= \sum_{k=0}^{m} \left(\sum_{t=0}^{m} c_{ut} p_{it}^k \right) A_k = \sum_{k=0}^{m} \left(\sum_{t=0}^{m} c_{ut} p_{it}^k \right) \sum_{l=0}^{m} c^{kl} A^*_l$$

$$= \sum_{l=0}^{m} \left(\sum_{k=0}^{m} \sum_{t=0}^{m} c_{ut} p_{it}^k c^{kl} \right) A_l^*.$$

Since

$$(4.10.7) \qquad \sum_{k=0}^{m} \sum_{t=0}^{m} c_{ut} p_{ti}^k c^{kl} = z_{ui} \delta_{ul},$$

$$(4.10.8) \qquad A_i A_u^* = \sum_{l=0}^{m} z_{ul} \delta_{ul} A_l^* = z_{ui} A_u^*.$$

Multiplying (4.10.8) by c_{wi} and summing up from 0 to m with respect to i, we get

$$(4.10.9) \qquad A_w^* A_u^* = \left(\sum_{i=0}^{m} c_{wi} z_{ui} \right) A_u^*.$$

In a similar manner we can get

$$(4.10.10) \qquad A_u^* A_w^* = \left(\sum_{i=0}^{m} c_{ui} z_{wi} \right) A_w^*.$$

Thus we get

$$(4.10.11) \qquad \sum_{i=0}^{m} c_{ui} z_{wi} = 0 \qquad (u \neq w,)$$

and

$$(4.10.12) \qquad A_u^{\#} = \left(\sum_{i=0}^{m} c_{ui} z_{ui} \right)^{-1} A_u^*, \qquad (u = 0,1,\ldots,n),$$

are mutually orthogonal idempotents.

The matrix A_u^* gives rise to a linear representation

(4.10.13) $$(u) : A_u^* A_i = z_{ui} A_u^*.$$

We would like to determine the multiplicity of each linear representation in the matrix algebra $\mathfrak{A}$.

Let $\alpha_0, \alpha_1, \ldots, \alpha_m$ be the respective multiplicities. First, by considering the trace of G_v, we find that

(4.10.14) $$\alpha_0 = 1.$$

Next considering the traces of $A_0, A_1, \ldots, A_m$, we also have

(4.10.15)
$$\begin{aligned}
\alpha_1 + \alpha_2 + \cdots + \alpha_m &= v - 1, \\
\alpha_1 z_{11} + \alpha_2 z_{21} + \cdots + \alpha_m z_{m} &= -n_1, \\
\ldots\ldots\ldots\ldots\ldots\ldots &\ldots, \\
\alpha_1 z_{1m} + \alpha_2 z_{2m} + \cdots + \alpha_m z_{mm} &= -n_m,
\end{aligned}$$

or

(4.10.16)
$$\begin{aligned}
\alpha_0 + \alpha_1 + \alpha_2 + \cdots + \alpha_m &= v, \\
\alpha_0 z_{01} + \alpha_1 z_{11} + \alpha_2 z_{21} + \cdots + \alpha_m z_{m1} &= 0, \\
\ldots\ldots\ldots\ldots\ldots\ldots &\ldots, \\
\alpha_0 z_{0m} + \alpha_1 z_{1m} + \alpha_2 z_{2m} + \cdots + \alpha_m z_{mm} &= 0.
\end{aligned}$$

Since the matrix $\|z_{ij}\|$ is nonsingular, $\alpha_0, \alpha_1, \ldots, \alpha_m$ are determined uniquely by (4.10.16).

Group Divisible Design. The number of elements is $v = mn$. They can be divided into m groups of n elements each, such that any two elements in the same group are first associates and two elements in different groups are second associates.

If the elements are numbered in the dictionary order with respect to groups, that is, the jth element in the ith group has the number $(j - 1)n + j$, then the association matrices have simple forms :

$$
(4.10.17) \qquad A_0 + A_1 = \left\| \begin{matrix} G_n & & & 0 \\ & G_n & & \\ & & \ddots & \\ 0 & & & G_n \end{matrix} \right\| \Bigg\}\, m \qquad \text{and} \quad A_2 = G_v - A_0 - A_1,
$$

$$
n_1 = n - 1, \quad n_2 = n(m - 1).
$$

The regular representation is as follows :

$$
A_0 \to P_0 = I_3,
$$

$$
A_1 \to P_1 = \left\| \begin{matrix} 0 & 1 & 0 \\ n - 1 & n - 2 & 0 \\ 0 & 0 & n - 1 \end{matrix} \right\| = C^{-1} \cdot \left\| \begin{matrix} n - 1 & & 0 \\ & n - 1 & \\ 0 & & -1 \end{matrix} \right\| \cdot C,
$$

(4.10.18)

$$
A_2 \to P_2 = \left\| \begin{matrix} 0 & 0 & 1 \\ 0 & 0 & n - 1 \\ n(m - 1) & n(m - 1) & n(m - 2) \end{matrix} \right\| = C^{-1} \cdot \left\| \begin{matrix} n(m - 1) & & 0 \\ & -n & \\ 0 & & 0 \end{matrix} \right\| \cdot C,
$$

and

$$
C = \left\| \begin{matrix} 1 & 1 & 1 \\ 1 & 1 & \dfrac{-1}{m - 1} \\ 1 & \dfrac{-1}{n - 1} & 0 \end{matrix} \right\|,
$$

$$
\sum_{j=0}^{2} c_{1j} z_{1j} = 1 \cdot 1 + 1 \cdot (n - 1) - \frac{1}{m - 1} (-n) = \frac{mn}{m - 1},
$$

$$\sum_{j=0}^{2} c_{2j}z_{2j} = 1\cdot 1 - \frac{1}{n-1}(-1) + 0\cdot 0 = \frac{n}{n-1}.$$

Thus three mutually orthogonal idempotents are

$$A_0^{\#} = \frac{1}{v}(A_0 + A_1 + A_2),$$

$$A_1^{\#} = \frac{m-1}{mn}\left(A_0 + A_1 - \frac{1}{m-1}A_2\right), \tag{4.10.19}$$

$$A_2^{\#} = \frac{n-1}{n}\left(A_0 - \frac{1}{n-1}A_1\right).$$

Indeed, for instance,

$$A_1A_1^{\#} = \frac{m-1}{mn}\left(A_1 + A_1^2 - \frac{1}{m-1}A_1A_2\right)$$

$$= \frac{m-1}{mn}\left[A_1 + n_1A_0 + (n-2)A_1 - \frac{1}{m-1}(n-1)A_2\right]$$

$$= \frac{(n-1)(m-1)}{mn}\left[A_0 + A_1 - \frac{1}{m-1}A_2\right] = (n-1)A_1^{\#}.$$

The multiplicities of the irreducible constituents are

$$\alpha_0 = 1, \quad \alpha_1 = m-1, \quad \alpha_2 = (n-1)m, \tag{4.10.20}$$

and these coincide with the ranks of idempotents $A_0^{\#}, A_1^{\#}, A_2^{\#}$, respectively. Indeed,

$$tr\left(A_0^{\#}\right) = 1, \quad tr\left(A_1^{\#}\right) = m-1, \quad tr\left(A_2^{\#}\right) = (n-1)m. \tag{4.10.21}$$

It is known that in this case the association scheme is determined uniquely by the parameters.

Triangular Design. The number of elements is $v = n(n-1)/2$. We take an $n \times n$ square and fill the $n(n-1)/2$ positions above the main diagonal by the different elements, taken in order. The positions in the main diagonal are left blank while the positions below the main diagonal are filled, so that the scheme is symmetrical with respect to the main diagonal. Two elements in the same column are first associates, whereas two elements that do not occur in the same column are second associates,

$$n_1 = 2n - 4, \quad n_2 = \frac{(n-2)(n-3)}{2}.$$

The regular representation of the association algebra is given by

$$A_0 \to P_0 = I_3,$$

$$A_1 \to P_1 = \begin{Vmatrix} 0 & 1 & 0 \\ 2n-4 & n-2 & 4 \\ 0 & n-3 & 2n-8 \end{Vmatrix} = C^{-1}\cdot \begin{Vmatrix} 2n-4 & & 0 \\ & n-4 & \\ 0 & & -2 \end{Vmatrix} \cdot C,$$

(4.10.22)

$$A_2 \to P_2 = \begin{Vmatrix} 0 & 0 & 1 \\ 0 & n-3 & 2n-8 \\ \frac{(n-2)(n-3)}{2} & \frac{(n-3)(n-4)}{2} & \frac{(n-4)(n-5)}{2} \end{Vmatrix}$$

$$= C^{-1}\cdot \begin{Vmatrix} \frac{(n-2)(n-3)}{2} & & 0 \\ & -(n-3) & \\ 0 & & 1 \end{Vmatrix} \cdot C,$$

and

$$C = \begin{Vmatrix} 1 & 1 & 1 \\ 2n-4 & n-4 & -4 \\ -(n-2)(n-3) & n-3 & -2 \end{Vmatrix},$$

$$\sum_{j=0}^{2} c_{1j} z_{1j} = 2n - 4 + (n-4)^2 + 4(n-3) = n(n-2),$$

$$\sum_{j=0}^{2} c_{2j} z_{2j} = -(n - 2)(n - 3) - 2(n - 3) - 2 = -(n - 1)(n - 2),$$

$$A_0^{\#} = \frac{1}{v}(A_0 + A_1 + A_2),$$

(4.10.23)
$$A_1^{\#} = \frac{1}{n(n - 2)}[(2n - 4)A_0 + (n - 4)A_1 - 4A_2],$$

$$A_2^{\#} = -\frac{1}{(n - 1)(n - 2)}[-(n - 2)(n - 3)A_0 + (n - 3)A_1 - 2A_2].$$

Thus the multiplicities are

(4.10.24)
$$\alpha_0 = tr\left(A_0^{\#}\right) = 1, \quad \alpha_1 = tr\left(A_1^{\#}\right) = n - 1,$$

$$\alpha_2 = tr\left(A_2^{\#}\right) = \frac{n(n - 3)}{2}.$$

We now define the relationship matrices of a PBIBD.

1. Identity relation. Corresponding to this relation we take $I = I_n$, the unit matrix of order n.

2. Universal relation. This relation should be represented by $G = G_n$, where G_n stands for the matrix of order n whose elements are all unity.

3. Block relation. This is represented by a matrix $B = \psi\psi'$.

4. Treatment relation. This should be represented by $m + 1$ matrices of order n

$$T_0, T_1, \ldots, T_m,$$

where

$$T_i = \left\| t_{fg}^{i} \right\| = \Phi A_i \Phi' \qquad (i = 0,1,\ldots,m)$$

It is seen immediately that

$$\sum_{i=0}^{m} T_i = G. \tag{4.10.26}$$

Also

$$G^2 = nG, \quad BG = GB = kG, \quad B^2 = kB, \tag{4.10.27}$$

and

$$T_i G = GT_i = rn_i G \qquad (i = 0,1,\ldots,m). \tag{4.10.28}$$

Since

$$NN' = \Phi' B \Phi = \sum_{i=0}^{m} \lambda_i A_i,$$

it follows that

$$T_0 B T_0 = \sum_{i=0}^{m} \lambda_i T_i. \tag{4.10.29}$$

Furthermore

$$T_i B T_j = \sum_{t=0}^{m} \left\{ \sum_{k,l=0}^{m} \lambda_k p_{ik}^{l} p_{lj}^{t} \right\} T_t. \tag{4.10.30}$$

Also

$$T_i T_j = r \sum_{t=0}^{m} p_{ij}^{t} T_t. \tag{4.10.31}$$

Hence the linear closure of the set of $4m + 3$ matrices

$$I, \quad G, \quad B, \quad T_u, \quad T_u B, \quad BT_u, \quad BT_u B, \quad (u = 1,2,\ldots,m) \tag{4.10.32}$$

is a linear associative algebra $\mathcal{R}$, which is called *the relationship algebra* of the PBIBD.

The relationshi algebra $\mathcal{R}$ contains a subalgebra $\mathcal{A}^*$ generated by I, G, T_u, $u = 1,\ldots,m$. which is isomorphic with the association algebra $\mathcal{A}$.

As a special case, the relationship algebra of a BIBD is the linear closure

(4.10.33) $[I,\ G,\ B,\ T_1,\ BT_1,\ T_1B,\ BT_1B]$.

By changing the basis by means of the relation

$$T + T_1 = G,$$

where $T \equiv T_0$, (4.10.33) is expressed as

(4.10.34) $[I,\ G,\ B,\ T,\ BT,\ TB,\ BTB]$.

This was treated by James [4] in some detail.

The relationship algebra $\mathscr{R}$ of a PBIBD is noncommutative in general. Since $\mathscr{R}$ is generated by symmetric matrices, it is completely reducible. Hence all irreducible representation of $\mathscr{R}$ are obtained by reducing its regular representation of ($\mathscr{R}$).

The totality of the multiples of G, denoted by $[G]$, is a one-dimensional two-sided ideal of $\mathscr{R}$, and

$$G^2 = nG, \quad BG = GB = kG, \quad T_iG = GT_i = rn_iG.$$

Hence we obtain a linear representation of $\mathscr{R}$:

(4.10.35) $\mathscr{R}_G^{(1)} : G \to n,\ \ B \to k,\ \ T_i \to rn_i \qquad (i = 1,\ldots,m).$

Next we shall consider the factor algebra $\mathscr{R}$ mod $[G]$. To this end, it will be convenient to change the basis of $\mathscr{R}$. Put

(4.10.36) $T_u^* = \Phi A_u^* \Phi' = \sum_{j=0}^{m} c_{uj} T_j \qquad (u = 0,1,\ldots,m).$

Then it is easy to see that

(4.10.37) $T_u^* T_v^* = \Phi A_u^* \Phi' \Phi A_v^* \Phi = r \Phi A_u^* A_v^* \Phi = 0 \quad \text{if} \quad u \neq v$

and

$$T_u^* T_u^* = r\Phi A_u^* A_u^* \Phi = r \sum_{i=0}^{m} c_{ui} z_{ui} T_u^* . \tag{4.10.38}$$

Thus m sets

$$[T_u^*, \quad BT_u^*, \quad T_u^* B, \quad BT_u^* B] \qquad (u = 1,2,\ldots,m) \tag{4.10.39}$$

are two-sided ideals of $\mathscr{R}$ mod $[G]$, and they annihilate each other. Hence there are m inequivalent, irreducible representations of the second degree each of which is derivable through a one-sided ideal of each of the two-sided ideals, provided $T_u^* B \neq BT_u^*$.

Now by direct calculation

(4.10.40)

$$\begin{bmatrix} T_u^* \\ T_u^* B \\ BT_u^* \\ BT_u^* B \end{bmatrix} B = \begin{Vmatrix} 0 & 1 & 0 & 0 \\ 0 & k & 0 & 0 \\ 0 & 0 & 0 & 1 \\ 0 & 0 & 0 & k \end{Vmatrix} \begin{bmatrix} T_u^* \\ T_u^* B \\ BT_u^* \\ BT_u^* B \end{bmatrix},$$

$$\begin{bmatrix} T_u^* \\ T_u^* B \\ BT_u^* \\ BT_u^* B \end{bmatrix} T_i = \begin{Vmatrix} & rz_{ui} & 0 & 0 & 0 \\ \sum_{k,l=0}^{m} & \lambda_k p_{ki}^l z_{ul} & 0 & 0 & 0 \\ & 0 & 0 & rz_{ui} & 0 \\ & 0 & 0 & \sum \lambda_k p_{ki}^l z_{ul} & 0 \end{Vmatrix} \begin{bmatrix} T_u^* \\ T_u^* B \\ BT_u^* \\ BT_u^* B \end{bmatrix}.$$

Thus we obtain m inequivalent representations of the second degree.

$$\mathscr{R}_u^{(2)} : \begin{cases} I \to \begin{Vmatrix} 1 & 0 \\ 0 & 1 \end{Vmatrix}, \quad G \to \begin{Vmatrix} 0 & 0 \\ 0 & 0 \end{Vmatrix}, \quad B \to \begin{Vmatrix} 0 & 1 \\ 0 & k \end{Vmatrix} \\ T_i \to \begin{Vmatrix} rz_{ui} & 0 \\ \sum_{k,l} z_{ul} p_{ki}^l \lambda_k & 0 \end{Vmatrix}, \end{cases} \qquad (u = 1,2,\ldots,m), \tag{4.10.41}$$

Other irreducible representations of $\mathcal{R}$ are obtained by considering the factor algebra $\mathcal{R}$ mod $\mathfrak{A}^*$. These are linear.

$$\text{(4.10.42)} \qquad \begin{aligned} &\mathcal{R}_0^{(1)} : I \to 1, \quad G \to 0, \quad B \to 0, \quad T_i \to 0; \\ &\mathcal{R}_1^{(1)} : I \to 1, \quad G \to 0, \quad B \to k, \quad T_i \to 0. \end{aligned}$$

We shall show that

$$\text{(4.10.43)} \qquad \begin{aligned} \mathcal{R} \sim (n - b + v - 1)\, \mathcal{R}_0^{(1)} + (b - v)\, \mathcal{R}_1^{(1)} \\ + \sum_{u=1}^{m} \alpha_u \mathcal{R}_u^{(2)}. \end{aligned}$$

Let

$$c_u^i = \frac{c_{ui}}{\sum_{j=0}^{m} c_{uj} z_{uj}}$$

Then

$$A_u^{\#} = \sum_{i=0}^{m} c_u^i A_i,$$

and therefore

$$\text{(4.10.44)} \qquad T_u^{\#} = \sum_{i=0}^{m} c_u^i T_i.$$

Let us consider m matrices

$$\text{(4.10.45)} \qquad V_u^{\#} = (T_u^{\#} - \frac{1}{k} B T_u^{\#})(T_u^{\#} - \frac{1}{k} T_u^{\#} B) \qquad (u = 1,\ldots,m),$$

It is seen that

$$V_u^{\#2} = \frac{rk - \rho_u}{k} V_u^{\#}, \qquad \rho_u = \sum_{i=0}^{m} z_{ui} \lambda_i \qquad (u = 1,2,\ldots,m).$$

In other words

(4.10.46) $$\frac{k}{r(rk - \rho_u)} (T_u^{\#} - \frac{1}{k}BT_u^{\#})(T_u^{\#} - \frac{1}{k}T_u^{\#}B)$$

are mutually orthogonal idempotents having the traces

(4.10.47) $$\alpha_1, \alpha_2, \ldots, \alpha_m$$

respectively.

(4.10.48) $$\frac{k}{r(rk - \rho_u)} \underline{x}'(T_u^{\#} - \frac{1}{k}BT_u^{\#})(T_u^{\#} - \frac{1}{k}T_u^{\#}B)\underline{x} = \frac{k}{rk - \rho_u} \underline{Q}'A_u^{\#}\underline{Q}.$$

From $A_u^{\#} = \sum_{i=0}^{m} c_u^i A_i$ it follows that

$$\underline{R}_u = A_u^{\#}\underline{Q} = \sum_{i=0}^{m} c_u^i A_i \underline{Q} = \sum_{i=0}^{m} c_u^i \begin{vmatrix} S_i(Q_1) \\ S_i(Q_2) \\ \vdots \\ S_i(Q_v) \end{vmatrix}.$$

Hence the αth component of $A_u^{\#}\underline{Q}$ is

(4.10.49) $$R_{u\alpha} = \sum_{i=0}^{m} c_u^i S_i(Q_\alpha).$$

Thus the sum of squares due to treatment is

(4.10.50) $$\begin{aligned} &\sum_{u=1}^{m} \frac{k}{r(rk - \rho_u)} \underline{x}'(T_u^{\#} - \frac{1}{k}BT_u^{\#})(T_u^{\#} - \frac{1}{k}T_u^{\#}B)\underline{x} \\ &= \sum_{u=1}^{m} \frac{k}{rk - \rho_u} \underline{R}_u'\underline{R}_u. \end{aligned}$$

This should coincide with $\sum_{\alpha=1}^{v} t_\alpha Q_\alpha$ given by (4.9.14). Indeed, from (4.9.3) one obtains

$$\sum_{u=1}^{m} \frac{rk - \rho_u}{k} A_u^{\#} \underline{t} = \underline{Q},$$

hence

$$\underline{t} = \sum_{u=1}^{m} \frac{k}{rk - \rho_u} A_u^{\#}\underline{Q},$$

and therefore

$$\underline{t}'\underline{Q} = \sum_{u=1}^{m} \frac{k}{rk - \rho_u} \underline{Q}'A_u^{\#}\underline{Q} = \sum_{u=1}^{m} \frac{k}{rk - \rho_u} \underline{R}_u'\underline{R}_u.$$

4.11. Partially Balanced Incomplete Block Design for the Factorial Combination of Two Factors

Let us denote by (i_1, i_2) the treatment in which the factors F_1 and F_2 appear at their i_1th and i_2th level, respectively. The total number of treatments is $v = s_1 s_2$, where s_i is the number of levels of the factor F_i, $i = 1, 2$.

One can define the following three associate-class association schemes as follows : two treatments (i_1, i_2) and (j_1, j_2) are said to be

first associates if $i_1 = j_1$, $i_2 \neq j_2$;

second associates if $i_1 \neq j_1$, $i_2 = j_2$;

third associates if $i_1 \neq j_1$, $i_2 \neq j_2$.

If the first and third or the second and third associates are amalgamated into one associate, then this scheme reduces to a *GD*-association scheme.

The parameters of the association are given as follows :

$$n_0 = 1, \quad n_1 = s_2 - 1, \quad n_2 = s_1 - 1, \quad n_3 = (s_1 - 1)(s_2 - 1),$$

$$\mathbf{P}^1 = \left\| \begin{matrix} p^1_{11} = s_2 - 2 & p^1_{12} = 0 & p^1_{12} = 0 \\ p^1_{21} & p^1_{22} = 0 & p^1_{23} = s_1 - 1 \\ p^1_{31} & p^1_{32} & p^1_{33} = (s_1 - 1)(s_2 - 2) \end{matrix} \right\|,$$

$$P^2 = \begin{Vmatrix} p_{11}^2 = 0 & p_{12}^2 = 0 & p_{13}^2 = s_2 - 1 \\ p_{21}^2 & p_{22}^2 = s_1 - 2 & p_{23}^2 = 0 \\ p_{31}^2 & p_{32}^2 & p_{33}^2 = (s_1 - 2)(s_2 - 1) \end{Vmatrix},$$

$$P^3 = \begin{Vmatrix} p_{11}^3 = 0 & p_{12}^3 = 1 & p_{12}^3 = s_2 - 2 \\ p_{21}^3 & p_{22}^3 = 0 & p_{23}^3 = s_1 - 2 \\ p_{31}^3 & p_{32}^3 & p_{33}^3 = (s_1 - 2)(s_2 - 2) \end{Vmatrix}.$$

Let the association matrices be A_0, A_1, A_2, and A_3. Then their regular representations are given by 4×4 matrices:

$$A_0 \to P_0 = I_4,$$

$$A_1 \to P_1 = \begin{Vmatrix} 0 & 1 & 0 & 0 \\ s_2 - 1 & s_2 - 2 & 0 & 0 \\ 0 & 0 & 0 & 1 \\ 0 & 0 & s_2 - 1 & s_2 - 2 \end{Vmatrix},$$

$$A_2 \to P_2 = \begin{Vmatrix} 0 & 0 & 1 & 0 \\ 0 & 0 & 0 & 1 \\ s_1 - 1 & 0 & s_1 - 2 & 0 \\ 0 & s_1 - 1 & 0 & s_1 - 2 \end{Vmatrix},$$

$$A_3 \to P_3 = \begin{Vmatrix} 0 & 0 & 0 & 1 \\ 0 & 0 & s_2 - 1 & s_2 - 2 \\ 0 & s_1 - 1 & 0 & s_1 - 2 \\ (s_1-1)(s_2-1) & (s_1-1)(s_2-2) & (s_1-2)(s_2-1) & (s_1-2)(s_2-2) \end{Vmatrix}.$$

By the general theory, there should be a nonsingular matrix C

$$C = \begin{Vmatrix} c_{00} & c_{01} & c_{02} & c_{03} \\ c_{10} & c_{11} & c_{12} & c_{13} \\ c_{20} & c_{21} & c_{22} & c_{23} \\ c_{30} & c_{31} & c_{32} & c_{33} \end{Vmatrix}, \quad c_{00} = c_{01} = c_{02} = c_{03} = 1,$$

such that

$$CP_i = \begin{Vmatrix} n_i & & & 0 \\ & n_{1i} & & \\ & & z_{2i} & \\ 0 & & & z_{3i} \end{Vmatrix} C \qquad (i = 0,1,2,3).$$

One can determine the eigenvalues of P_i as follows :

$n_0 = 1, \quad z_{10} = z_{20} = z_{30} = 1;$

$n_1 = s_2 - 1, \quad z_{11} = s_2 - 1, \quad z_{21} = -1, \quad z_{31} = -1;$

$$n_2 = s_1 - 1, \quad z_{12} = s_1 - 1, \quad z_{22} = -1, \quad z_{32} = +1;$$

$$n_3 = (s_1 - 1)(s_2 - 1), \quad z_{13} = -(s_1 + s_2 - 1), \quad z_{23} = 1, \quad z_{33} = -1;$$

$$\alpha_1 = s_2 - 1, \quad \alpha_2 = s_1 - 1, \quad \alpha_3 = (s_1 - 1)(s_2 - 1) - (s_2 - 1);$$

and

$$A_i^{\#} = \left[\sum_{j=0}^{3} c_{ij} z_{ij} \right]^{-1} \left[c_{i0} A_0 + c_{i1} A_1 + c_{i2} A_2 + c_{i3} A_3 \right] (i = 0,1,2,3).$$

Since

$$tr\left(A_i^{\#}\right) = v c_{i0} \left(\sum c_{ij} z_{ij} \right)^{-1} = \alpha_i,$$

we have

$$c_{10} \propto s_1 - 1, \quad c_{20} \propto s_2 - 1, \quad c_{30} \propto (s_1 - 1)(s_2 - 1).$$

Thus one can write down four orthogonal idempotents as follows:

$$A_0^{\#} = \frac{1}{v}[A_0 + A_1 + A_2 + A_3],$$

$$A_1^{\#} = \frac{1}{v}[(s_1 - 1)A_0 + (s_1 - 1)A_1 - A_2 - A_3],$$

$$A_2^{\#} = \frac{1}{v}[(s_2 - 1)A_0 - A_1 + (s_2 - 1)A_2 - A_3],$$

$$A_3^{\#} = \frac{1}{v}[(s_1 - 1)(s_2 - 1)A_0 - (s_1 - 1)A_1 - (s_2 - 1)A_2 + A_3].$$

The sum of squares due to treatments adjusted $\underline{t}'\underline{Q}$ can be written

$$S_t^2 = \underline{t}'\underline{Q} = \sum_{u=1}^{3} \frac{rk-\rho_u}{k} \underline{x}'(T_u^{\#} - \frac{1}{k}BT_u^{\#})(T_u^{\#} - \frac{1}{k}T_u^{\#}B)\underline{x}.$$

Since

$$\underline{x}'(T_u^{\#} - \frac{1}{k}BT_u^{\#})(T_u^{\#} - \frac{1}{k}T_u^{\#}B)\underline{x} = r\underline{Q}'A_u^{\#}\underline{Q},$$

the partition of S_t^2 is

$$S_t^2 = \frac{rk - \rho_1}{k} \underline{Q}'A_1\underline{Q} + \frac{rk - \rho_2}{k} \underline{Q}'A_2\underline{Q} + \frac{rk - \rho_3}{k} \underline{Q}'A_3\underline{Q},$$

and under the linear model,

$$\left\{x(i_1, i_2)\right\} = \tau(i_1, i_2) = \gamma + \alpha_{i_1} + \gamma_{i_2} + \delta_{i_1 i_2},$$

one can see that

$$A_1^{\#}\underline{\tau} = \begin{bmatrix} \alpha_1 \\ \vdots \\ \alpha_1 \\ \alpha_2 \\ \vdots \\ \alpha_2 \\ \vdots \\ \alpha_{s_1} \\ \vdots \\ \alpha_{s_1} \end{bmatrix} \begin{matrix} \left.\begin{matrix} \\ \\ \\ \end{matrix}\right\} s_2 \\ \left.\begin{matrix} \\ \\ \\ \end{matrix}\right\} s_2 \\ \\ \left.\begin{matrix} \\ \\ \\ \end{matrix}\right\} s_2 \end{matrix}, \quad A_2^{\#}\underline{\tau} = \begin{bmatrix} \gamma_1 \\ \gamma_2 \\ \vdots \\ \gamma_{s_2} \\ \gamma_1 \\ \vdots \\ \gamma_{s_2} \\ \vdots \\ \gamma_1 \\ \vdots \\ \gamma_{s_2} \end{bmatrix}, \quad \text{and } A_3^{\#}\underline{\tau} = \begin{bmatrix} \delta_{11} \\ \vdots \\ \delta_{1s_2} \\ \delta_{21} \\ \vdots \\ \delta_{2s} \\ \vdots \\ \delta_{s_1 1} \\ \vdots \\ \delta_{s_1 s_2} \end{bmatrix}.$$

In the special case $\lambda_2 = \lambda_3$,

$$\rho_1 = r + (s_2 - 1)\lambda_1 - s_2\lambda_2,$$

$$\rho_2 = r - \lambda_1 = \rho_3,$$

$$\begin{aligned} rk - \rho_1 &= r(k - 1) - (s_2 - 1)\lambda_1 - s_2\lambda_2 \\ &= (s_2 - 1)\lambda_1 + (s_1 - 1)s_2\lambda_2 - (s_2 - 1)\lambda_1 - s_2\lambda_2 \\ &= s_1 s_2 \lambda_2, \end{aligned}$$

$$rk - \rho_2 = r(k - 1) + \lambda_1,$$

$$\underline{Q}'A_1^{\#}\underline{Q} = s_2 \sum_{i=1}^{s_1} t_{i\cdot}^2, \quad \underline{Q}'A_2^{\#}\underline{Q} = s_1 \sum_{j=1}^{s_2} t_{\cdot j}^2, \quad \underline{Q}'A_3^{\#}\underline{Q} = \sum_{i,j} (t_{ij} - t_{i\cdot} - t_{\cdot j})^2,$$

$$\frac{rk - \rho_1}{k}\underline{Q}'A^{\#}\underline{Q} = \frac{vs_2\lambda_2}{k} \sum_{i=1}^{s} t_{i\cdot}^2, \quad \frac{rk - \rho_2}{k}\underline{Q}'A^{\#}\underline{Q} = \frac{s_1(rk - r + \lambda_1)}{k} \sum_{j=1}^{s_2} t_{\cdot j}^2,$$

$$\frac{rk - \rho_3}{k}\underline{Q}'A_3^{\#}\underline{Q} = \frac{rk - r + \lambda_1}{k}\sum_i \sum_j t_{ij}^2.$$

Thus one confirms the Bradley-Kramer results described in Section 4.9.

We shall present the analysis of variance of a PBIBD of triangular type in detail with a numerical example.

In this case we have

$$A_0^{\#} = \frac{2}{n(n-1)} [A_0 + A_1 + A_2] \qquad (\alpha_0 = 1),$$

$$A_1^{\#} = \frac{1}{n(n-1)} \left[(2n - 4)A_0 + (n - 4)A_1 - 4A_2\right] \qquad (\alpha_1 = n - 1),$$

$$A_2^{\#} = \frac{1}{(n-1)(n-2)} \left[(n-2)(n-3)A_0 - (n-3)A_1 + 2A_2\right]$$

$$\left[\alpha_2 = n(n-3)/2\right].$$

We assure the following inner structure of the treatment effects:

$$n = 5, \qquad v = n(n-1)/2 = 10$$

	$\theta_1+\theta_2+\pi_{12}$	$\theta_1+\theta_3+\pi_{13}$	$\theta_1+\theta_4+\pi_{14}$	$\theta_1+\theta_5+\pi_{15}$
$\theta_2+\theta_1+\pi_{21}$		$\theta_2+\theta_3+\pi_{23}$	$\theta_2+\theta_4+\pi_{24}$	$\theta_2+\theta_5+\pi_{25}$
$\theta_3+\theta_1+\pi_{31}$	$\theta_3+\theta_2+\pi_{32}$		$\theta_3+\theta_4+\pi_{34}$	$\theta_3+\theta_5+\pi_{35}$
$\theta_4+\theta_1+\pi_{41}$	$\theta_4+\theta_2+\pi_{42}$	$\theta_4+\theta_3+\pi_{43}$		$\theta_4+\theta_5+\pi_{45}$
$\theta_5+\theta_1+\pi_{51}$	$\theta_5+\theta_2+\pi_{52}$	$\theta_5+\theta_3+\pi_{53}$	$\theta_5+\theta_4+\pi_{54}$	

The inner parameters are subjected to the restrictions

$$\sum_{i=1}^{n} \theta_i = 0, \qquad \pi_{ij} = \pi_{ji}, \qquad \sum_{j(\neq i)} \pi_{ij} = 0 \qquad (i = 1,2,\ldots,n),$$

so that these are just $v - 1$ independent parameters.

By direct calculations, we obtain the following expression :

$$A_1^{\#}\underline{\tau} = n^{-1}$$

$$\begin{bmatrix}
(n-2)\theta_1 + (n-2)\theta_2 - 2\theta_3 - \cdots - 2\theta_{n-1} - 2\theta_n \\
(n-2)\theta_1 - 2\theta_2 + (n-2)\theta_3 - \cdots - 2\theta_{n-1} - 2\theta_n \\
\cdots \\
(n-2)\theta_1 - 2\theta_2 - 2\theta_3 - \cdots + (n-2)\theta_{n-1} - 2\theta_n \\
(n-2)\theta_1 - 2\theta_2 - 2\theta_3 - \cdots - 2\theta_{n-1} + (n-2)\theta_n \\
-2\theta_1 + (n-2)\theta_2 + (n-2)\theta_3 - \cdots - 2\theta_{n-1} - 2\theta_n \\
-2\theta_1 + (n-2)\theta_2 - 2\theta_3 + \cdots - 2\theta_{n-1} - 2\theta_n \\
\cdots \\
-2\theta_1 + (n-2)\theta_2 - 2\theta_3 - \cdots - 2\theta_{n-1} + (n-2)\theta_n \\
\cdots \\
-2\theta_1 - 2\theta_2 - 2\theta_3 - \cdots - 2\theta_{n-1} + (n-2)\theta_n
\end{bmatrix}$$

and

the α-th component of $A_2^{\#}\underline{\tau} = [(n-1)(n-2)]^{-1}[(n-1)(n-3)$
$\cdot$interaction $\pi_{ij}(i < j)$ of τ_α
$- (n-3)\cdot\sum$ interactions corresponding
to the first associates of ϕ_α
$+ 2\cdot\sum$ interactions corresponding to
the second associates of ϕ_α].

Thus the vector $A_1^{\#}\underline{\tau}$ represents the contrasts between the main effects θ, and the vector $A_2^{\#}\underline{\tau}$ represents the contrasts between interactions π_{ij}.

Now we show a numerical example of the analysis of variance of a PBIBD of triangular type (Table 4.4).

TABLE 4.4

A Design of the Triangular Type

Block treatments	1	2	3	4	5	6	7	8	9	10	Treatment total
1=(1,2)			2.31		2.81	1.65			2.58		9.40
2=(1,3)		2.51					1.41	1.90	3.06		8.88
3=(1,4)	2.89		2.29					1.95		2.04	9.16
4=(1,5)				2.54		2.09	2.36			2.03	9.02
5=(2,3)	2.28			2.81					2.20	2.07	9.36
6=(2,4)		1.77	2.49	2.31			3.02				9.59
7=(2,5)	2.72	2.29				1.57		2.60			9.18
8=(3,4)				2.81	2.99	2.28		2.44			10.52
9=(3,5)	2.54		2.44		2.23		2.12				9.33
10=(4,5)		1.54			2.87				2.77	2.09	9.27
Block totals	10.43	8.11	9.53	10.47	10.95	7.59	8.91	8.89	10.61	8.22	93.71

There are n kind ingredients $I_1, I_2, \ldots, I_n$ that are known to be efficient in producing weight gain in hogs if added to the feedstuff. We want to know whether there are interactions between any two of the ingredients when the mixtures of the two are added to the feedstuff.

We make $v = n(n - 1)/2$ mixtures of the possible pairs (I_i, I_j), $i \neq j$. The main effects of the n original ingredients are denoted by θ_i, $i = 1,2,\ldots,n$, and the interactions between I_i and I_j are denoted by π_{ij}. Then the inner-parametric presentations of the mixture are given by $\theta_\alpha = \theta_i + \theta_j + \pi_{ij}$ if the α-th treatment is the mixture of I_i and I_j for $\alpha = 1,2,\ldots,v = n(n - 1)/2$. Hence in this situation the association scheme of the triangular type is naturally defined among the v treatments (Table 4.5).

TABLE 4.5

Association

Treatment	First associates	Second associates
1	2, 3, 4, 5, 6, 7	8, 9, 10
2	1, 3, 4, 5, 8, 9	6, 7, 10
3	1, 2, 4, 6, 8, 10	5, 7, 9
4	1, 2, 3, 7, 9, 10	5, 6, 8
5	1, 2, 6, 7, 8, 9	3, 4, 10
6	1, 3, 5, 7, 8, 10	2, 4, 9
7	1, 4, 5, 6, 9, 10	2, 3, 8
8	2, 3, 5, 6, 9, 10	1, 4, 7
9	2, 4, 5, 7, 8, 10	1, 3, 6
10	3, 4, 6, 7, 8, 9	1, 2, 5

Suppose by taking 10 litters of four hogs each as blocks, a PBIBD of triangular type with the parameters $n = 5$, $v = b = 10$, $r = k = 4$, $\lambda_1 = 1$, and $\lambda_2 = 2$ is adopted. Observations are the weight gains (in pounds) in hogs that had been fed the mixtures of ingredients for 3 months. This experiment is a hypothetical one, and the data are borrowed from Bose and Shimamoto [5] therefore this example should be regarded as a purely illustrative one.

The adjusted treatment tables and related sums are given in Table 4.6, and the analysis of variance is shown in Table 4.7.

TABLE 4.6

Adjusted Treatment Totals and Related Sums

	Q	A_1Q	A_2Q	$A_1^{\#}Q$	$A_2^{\#}Q$
1	-0.2700	0.0525	0.2175	-0.1625	-0.1075
2	-0.2500	-0.3075	0.5575	-0.2692	0.0192
3	-0.1075	0.8800	-0.7725	0.2217	-0.3292
4	0.2225	-1.0300	0.8075	-0.1950	0.4175
5	-0.5725	0.6600	-0.0875	-0.1617	-0.4108
6	0.3350	0.3175	-0.6505	0.3292	0.0058
7	0.4250	-1.1125	0.6875	-0.0875	0.5125
8	1.0450	-1.4225	0.3775	-0.2225	0.8225
9	-0.6250	0.6675	-0.0425	-0.1992	-0.4308
10	-0.2025	1.2950	-1.0925	0.2967	-0.4992
Total	0.0000	0.0000	0.0000	0.0000	0.0000

TABLE 4.7

Analysis of Variance

Sources of variation	Sum of squares	Degrees of Freedom	Mean sum of squares	Variance ratio
Blocks	3.2284	9		
Treatment eliminating blocks	0.7467	9	0.08295	
Main effects	0.1343	4	0.03357	0.230
Interactions	0.6124	5	0.12248	0.841
Errors	3.0585	21	0.14550	
Total	7.0336	39		

Now in this case, since

$$\rho_1 = z_{10}\lambda_0 + z_{11}\lambda_1 + z_{12}\lambda_2 = 4 + 1\cdot 1 - 2\cdot 2 = 1,$$

$$\rho_2 = z_{20}\lambda_0 + z_{21}\lambda_1 + z_{22}\lambda_2 = 4 - 2\cdot 1 + 2\cdot 2 = 4,$$

and

$$A_1^{\#} = \frac{1}{15}(6A_0 + A_1 - 4A_2), \qquad \alpha_1 = 4,$$

$$A_2^{\#} = \frac{1}{12}(6A_0 - 2A_1 + 2A_2), \qquad \alpha_2 = 5,$$

it follows that

$$A_1^{\#}\underline{Q} = \frac{1}{15}(6\underline{Q} + A_1\underline{Q} - 4A_2\underline{Q}),$$

$$A_2^{\#}\underline{Q} = \frac{1}{6}(3\underline{Q} - A_1\underline{Q} + A_2\underline{Q}).$$

There is a relation $A_1^{\#}\underline{Q} + A_2^{\#}\underline{Q} = \underline{Q}$. Finally the sum of squares due to main effects and interactions is given by

$$\underline{x}'V_1^{\#}\underline{x} = \frac{k}{rk - \rho_1}\,\underline{Q}'A_1^{\#}\underline{Q} = \frac{4}{15}\,\underline{Q}'A_1^{\#}\underline{Q}$$

and

$$\underline{x}'V_2^{\#}\underline{x} = \frac{k}{rk - \rho_2}\,\underline{Q}'A_2^{\#}\underline{Q} = \frac{1}{3}\underline{Q}'A_2^{\#}\underline{Q},$$

respectively, satisfying the relation

$$\frac{4}{15}\underline{Q}'A_1^{\#}\underline{Q} + \frac{1}{3}\underline{Q}'A_2^{\#}\underline{Q} = \underline{t}'\underline{Q}.$$

Thus we get Table 4.7 of the analysis of variance by using the auxiliary Tables 4.4, 4.5, and 4.6.

REFERENCES

1. W. G. Youden, *Biometrics, 7,* 124 (1951).
2. C. Y. Kramer and R. A. Bradley, *Biometrics, 13,* 197 (1957).
3. C. Y. Kramer and R. A. Bradley, *Ann. Math. Statist., 28,* 349 (1957).
4. A. T. James, *Ann. Math. Statist., 28,* 993 (1957).
5. R. C. Bose and T. Shimamoto, *J. Amer. Stat. Assoc., 27,* 151 (1952).

Chapter 5

THEORY OF BLOCK DESIGNS -- INTERBLOCK ANALYSIS

The block effect $\underline{\beta}$ should be regarded as a random vector.

5.1. Preliminaries

The Linear Mixed Model with Independent Components. Let $y_1, y_2, \ldots, y_n$ be random variables such that

$$y_1 = e_1 + a_{11}p_1 + a_{12}p_2 + \qquad + a_{m1}p_m + l_{11}x_1 + \qquad + l_{b1}x_b,$$

$$y_2 = e_2 + a_{12}p_1 + a_{22}p_2 + \qquad + a_{m2}p_m + l_{12}x_1 + \qquad + l_{b2}x_b,$$

$$\cdots\cdots\cdots\cdots\cdots\cdots\cdots\cdots\cdots\cdots\cdots\cdots\cdots\cdots$$

$$y_n = e_n + a_{1n}p_1 + p_{2n}p_2 + \qquad + a_{mn}p_m + l_{1n}x_1 + \qquad + l_{bn}x_b,$$

where $e_1, \ldots, e_n$ are independently distributed as $N(0,\sigma^2)$, $x_1, \ldots, x_b$ are independently distributed as $N(0,\sigma_1^2)$, and $\underline{e}$ and $\underline{x}$ are mutually independent. The $p_1, \ldots, p_m$ are unknown parameters, and a_{ij} and l_{ij} are known constants. In vector notation this can be expressed as

$$\text{(5.1.1)} \qquad \underline{y} = \underline{e} + A'\underline{p} + L'\underline{x},$$

where

$$\text{(5.1.2)} \qquad A' = \left\| \begin{matrix} a_{11} & a_{21} & \cdots & a_{m1} \\ a_{12} & a_{22} & \cdots & a_{m2} \\ \cdot & \cdot & \cdot & \cdot \\ a_{1n} & a_{2n} & \cdots & a_{mn} \end{matrix} \right\|, \qquad L' = \left\| \begin{matrix} l_{11} & l_{21} & \cdots & l_{b1} \\ l_{12} & l_{22} & \cdots & l_{b2} \\ \cdot & \cdot & \cdot & \cdot \\ l_{1n} & l_{2n} & \cdots & l_{bn} \end{matrix} \right\|.$$

Hence

$$E(\underline{y}) = A'\underline{p}, \tag{5.1.3}$$

$$\begin{aligned} Var(y_u) &= \sigma^2 + (l_{1u}^2 + l_{2u}^2 + \cdots + l_{bu}^2)\sigma^2{}_1, \\ cov(y_u, y_u) &= (l_{1u}l_{1v} + l_{2u}l_{2v} + \cdots + l_{bu}l_{bv})\sigma_1^2, \end{aligned} \tag{5.1.4}$$

that is,

$$Var(y) = \sigma^2 I + L'L\sigma_1^2 \equiv \Sigma \text{ (say)}. \tag{5.1.5}$$

For a linear function $\underline{c}'\underline{y}$, we have

$$E(\underline{c}'\underline{y}) = \underline{c}'A'\underline{p},$$

$$\begin{aligned} Var(\underline{c}'\underline{y}) &= \underline{c}'(\sigma^2 I + L'L\sigma_1^2)\underline{c}, \\ Cov(\underline{c}'\underline{y}, \underline{d}'\underline{y}) &= \underline{c}'(\sigma^2 I + L'L\sigma_1^2)\underline{d}. \end{aligned} \tag{5.1.6}$$

The necessary and sufficient condition for $\underline{c}'\underline{y}$ and $\underline{d}'\underline{y}$ to be uncorrelated is that

$$\underline{c}'(\sigma^2 I + L'L\sigma_1^2)\underline{d} = 0. \tag{5.1.7}$$

A linear function $\underline{c}'\underline{y}$ is said to *belong to the error* if the expectation vanishes independently of parameters. Hence the necessary and sufficient condition for $\underline{c}'\underline{y}$ to belong to the error is

$$\underline{c}'\underline{A}' = 0 \quad \text{or} \quad A\underline{c} = 0. \tag{5.1.8}$$

Now (5.1.8) can be regarded as a set of homogeneous linear equations for determining $c_1, c_2, \ldots, c_n$.

We shall suppose that $m < n$. Let

$$\text{rank } A \equiv n_0 \leqq m.$$

Then (5.1.8) has $n - n_0$ linearly independent solutions. We thus get $n_e \equiv n - n_0$ linearly independent linear functions belonging to the error

Let V_e denote the set of all the linear functions of the y's belonging to the error and let V_0 be the set of all the linear functions of the y's that are uncorrelated to every linear function belonging to the error. Let $\underline{e}_1'\underline{y}$, $\underline{e}_2'\underline{y}, \ldots, \underline{e}_{n_e}'\underline{y}$ be a set of n_e linearly independent linear functions belonging to the error and being mutually uncorrelated. Any non-null linear function belonging to the error is then of the form

$$\underline{e}'\underline{y} = \lambda_1\underline{e}_1'\underline{y} + \cdots + \lambda_{n_e}\underline{e}_{n_e}'\underline{y}. \tag{5.1.9}$$

If $\underline{e}'\underline{y}$ belongs to V_0, then

$$0 = Cov(\underline{e}'\underline{y}, \underline{e}_i'\underline{y}) = \lambda_i\ Var(\underline{e}_i'\underline{y})$$

$$\rightarrow \lambda_i = 0, \quad i = 1,2,\ldots,n_e.$$

This makes $\underline{e}'\underline{y} = 0$ which contradicts to our starting assumption. Hence $\underline{e}'\underline{y}$ of the form of (5.1.9) cannot belong to V_0.

Lemma 5.1.1. No nonnull linear function of the y's can simultaneously belong to both the sets V_0 and V_e.

If $\underline{c}'\underline{y}$ belongs to V_0, then

$$\underline{e}_i'(\sigma^2 I + L'L\sigma_1^2)\underline{c} = 0 \qquad (i = 1,2,\ldots,n_e),$$

or

$$M_e(\sigma^2 I + L'L\sigma_1^2)\underline{c} = 0, \tag{5.1.10}$$

where

$$M_e = \left\| \begin{matrix} \underline{e}_1' \\ \underline{e}_2' \\ \vdots \\ \underline{e}_{n_e}' \end{matrix} \right\|.$$

Equation (5.1.10) can be regarded as a set of n_e independent homogeneous linear equations for determining $\underline{c}$. Hence the number of linearly independent solutions for $\underline{c}$ is

$$n_0 \equiv n - n_e.$$

Let these linearly independent linear functions be

$$\underline{c}_1'\underline{y},\ \underline{c}_2'\underline{y}, \ldots,\ \underline{c}_{n_0}'\underline{y}.$$

Then any linear function belonging to V_0 must be of the form

$$\underline{c}'\underline{y} = \mu_1\underline{c}_1'\underline{y} + \cdots + \mu_{n_0}\underline{c}_{n_0}\underline{y}.$$

Lemma 5.1.2. We can choose n_e linear functions belonging to the error in such a way that any two of them are uncorrelated to each other.

Proof. Let $\underline{d}'\underline{y}, \ldots, \underline{d}_{n_e}'\underline{y}$ be any set of n_e linearly independent nonnull linear functions belonging to the error. We can write this set as $D\underline{y}$, where

$$D = \begin{Vmatrix} \underline{d}_1' \\ \vdots \\ \underline{d}_{n_e}' \end{Vmatrix}$$

$$Var[D\underline{y}] = D\Sigma D'.$$

Let $\underline{z} = HD\underline{y}$, $H(n_e \times n_e)$. Then

$$Var[\underline{z}] = HD\Sigma D'H'.$$

We can choose H to be an orthogonal matrix such that $HD\Sigma D'H'$ is a diagonal form. Hence if we set

$$M_e = HD,$$

then $M_e\underline{y}$ is a set of n_e mutually uncorrelated linear functions belonging to the error. A necessary and sufficient condition for $\underline{c}'\underline{y}$ to belong to V_0 is

$$M_e(\sigma^2 I + L'L\sigma_1^2)\underline{c} = 0.$$

The function $\underline{c}'\underline{y}$ is said to be an unbiased linear estimate of $\underline{l}'\underline{p}$ if

$$E(\underline{c}'\underline{y}) = \underline{l}'\underline{p},$$

that is,

$$\underline{c}'A' = \underline{l}'$$

or

$$A\underline{c} = \underline{l}.$$

The function $\underline{l}'\underline{p}$ is said to be *estimable* if a corresponding linear function $\underline{c}'\underline{y}$ exists as its unbiased linear estimate. The condition for this is

$$\text{Rank } A = \text{Rank } (A,\underline{l}).$$

If $\underline{l}'\underline{p}$ is estimable, then the best linear unbiased estimate (BLUE) is the unbiased linear estimate with minimum variance.

Theorem 5.1.1. If $\underline{l}'\underline{p}$ is estimable, then

(i) there exists a unique linear function $\underline{c}'\underline{y}$ belonging to V_0 for which $E(\underline{c}'\underline{y}) = \underline{l}'\underline{p}$.

(ii) $\underline{c}'\underline{y}$ is the best linear unbiased estimate of $\underline{l}'\underline{p}$.

Proof. Since $\underline{l}'\underline{p}$ is estimable, there exists at least one linear function $\underline{d}'\underline{y}$ such that

$$E(\underline{d}'\underline{y}) = \underline{l}'\underline{p}.$$

Now $\underline{d}'\underline{y}$ can be written as

$$\underline{d}'\underline{y} = \underline{c}'\underline{y} + \underline{e}'\underline{y},$$

where $\underline{c}'\underline{y}$ belongs to V_0 and $\underline{e}'\underline{y}$ belongs to V_e, and this decomposition being unique,

$$E(\underline{d}'\underline{y}) = E(\underline{c}'\underline{y}) + E(\underline{e}'\underline{y}) = E(\underline{c}'\underline{y})$$

and therefore

$$E(\underline{c}'\underline{y}) = \underline{l}'\underline{p}.$$

If possible, let there exist a $\underline{c}_1'\underline{y}$ belonging to V_0 such that

$$E(\underline{c}_1'\underline{y}) = \underline{l}'\underline{p}.$$

Hence

$$E[(\underline{c}_1' - \underline{c}')\underline{y}] = 0.$$

Thus $(\underline{c}_1' - \underline{c}')y$ belongs to both V_0 and V_e simultaneously, and therefore this must be the null form. Hence

$$\underline{c}_1'\underline{y} = \underline{c}'\underline{y}.$$

This proves part (i) of the theorem.

Now

$$Var(\underline{d}'\underline{y}) = Var(\underline{c}'\underline{y}) + Var(\underline{e}'\underline{y}) + 2Cov(\underline{d}'\underline{y}, \underline{e}'\underline{y})$$

$$= Var(\underline{e}'\underline{y}) + Var(\underline{e}'\underline{y}) \geqq Var(\underline{c}'\underline{y}).$$

Equality holds only when $Var(\underline{e}'\underline{y}) = 0$. This shows that $\underline{c}'\underline{y}$ is of the minimum variance. Hence $\underline{c}'\underline{y}$ is the BLUE of $\underline{l}'\underline{p}$.

We may call V_0 the estimation set. If $\underline{l}'\underline{p}$ is estimable, then there exists *a unique linear function in the estimation set* that is an unbiased linear estimate with minimum variance.

Let $\underline{a}_1', \underline{a}_2', \ldots, \underline{a}_m'$ be the m row vectors of A. Then if $\underline{e}'\underline{y}$ belongs to error,

$$\underline{e}'\underline{A} = 0,$$

which was already shown. Therefore

$$\underline{e}'\underline{a}_j = 0 \qquad (j = 1,2,\ldots,m)$$

Let $\underline{c}_j$ be determined by

$$(\sigma^2 I + L'L\sigma_1^2)\underline{c}_j = \underline{a}_j \qquad (j = 1,2,\ldots,m),$$

or

$$\underline{c}_j = \Sigma^{-1}\underline{a}_j.$$

Then

$$\underline{e}'\Sigma\underline{c}_j = \underline{e}'\underline{a}_j = 0.$$

Hence $\underline{c}_j$ belongs to V_0. Among the vectors $\underline{c}_1, \underline{c}_2, \ldots, \underline{c}_m$ only n_0 are linearly independent. The vectors $\underline{c}_1'\underline{y}, \ldots, \underline{c}_m'\underline{y}$ will form a generating set for V_0 and they will be a basis if $m = n_0$. Let C be the matrix

$$C = \begin{Vmatrix} \underline{c}_1' \\ \vdots \\ \underline{c}_m' \end{Vmatrix}.$$

Then

$$\Sigma(\underline{c}_1 \cdots \underline{c}_m) = (\underline{a}_1 \cdots \underline{c}_m)$$

or

$$C' = \Sigma^{-1}A',$$

$$C = A\Sigma^{-1},$$

where

$$\Sigma = \sigma^2 I + L'L\sigma_1^2.$$

Any arbitrary linear function of the estimation set can be written as

$$q_1\underline{c}_1'\underline{y} + q_2\underline{c}_2'\underline{y} + \cdots + q_m\underline{c}_m'\underline{y}$$

$$= (q_1 q_2 \cdots q_m) \begin{Vmatrix} \underline{c}_1' \\ \underline{c}_2' \\ \vdots \\ \underline{c}_m' \end{Vmatrix} \underline{y} = \underline{q}'C\underline{y} = \underline{q}'A\Sigma^{-1}\underline{y},$$

and thus we have the following corollary :

Corollary 5.1.1. The estimation set is generated by the elements $A\Sigma^{-1}\underline{y}$. If $\underline{l}'\underline{p}$ is estimable, then its best estimate is given by

$$\underline{q}'A\Sigma^{-1}\underline{y},$$

where $\underline{q}$ is determined by

$$E(q'A\Sigma^{-1}\underline{y}) = \underline{l}'\underline{p},$$

$$\underline{q}'A\Sigma^{-1}A' = \underline{l}',$$

or

$$A\Sigma^{-1}\underline{A}'\underline{q} = \underline{l}.$$

This is a *conjugate normal equation* determining $\underline{q}$.

Let

$$\hat{\underline{p}} = \begin{bmatrix} \hat{p}_1 \\ \hat{p}_2 \\ \vdots \\ \hat{p}_m \end{bmatrix}$$

be determined by the *normal equation*

$$A\Sigma^{-1}A'\hat{\underline{p}} = A\Sigma^{-1}\underline{y}.$$

Then $\underline{l}'\hat{\underline{p}} = \underline{q}'A\Sigma^{-1}A'\hat{\underline{p}} = \underline{q}'A\Sigma^{-1}\underline{y}$ is the best estimate of $\underline{l}'\underline{p}$. The normal equation can also be written as

$$E(A\Sigma^{-1}\underline{y}) = A\Sigma^{-1}\underline{y}.$$

5.2. Interblock Analysis

Treatment sets are put at random to block. We consider that the block effects are random variables instead of being constants. We shall consider incomplete block designs in which the block size is constant and equal to k.

Let the ith treatment be replicated r_i times. The number of treatments is v, and the number of blocks is b; $\lambda_{i\alpha}$ is the number of times the ith and αth treatments occur together in a block. The incidence matrix is

$$N = \begin{Vmatrix} n_{11} & n_{12} & \cdots & n_{1b} \\ n_{21} & n_{22} & \cdots & n_{2b} \\ \cdot & \cdot & \cdots & \cdot \\ n_{v1} & n_{v2} & \cdots & n_{vb} \end{Vmatrix},$$

$$\sum_{i=1}^{v} r_i = bk = n \quad \text{the total number of experimental units.}$$

and

$$\sum_{\alpha \neq i} \lambda_{i\alpha} = r_i(k - 1).$$

In fact

$$r_i \left\{ \overbrace{\begin{matrix} \theta & * & * & \cdots & * \\ \theta & * & * & \cdots & * \\ \vdots & \vdots & \vdots & & \vdots \\ \theta & * & * & \cdots & * \end{matrix}}^{k} \right. \qquad \begin{aligned} &(\lambda_{ii} \text{ being omitted}), \\ &\lambda_{i1} + \lambda_{i2} + \cdots + \lambda_{ir} = r_i(k - 1), \\ &\sum_{\alpha=1}^{v} \lambda_{i\alpha} = r_i k. \end{aligned}$$

Let the uth experimental unit occur in the jth block and suppose the ith treatment has been applied to it. Then we can write the response as

$$\begin{array}{ccccccccc} y_u = & e_u & + & \gamma & + & \tau_i & + & x_j \\ & \downarrow & & & & \downarrow & & \downarrow \\ & N(0,\sigma^2) & & & & \text{effect of the } i\text{th treatment} & & N(0,\sigma_1^2) \end{array}$$

We can write

$$\underline{y} = \underline{e} + \gamma \underline{j}_n + \Phi\underline{\tau} + \Psi\underline{x},$$

$$\underline{j}_n'\underline{y} = G = \text{grand total},$$

$$\Phi'\underline{y} = \underline{T} = \begin{bmatrix} T_1 \\ \vdots \\ T_v \end{bmatrix} : \text{treatment totals},$$

$$\Psi'\underline{y} = \underline{B} = \begin{bmatrix} B_1 \\ \vdots \\ B_b \end{bmatrix} : \text{ block totals.}$$

We shall consider the first k units to be units of the first block, the next k units to be units of the second block and so on. In this case

$$\Psi\Psi' = \left\| \begin{array}{c|c|c|c} \begin{matrix} 1 & \cdots & 1 \\ \vdots & & \vdots \\ 1 & \cdots & 1 \end{matrix} & 0 & 0 & 0 \\ \hline 0 & \begin{matrix} 1 & \cdots & 1 \\ \vdots & & \vdots \\ 1 & \cdots & 1 \end{matrix} & 0 & 0 \\ \hline 0 & 0 & \begin{matrix} 1 & \cdots & 1 \\ \vdots & & \vdots \\ 1 & \cdots & 1 \end{matrix} & 0 \\ \hline 0 & 0 & 0 & \begin{matrix} 1 & \cdots & 1 \\ \vdots & & \vdots \\ 1 & \cdots & 1 \end{matrix} \end{array} \right\|$$

Thus

$$\Sigma = I\sigma^2 + \Psi\Psi'\sigma_1^2 = K \times I_b,$$

where the diagonal elements of K are $\sigma^2 + \sigma_1^2$ and the off-diagonal elements are σ_1^2.

$$\Sigma^{-1} = K^{-1} \times I_b.$$

It is obvious that K^{-1} is given by the form

$$K^{-1} = \begin{vmatrix} x & y & y & \cdots & y \\ y & x & y & \cdots & y \\ \vdots & \vdots & \vdots & & \vdots \\ y & y & y & \cdots & x \end{vmatrix},$$

where x and y are determined by

$$(5.2.1)\qquad \begin{aligned} &(\sigma^2 + \sigma_1^2)x + (k - 1)\sigma_1^2 y = 1, \\ &\sigma_1^2 x + [(\sigma_1^2 + \sigma^2) + (k - 2)\sigma_1^2]y = 0. \end{aligned}$$

Solving (5.2.1), we get

$$x = \frac{\sigma^2 + (k - 1)\sigma_1^2}{\sigma^2\left(\sigma^2 + k\sigma_1^2\right)}, \qquad y = -\frac{\sigma_1^2}{\sigma^2\left(\sigma^2 + k\sigma_1^2\right)}.$$

Since

$$x - y = \frac{1}{\sigma^2},$$

we can write

$$K^{-1} = \frac{1}{\sigma^2} I_k - \frac{\sigma_1^2}{\sigma^2(\sigma^2 + k\sigma_1^2)} J_{k,k}.$$

The estimation set consists of the elements of the vectors

$$\begin{Vmatrix} \underline{j}_n' \\ \Phi' \end{Vmatrix} \Sigma^{-1}\underline{y}.$$

Put

$$\underline{z} = \begin{bmatrix} z_1 \\ \vdots \\ z_u \\ \vdots \\ z_n \end{bmatrix} \Sigma^{1}\underline{y},$$

where

$$z_u = \frac{y_u}{\sigma^2} - \frac{B_{.j}\sigma_1^2}{\sigma^2(\sigma^2 + k\sigma_1^2)}$$

if y_u comes from the jth block. Hence, by our assumption, this holds for

$$k(j - 1) < u \leqq kj.$$

Thus

$$\underset{\sim}{j}_n'\Sigma^{-1}\underline{y} = \sum_u\left[\frac{y_u}{\sigma^2} - \frac{B_j\sigma_1^2}{\sigma^2(\sigma^2 + k\sigma_1^2)}\right] = \frac{G}{\sigma^2} - \frac{k\sigma_1^2}{\sigma^2(\sigma^2 + k\sigma_1^2)}(B_1 + \cdots + B_b)$$

$$= \frac{G}{\sigma^2 + k\sigma_1^2}.$$

Now let ith column of Φ be $\underline{\zeta}_i$. The ith element of $\Phi'\Sigma^{-1}\underline{y}$ is

$$\frac{T_i}{\sigma^2} - \frac{\sigma_1^2}{\sigma^2(\sigma^2 + k\sigma_1^2)}(n_{i1}B_1 + n_{i2}B_2 + \cdots + n_{ib}B_b).$$

Hence we have

$$\Phi'\Sigma^{-1}\underline{y} = \frac{\underline{T}}{\sigma^2} - \frac{\sigma_1^2}{\sigma^2(\sigma^2 + k\sigma_1^2)} N\underline{B}.$$

Put

$$Q_i = T_i - \frac{1}{k}(n_{i1}B_1 + \cdots + n_{ib}B_b) \qquad (i = 1,2,\ldots,v),$$

or, in matrix notation,

$$\underline{Q} = \underline{T} - \frac{1}{k} N\underline{B}.$$

Thus we get the following result:

$$\Phi'\Sigma^{-1}\underline{y} = \frac{\underline{T}}{\sigma^2} - \frac{k\sigma_1^2(\underline{T} - \underline{Q})}{\sigma^2(\sigma^2 + k\sigma_1^2)} = \frac{1}{\sigma^2}\underline{Q} + \frac{\underline{T} - \underline{Q}}{\sigma^2 + k\sigma_1^2}.$$

Let us put

$$w = \frac{1}{\sigma^2}, \qquad w' = \frac{1}{(\sigma^2 + k\sigma_1^2)}.$$

Then we reach

$$\left\|\begin{matrix} \underset{\sim}{j}_n' \\ \\ \Phi' \end{matrix}\right\| \Sigma^{-1}\underline{y} = \left\|\begin{matrix} w'G \\ \\ w\underline{Q} + w'(\underline{T} - \underline{Q}) \end{matrix}\right\|.$$

Hence the estimation set is generated by

(5.2.2)
$$w'G,\ wQ_1 + w'(T_1 - Q_1),$$
$$wQ_2 + w'(T_2 - Q_2),\ldots,wQ_v + w'(T_v - Q_v);$$

that is, the best estimate of any linear function of the parameters

$$\gamma,\ \tau_1,\ \tau_2,\ldots,\tau_v$$

must be some linear combination of the above functions, that is,

$$q_0w'G + \sum_{i=1}^{v}\left[(wQ_i + w'(T_i - Q_i)\right]q_i. \tag{5.2.3}$$

If (5.2.3) estimates a linear function of the treatment effects only, the coefficient of γ in its expectation must vanish, that is,

$$nq_0w' + w'(r_1q_1 + \cdots + r_vq_v) = 0,$$
$$q_0 = -\frac{1}{n}(r_1q_1 + \cdots + r_vq_v).$$

This is the condition for the expectation of (5.2.3) to be free of γ. Substituting for q_0 in (5.2.3), we get

$$\sum_{j=1}^{v}\ [wQ_i + w'(T_i - Q_i - \frac{r_iG}{n})]q_i \tag{5.2.4}$$

as the general form of the best linear estimate of any linear function of the treatment effects only.

Suppose
$$\tilde{Q}_i = T_i - Q_i - \frac{r_iG}{n} = \sum_{j=1}^{b} n_{ij}\left(\frac{B_j}{k} - \frac{G}{n}\right),$$
where

$\frac{B_j}{k} - \frac{G}{n}$: difference of the average of the jth block and the general average.

Then the general form of the best estimate of a linear function of the τ's only is given by

$$(5.2.5) \qquad \sum_{i=1}^{v} (wQ_i + w'\tilde{Q}_i)q_i.$$

Since we have assumed the model

$$y_u = e_u + \gamma + \tau_i + x_j,$$

$$E(y_u) = \gamma + \tau_i,$$

$$E(G) = n\gamma + r_1\tau_1 + r_2\tau_2 + \cdots + r_v\tau_v = \underline{j}_n'\underline{j}_n\gamma + \underline{r}'\underline{\tau},$$

$$E(B_j) = k\gamma + n_{1j}\tau_1 + \cdots + n_{vj}\tau_v \qquad (j = 1,2,\ldots,b),$$

or

$$E(\underline{B}) = k\gamma\underline{j}_b + N'\underline{\tau},$$

$$E(T_i) = r_i\tau_i + r_i\gamma \qquad (i = 1,2,\ldots,v),$$

or

$$E(\underline{T}) = D_r\underline{\tau} + \gamma\underline{r}.$$

Since $\underline{Q} = \underline{T} - (1/k)N\underline{B}$, we have

$$E(\underline{Q}) = D_r\underline{\tau} + \gamma\underline{r} - \frac{1}{k}N\,(k\gamma\underline{j}_b + N'\underline{\tau})$$

whereas

$$N\underline{j}_b = \underline{r}, \qquad NN' = \begin{Vmatrix} r_1 & \lambda_{12} & \cdots & \lambda_{1v} \\ \lambda_{21} & r_2 & \cdots & \lambda_{2v} \\ \vdots & \vdots & & \vdots \\ \lambda_{v1} & \lambda_{v2} & \cdots & r_v \end{Vmatrix},$$

$$E(\underline{Q}) = D_r\underline{\tau} - \frac{1}{k}NN'\underline{\tau} = \left(D_r - \frac{1}{k}NN'\right)\underline{\tau} = C\underline{\tau},$$

$$c_{ii} = r_i\left(1 - \frac{1}{k}\right), \qquad c_{i\alpha} = -\frac{\lambda_{i\alpha}}{k},$$

$$E(\tilde{Q}_i) = E\left(T_i - Q_i - \frac{r_iG}{n}\right) = \tilde{c}_{i1}\tau_1 + \tilde{c}_{i2}\tau_2 + \cdots + \tilde{c}_{iv}\tau_v,$$

where

$$\tilde{c}_{ii} = \frac{r_{i\cdot}}{k} - \frac{r_{i\cdot}^2}{n},$$

$$\tilde{c}_{i\alpha} = \frac{\lambda_{i\alpha}}{k} - \frac{r_{i\cdot}r_{\alpha}}{n} \quad \text{if} \quad i \neq \alpha.$$

In matrix notation, we have

$$E(\underline{\tilde{Q}}) = D_{r}\underline{\tau} + \gamma\underline{r} - C\underline{\tau} - \frac{1}{n}\underline{r}(\underline{j}_n'\underline{j}_n\gamma + \underline{r}'\underline{\tau})$$

$$= \left(D_r - C - \frac{1}{n}\underline{r}\,\underline{r}'\right)\underline{\tau} \equiv \tilde{C}\underline{\tau},$$

where

$$\tilde{C} = D_r - C - \frac{1}{n}\underline{r}\,\underline{r}'.$$

Now

$$\sum_{\alpha=1}^{v} \tilde{c}_{i\alpha} = 0.$$

Putting $\lambda_{ii} = r_{i\cdot}$, we get

$$\tilde{c}_{i\alpha} = \frac{\lambda_{i\alpha}}{k} - \frac{r_{i\cdot}\lambda_{\alpha}}{n},$$

$$\sum_{\alpha=1}^{v} \tilde{c}_{i\alpha} = \sum_{\alpha=1}^{v} \left(\frac{\lambda_{i\alpha}}{k} - \frac{r_{i\cdot}r_{\alpha}}{n}\right) = r_{i\cdot} - r_{i\cdot} = 0,$$

$$\sum_{\alpha=1}^{v} \lambda_{i\alpha} = kr_{i\cdot}, \qquad \sum_{\alpha=1}^{v} r_{\alpha} = n.$$

The sum of the elements of any row or column of C *vanishes.*

In (5.2.5) henceforth we shall use $\tilde{\omega}$ instead of w'. Then

$$E\left[\underline{q}'(\omega\underline{Q} + \tilde{\omega}\underline{\tilde{Q}})\right] = q'(\omega C + \tilde{\omega}\tilde{C})\underline{\tau}.$$

The sum of the coefficients of $\underline{\tau}$ vanishes in the above expectation, in fact

$$\underline{q}'(\omega C + \tilde{\omega}\tilde{C})\underline{j}_v = 0.$$

Hence only treatment contrasts can be estimated.

Suppose we want to estimate

$$\underline{l}'\underline{\tau} = l_1\tau_1 + \cdots + l_v\tau_v,$$

where $\underline{l}'\underline{j}_v = 0$. Let the best estimate be

$$\underline{q}'(\omega\underline{Q} + \tilde{\omega}\underline{\tilde{Q}}).$$

Then $\underline{q}$ should be determined by

$$\underline{q}'(\omega C + \tilde{\omega}\tilde{C}) = \underline{l}',$$

or

(5.2.6) $$(\omega C + \tilde{\omega}\tilde{C})\underline{q} = \underline{l}.$$

This is called the *conjugate normal equation* determining $\underline{q}$.

Theorem 5.2.1. If the treatment contrast $\underline{l}'\underline{\tau}$ is estimable, its best estimate is obtained by substituting in $\underline{l}'\underline{\tau}$ any solution $\hat{\underline{\tau}}$ of the adjusted normal equation

$$(\omega C + \tilde{\omega}\tilde{C})\hat{\underline{\tau}} = \omega\underline{Q} + \tilde{\omega}\underline{\tilde{Q}}.$$

Proof. Let $\underline{q}$ be determined by (5.2.6). Then

$$\underline{l}'\hat{\underline{\tau}} = \underline{q}'(\omega C + \tilde{\omega}\tilde{C})\hat{\underline{\tau}} = \underline{q}'(\omega\underline{Q} + \tilde{\omega}\underline{\tilde{Q}}),$$

that is, this is the best estimate of $\underline{l}'\underline{\tau}$ [see (5.2.5)].

Next we consider variances :

$$\underline{y} = \underline{e} + \gamma\underline{j} + \Phi\underline{\tau} + \Psi\underline{x},$$

$$Var(\underline{y}) = \sigma^2 I + \Psi\Psi'\sigma_1^2$$

$$\underline{T} = \begin{bmatrix} T_1 \\ T_2 \\ \vdots \\ T_v \end{bmatrix} = \Phi'\underline{y}, \qquad \underline{B} = \begin{bmatrix} B_1 \\ B_2 \\ \vdots \\ B_b \end{bmatrix} = \Psi'\underline{y}.$$

In the first place, we shall calculate

$$Var \begin{bmatrix} \underline{T} \\ \\ \underline{B} \end{bmatrix} = Var \begin{bmatrix} \Phi' \\ \\ \Psi' \end{bmatrix} \underline{y} = \begin{bmatrix} \Phi' \\ \\ \Psi' \end{bmatrix} (\sigma^2 I + \Psi\Psi'\sigma_1^2)(\Phi\bar{\Psi}) = \begin{Vmatrix} \Phi'\Sigma\Phi & \Phi'\Sigma\Psi \\ \\ \Psi'\Sigma\Phi & \Psi'\Phi\Psi \end{Vmatrix}.$$

Since $\Phi'\Phi = D_r$ and $\Phi'\Psi = N$, we have

$$\Phi'\Sigma\Phi = \Phi'(\sigma^2 I + \Psi\Psi'\sigma_1^2)\Phi = D_r\sigma^2 + NN'\sigma_1^2.$$

Thus

$$\text{(5.2.7)} \qquad Var\ \underline{T} = D_r\sigma^2 + NN'\sigma_1^2,$$

that is

$$Var(T_i) = r_i\sigma^2 + r_i\sigma_1^2 = r_i(\sigma^2 + \sigma_1^2) \quad .$$

Since

$$\Psi'\Sigma\Psi = \sigma^2\Psi'\Psi + \Psi\Psi'(\Psi\Psi')'\sigma^2$$

we have

$$Var(B_j) = k\sigma^2 + k^2\sigma_1^2 = k(\sigma^2 + k\sigma_1^2) = \frac{k}{\tilde{\omega}},$$

$$\text{(5.2.8)} \qquad Cov(B_i, B_j) = 0 \quad \text{for} \quad i \neq j.$$

Since

$$\Phi'\Sigma\Psi = \Phi'(\sigma^2 I + \Psi\Psi'\sigma_1^2)\Psi = \sigma^2 N + kN\sigma_1^2 = \frac{1}{\tilde{\omega}}N,$$

we have

$$Cov(T_i, B_i) = \frac{n_{ij}}{\tilde{\omega}} = \begin{cases} \frac{1}{\tilde{\omega}} & \text{if the } i\text{th treatment occurs in the } j\text{th block,} \\ 0 & \text{otherwise.} \end{cases}$$

Since

$$C = D_r - \frac{1}{k} NN',$$

$$D_r\sigma^2 + NN'\sigma_1^2 = D_r\sigma^2 + k(D_r - C)\sigma_1^2 = C\cdot\sigma^2 + (D_r - C)(\sigma^2 + k\sigma^2)$$
$$= \frac{C}{\omega} + \frac{D_r - C}{\tilde{\omega}},$$

we have

$$Var\begin{bmatrix} \underline{T} \\ \\ \underline{B} \end{bmatrix} = \begin{Vmatrix} \frac{C}{\omega} + \frac{D_r - C}{\tilde{\omega}} & \frac{N}{\tilde{\omega}} \\ \\ \frac{N}{\omega} & \frac{k}{\tilde{\omega}} I \end{Vmatrix}.$$

Next

$$\underline{Q} = \underline{T} - \frac{1}{k} N\underline{B},$$

$$\underline{B} = \qquad \underline{B},$$

that is

$$Var\begin{bmatrix} \underline{Q} \\ \\ \underline{B} \end{bmatrix} = \begin{Vmatrix} I_v & -\frac{1}{k} N \\ \\ 0 & I_b \end{Vmatrix} \cdot \begin{Vmatrix} \frac{C}{\omega} + \frac{D_r - C}{\tilde{\omega}} & \frac{N}{\tilde{\omega}} \\ \\ \frac{N}{\omega} & \frac{k}{\tilde{\omega}} I \end{Vmatrix} \cdot \begin{Vmatrix} I_v & 0 \\ \\ -\frac{1}{k} N' & I_b \end{Vmatrix} = \begin{Vmatrix} \frac{C}{\omega} & 0 \\ \\ 0 & \frac{k}{\tilde{\omega}} I \end{Vmatrix},$$

that is

$$Var(\underline{Q}) = \frac{C}{\omega} \quad \text{or} \quad \sigma^2 C \rightarrow Var(Q_i) = c_{ii}\sigma^2 = r_i(1 - \frac{1}{k})\sigma^2,$$

$$Cov(Q_i, Q_\alpha) = c_{i_\alpha}\sigma^2 = -\frac{\lambda_{i\alpha}}{k}\sigma^2,$$

$$Cov(\underline{Q}, \underline{B}) = 0 \rightarrow Cov(Q_i, B_i) = 0,$$

$$Q_i = T_i - \frac{1}{k}(n_{i1}B_1 + \cdots + n_{ib}B_b),$$

$$\tilde{Q}_i = T_i - Q_i - \frac{r_i G}{n},$$

$$E(\underline{Q}) = C\underline{t}, \qquad C = D_r - \frac{1}{k}NN',$$

$$E(\tilde{\underline{Q}}) = \tilde{C}\underline{t}, \qquad C = \frac{1}{k}NN' - \frac{1}{n}\underline{r}\,\underline{r}',$$

$$Var(\underline{Q}) = \frac{C}{\omega},$$

$$Var(\underline{\tilde{Q}}) = \frac{\tilde{C}}{\tilde{\omega}}.$$

Proof of $Var(\underline{\tilde{Q}}) = \tilde{C}/\tilde{\omega}$:

$$\underline{\tilde{Q}} = \underline{T} - \underline{Q} - \frac{G}{n}\underline{r} = \underline{T} - \underline{Q} - \frac{\underline{r}}{n}\underline{j}_b'\underline{B} = \underline{T} - \left(\underline{T} - \frac{1}{k}N\underline{B}\right) - \frac{\underline{r}}{n}\underline{j}_b'\underline{B}$$

$$= \left(\frac{1}{k}N - \frac{\underline{r}}{n}\underline{j}_b'\right)\underline{B},$$

$$Var(\underline{\tilde{Q}}) = \left(\frac{1}{k}N - \frac{\underline{r}}{n}\underline{j}_b'\right)\frac{k}{\tilde{\omega}}I_b\left(\frac{1}{k}N' - \frac{1}{n}\underline{j}_b\underline{r}'\right)$$

$$= \frac{k}{\tilde{\omega}}\left(\frac{NN'}{k^2} - \frac{1}{nk}N\underline{j}_b\underline{r}' - \frac{1}{nk}\underline{r}\underline{j}_b'N' + \frac{1}{n^2}\underline{r}\underline{j}_b'\underline{j}_b\underline{r}'\right)$$

$$= \frac{1}{\tilde{\omega}}\left(\frac{1}{k}NN' - \frac{2}{n}\underline{r}\,\underline{r}' + \frac{1}{n}\underline{r}\,\underline{r}'\right) = \frac{1}{\tilde{\omega}}\left(\frac{NN'}{k} - \frac{1}{n}\underline{r}\,\underline{r}'\right) = \frac{1}{\tilde{\omega}}\tilde{C}.$$

Q.E.D.

In other words

$$(5.2.13) \qquad Var(\tilde{Q}_{i.}) = \frac{\tilde{c}_{ii}}{\tilde{\omega}} = (\sigma^2 + k\sigma_1^2)\left(\frac{r_{i.}}{k} - \frac{r_{i.}^2}{n}\right),$$

$$Cov(\tilde{Q}_{i.},\tilde{Q}_{\alpha}) = \frac{\tilde{c}_{i\alpha}}{\tilde{\omega}} = (\sigma^2 + k\sigma_1^2)\left(\frac{\lambda_{i\alpha}}{k} - \frac{r_{i.}r_{\alpha}}{n}\right).$$

Recall the relation

$$Q_{i.} = \sum_{j=1}^{b} n_{ij}\left(\frac{B_{.j}}{k} - \frac{G}{n}\right).$$

This is the sum of the deviations of block averages from the grand average! Finally, we want to show that

$$(5.2.14) \qquad Cov(Q_{i.},\tilde{Q}_{\alpha}) = 0.$$

Since

$$\underline{Q} = \left(\frac{N}{k} - \frac{r}{n}\, \underline{j}_b'\right)\underline{B}, \qquad \underline{Q} = \underline{T} - \frac{1}{k} N\underline{B},$$

$$Var\begin{bmatrix} \underline{Q} \\ \tilde{\underline{Q}} \end{bmatrix} = Var\left\{\begin{Vmatrix} I & -\frac{1}{k} N \\ 0 & \frac{N}{k} - \frac{r}{n}\, \underline{j}_b' \end{Vmatrix} \begin{bmatrix} \underline{T} \\ \underline{B} \end{bmatrix}\right\}$$

$$= \begin{Vmatrix} I & -\frac{1}{k} N \\ 0 & \frac{1}{k}N - \frac{r}{n}\, \underline{j}_b' \end{Vmatrix} \cdot \begin{Vmatrix} \frac{C}{\tilde{\omega}} + \frac{\tilde{C}}{\tilde{\omega}} & \frac{N}{\tilde{\omega}} \\ \frac{N'}{\tilde{\omega}} & \frac{k}{\tilde{\omega}} I \end{Vmatrix} \cdot \begin{Vmatrix} I & 0 \\ -\frac{1}{k} N' & \frac{1}{k} N' - \frac{1}{n}\, \underline{j}_b \underline{r} \end{Vmatrix}$$

$$= \begin{Vmatrix} \frac{C}{\tilde{\omega}} + \frac{\tilde{C}}{\tilde{\omega}} - \frac{NN'}{k\tilde{\omega}} & 0 \\ \frac{NN'}{k\tilde{\omega}} - \frac{r}{n}\, \underline{j}_b' N' & \frac{N}{\tilde{\omega}} - \underline{r}\, \underline{j}_b' \end{Vmatrix} \cdot \begin{Vmatrix} I & 0 \\ -\frac{N'}{k} & \frac{N'}{k} - \frac{1}{n}\, \underline{j}_b \underline{r} \end{Vmatrix} = \begin{Vmatrix} \frac{\tilde{C}}{\tilde{\omega}} & 0 \\ 0 & \frac{\tilde{C}}{\tilde{\omega}} \end{Vmatrix},$$

$$(5.2.15) \qquad (\omega C + \tilde{\omega}\tilde{C})\underline{\tilde{t}} = \omega\underline{Q} + \tilde{\omega}\underline{\tilde{Q}},$$

$$Var(\omega Q_i + \tilde{\omega}\tilde{Q}_i) = \omega^2\ Var\ Q_i + \tilde{\omega}^2\ Var\ \tilde{Q}_i = \omega c_{ii} + \tilde{\omega}\tilde{c}_{ii}$$

$$Cov(\omega Q_i + \tilde{\omega}\tilde{Q}_i,\ \omega Q_\alpha + \tilde{\omega}\tilde{Q}_\alpha) = \omega^2\ Cov(Q_i, Q_\alpha) + \tilde{\omega}^2\ Cov(\tilde{Q}_i, \tilde{Q}_\alpha) = \omega c_{i\alpha} + \tilde{\omega}\tilde{c}_{i\alpha}$$

Hence we get

$$(5.2.16) \qquad Var(\omega\underline{Q} + \tilde{\omega}\underline{\tilde{Q}}) = \omega C + \tilde{\omega}\tilde{C}.$$

As $1/\tilde{\omega}$ is the variance of the block total,

$$\text{information } I(B_j) = \frac{\tilde{\omega}}{k},$$

and hence the physical meaning of $\tilde{\omega}$ is

$$\tilde{\omega} = \text{information of block total.}$$

We had shown that the best estimate of the treatment contrasts can be written in the form

$$\underline{q}'(\omega\underline{Q} + \tilde{\omega}\underline{\tilde{Q}}),$$

where $\underline{q}$ is determined by the conjugate normal equation

$$(\omega C + \tilde{\omega}\tilde{C})\underline{q} = \underline{l}.$$

Therefore suppose we want to estimate the contrast

$$\underline{l}'\underline{\tau} = l_1\tau_1 + \cdots + l_v\tau_v.$$

Then the best estimate of $\underline{l}'\underline{\tau}$ is given by

$$\underline{q}'(\omega\underline{Q} + \tilde{\omega}\underline{\tilde{Q}}),$$

where

(5.2.17) $$(\omega C + \tilde{\omega}\tilde{C})\underline{q} = \underline{l}.$$

Thus

$$\text{the variance of the best estimate} = \underline{q}'(\omega C + \tilde{\omega}\tilde{C})\underline{q}$$

(5.2.18) $$= \underline{q}'\underline{l} = l_1q_1 + \cdots + l_vq_v$$

Suppose a solution of (5.2.15) is

$$\underline{\hat{t}} = F^*(\omega\underline{Q} + \tilde{\omega}\underline{\tilde{Q}}),$$

where F^* is a conditional inverse of the matrix

$$F = \omega C + \tilde{\omega}\tilde{C}.$$

Then a solution of (5.2.17) would be

$$\underline{q} = F^*\underline{l},$$

and the variance of the best estimate = $Var(\underline{l}'\underline{\hat{t}}) = \underline{l}'\underline{q} = \underline{l}'F^*\underline{l}$.

For example,

$$Var(t_i - t_\alpha) = f^*_{ii} - f^*_{i\alpha} - f^*_{\alpha i} + f^*_{\alpha\alpha}, \qquad F^* = \|f^*_{i\alpha}\|.$$

5.3. Application to Special Designs

Balanced Incomplete Block Design. In this case we have

$$c_{ii} = r(1 - \frac{1}{k}), \qquad c_{i\alpha} = -\frac{\lambda}{k}$$

$$\tilde{c}_{ii} = \frac{r}{k} - \frac{r^2}{k} = \frac{r}{k}(1 - \frac{r}{b}), \qquad \tilde{c}_{i\alpha} = \frac{\lambda}{k} - \frac{r^2}{n} = \frac{\lambda}{k} - \frac{r}{v},$$

$$c_{ii} - c_{i\alpha} = r - \frac{r}{k} + \frac{\lambda}{k} = \frac{r(k-1) + \lambda}{k} = \frac{\lambda_v}{k},$$

$$\tilde{c}_{ii} - \tilde{c}_{i\alpha} = \frac{r}{k} - \frac{r^2}{kb} - \frac{\lambda}{k} + \frac{r}{v} = \frac{r - \lambda}{k} = \frac{v\lambda}{k}.$$

The ith equation of the normal equation is

$$(5.3.1) \qquad \begin{aligned}(\omega c_{i1} + \tilde{\omega}\tilde{c}_{i1})t_1 + (\omega c_{i2} + \tilde{\omega}\tilde{c}_{i2})t_2 \\ + \cdots + (\omega c_{iv} + \tilde{\omega}\tilde{c}_{iv})t_v = \omega Q_i + \tilde{\omega}\tilde{Q}_i .\end{aligned}$$

The easiest constraint is to put that

$$(5.3.2) \qquad t_1 + t_2 + \cdots + t_v = 0.$$

By multiplying (5.3.2) by $\omega c_{i\alpha} + \tilde{\omega}\tilde{c}_{i\alpha}$ and subtracting from (5.3.1), we get

$$(5.3.3) \qquad [\omega(c_{ii} - c_{i\alpha}) + (\tilde{c}_{ii} - \tilde{c}_{i\alpha})]t_i = \omega Q_i + \tilde{\omega}\tilde{Q}_i,$$

$$[\omega\frac{v\lambda}{k} + \tilde{\omega}(r - \frac{v\lambda}{k})]t_i = \omega Q_i + \tilde{\omega}\tilde{Q}_i,$$

Where

$$t_i = \frac{\omega Q_i + \tilde{\omega}\tilde{Q}_i}{(v\lambda/k)\omega + [r - (v\lambda/k)]\tilde{\omega}} = \frac{\omega Q_i + \tilde{\omega}\tilde{Q}_i}{r[\omega E + \tilde{\omega}(1 - E)]}$$

is called the *combined intrablock and interblock estimate*. The Intrablock estimate was given by

$$\frac{Q_i}{rE}.$$

The combined intrablock and interblock variance of $\hat{t}_i - \hat{t}_\alpha$ is

$$Var(t_i - t_\alpha) = \frac{1}{r[\omega E + \tilde{\omega}(1 - E)]} + \frac{1}{r[\omega E + \tilde{\omega}(1 - E)]}$$

$$= \frac{2}{r[\omega E + \tilde{\omega}(1 - E)]},$$

$$E = \frac{v\lambda}{kr} = \frac{v(k - 1)}{k(v - 1)} = \frac{1 - (1/k)}{1 - (1/v)} < 1$$

so long as $k < v$. The intrablock variance of $t_i - t_\alpha$ is

$$Var(t_i - t_\alpha) = \frac{2}{r\omega E}.$$

$$\frac{\text{information of combined estimate}}{\text{information of intrablock estimate}} = \frac{\omega E + \tilde{\omega}(1 - E)}{\omega E} = 1 + \frac{\tilde{\omega}}{\omega}\left(\frac{1}{E} - 1\right) > 1,$$

the information of the combined estimate is greater than that of the intrablock estimate!

If we use the estimates of ω and $\tilde{\omega}$, then the distribution problem arises.

Lattice Design with m Orthogonal Replications. In this case the number of treatments is $v = k^2$, we also have

row, column, latin squares $1, 2, \ldots, m - 2$.

$$b = km, \quad r = m,$$

$$c_{ii} = m\left(1 - \frac{1}{k}\right), \quad c_{i\alpha} = -\frac{\lambda_{i\alpha}}{k},$$

$$\tilde{c}_{ii} = \frac{m}{k}\left(1 - \frac{1}{k}\right), \quad \tilde{c}_{i\alpha} = \frac{\lambda_{i\alpha}}{k} - \frac{m}{k^2};$$

$$\lambda_{i\alpha} = \begin{cases} 1 \text{ if } i \text{ and } \alpha \text{ occur in the same row or column, or correspond to the same letter in one of the Latin squares,} \\ 0 \text{ otherwise.} \end{cases}$$

$$f_{ii} \equiv \omega c_{ii} + \tilde{\omega}\tilde{c}_{ii}, \quad P_i \equiv \omega Q_i + \tilde{\omega}\tilde{Q}_i, \text{ etc.}$$

$t_R(i)$ = sum of the treatment effects of all the treatments occuring in the same row as the ith treatment;

$Q_R(i)$ = sum of the adjusted treatment effects of all the treatments occuring in the same row as the ith treatment.

The parameters $t_C(i)$, $Q_C(i)$; $t_j(i)$, $Q_j(i)$, $j = 1,2,\ldots,m - 2$ are similarly defined. For example,

							(j)		
1	2	3	4	5	C	E	D	B	A
6	7	8	9	10	A	C	B	E	D
11	12	13	14	15	E	B	A	D	C
16	17	18	19	20	B	D	C	A	E
21	22	23	24	25	D	A	E	C	B

$$t_R(1) = t_1 + t_2 + t_3 + t_4 + t_5,$$

$$Q_R(1) = Q_1 + Q_2 + Q_3 + Q_4 + Q_5,$$

$$t_j(1) = t_1 + t_7 + t_{15} + t_{18} + t_{24},$$

$$Q_j(1) = Q_1 + Q_7 + Q_{15} + Q_{18} + Q_{24},$$

$$f_{ii} = m(1 - \frac{1}{k})\omega + \frac{m}{k}(1 - \frac{1}{k})\tilde{\omega} \equiv \alpha, \quad \text{say},$$

$$f_{i\alpha} = \frac{\omega}{k} + (\frac{1}{k} - \frac{m}{k^2})\tilde{\omega} \equiv \beta \text{ (say) if } \lambda_{i\alpha} = 1$$

$$(i \neq \alpha)$$

$$= -\frac{\tilde{\omega}m}{k^2} \qquad \equiv \gamma \text{ (say) if } \lambda_{i\alpha} = 0$$

The ith normal equation is

$$f_{i1}t_1 + f_{i2}t_2 + \cdots + f_{ii}t_i + \cdots + f_{iv}t_v = \omega Q_i + \tilde{\omega}\tilde{Q}_i.$$

or

$$(\alpha - m\beta)t_i + \beta[t_R(i) + t_C(i) + t_1(i) + \cdots + t_{m-2}(i)]$$

(5.3.5)

$$+ \gamma t^*(i) = \omega Q_i + \tilde{\omega}\tilde{Q}_i,$$

where $t^*(i)$ is the sum of the treatment effects that do not occur together with the ith treatment in any block.

We add the restriction

$$\tau_1 + \tau_2 + \cdots + \tau_v = 0,$$

so that

$$-(m - 1)t_i + t_R(i) + t_C(i) + t_1(i) + \cdots + t_{m-2}(i) + t^*(i) = 0. \tag{5.3.6}$$

Eliminating $t^*(i)$, we have

$$[\alpha - m\beta + (m - 1)\gamma]t_i + (\beta - \gamma)[t_R(i) + t_c(i) + t_1(i) + \cdots + t_{m-2}(i)] = \omega Q_i + \tilde{\omega}\tilde{Q}_i \equiv P_i. \tag{5.3.7}$$

If we write down equations like these for every treatment that occurs in the same row as i and add up, we get

$$[\alpha - m\beta + (m - 1)\gamma]t_R(i) + k(\beta - \gamma)t_R(i) = P_R(i), \tag{5.3.8}$$

where $P_R(i)$ is the sum of the P's for all treatments in the same row as i. This reduces to

$$[(m - 1)\omega + \tilde{\omega}]t_R(i) = P_R(i) \tag{5.3.9}$$

if we note that

$$\alpha - m\beta + (m - 1)\gamma = m\omega$$

$$k(\beta - \gamma) = -\omega + \tilde{\omega}$$

Therefore

$$t_R(i) = \frac{P_R(i)}{(m - 1)\omega + \tilde{\omega}}. \tag{5.3.10}$$

Similarly we get

(5.3.11) $$t_C(i) = \frac{P_C(i)}{(m-1)\omega + \tilde{\omega}}.$$

and

(5.3.12) $$t_j(i) = \frac{P_j(i)}{(m-1)\omega + \tilde{\omega}} \qquad (j = 1,2,\ldots,m-2).$$

Substituting these equations into (5.3.7), we obtain

(5.3.13) $$m\omega\hat{t}_i = P_i + \frac{\omega - \tilde{\omega}}{k[(m-1)\omega + \tilde{\omega}]}[P_R(i) + P_C(i) + P_1(i) + \ldots + P_{m-2}(i)].$$

For actual computations it will be sufficient to know

$$P_1, P_2, \ldots, P_v,$$

$$P_R(i),\ P_C(i),\ P_1(i), \ldots, P_{m-2}(i) \qquad (i = 1,2,\ldots,v),$$

$$f^*_{ii} = \frac{2}{m\omega}\left\{1 + \frac{m(\omega - \tilde{\omega})}{k[(m-1)\omega + \tilde{\omega}]}\right\}$$

$$f^*_{i\alpha} = \frac{\omega - \tilde{\omega}}{m\omega k[(m-1)\omega + \tilde{\omega}]} \quad \text{or} \quad 0 \quad \text{according as } \lambda_{i\alpha} = 1 \text{ or } 0,$$

and

(5.3.15) $$Var(t_i - t_\alpha) = f^*_{ii} + f^*_{\alpha\alpha} - f^*_{i\alpha} - f^*_{\alpha i}.$$

Thus, if i and α occur together in a block,

(5.3.16) $$Var(t_i - t_\alpha) = \frac{2}{m\omega}\ 1 + \frac{(m-1)(\omega - \tilde{\omega})}{k[(m-1)\omega + \tilde{\omega}]},$$

and if i and α do not occur together in a block,

(5.3.17) $$Var(t_i - t_\alpha) = \frac{2}{m\omega}\left\{1 + \frac{m(\omega - \tilde{\omega})}{k[(m-1)\omega + \tilde{\omega}]}\right\}.$$

Combined Intrablock and Interblock Analysis for a PBIBD. The parameters in this case are as follows :

v:treatments with association,
b:blocks,
k:block size,
r:replications,

$$\alpha_0 \equiv f_{ii} = \omega C_{ii} + \tilde{\omega}\tilde{C}_{ii} = \omega r\left(1 - \frac{1}{k}\right) + \tilde{\omega}\left(\frac{r}{k} - \frac{r}{v}\right),$$
(5.3.18)
$$\alpha_j \equiv f_{i\alpha} = \omega C_{i\alpha} + \tilde{\omega}\tilde{C}_{i\alpha} = -\frac{\omega\lambda_j}{k} + \tilde{\omega}\left(\frac{\lambda_j}{k} - \frac{r}{v}\right)$$

if i and α are the jth associates. The ith normal equation is

$$f_{i1}t_1 + f_{i2}t_2 + \cdots + f_{iv}t_v = P_i. \tag{5.3.19}$$

The normal equation can be written as

$$(\alpha_0 B_0 + \alpha_1 B_1 + \cdots + \alpha_m B_m)\underline{t} = \underline{P}$$

or

$$\left(\sum_{k=0}^{m} \alpha_k B_k\right)\underline{t} = \underline{P}. \tag{5.3.20}$$

Multiplying (5.3.20) by B_j, we get

$$\left(\sum_{k=0}^{m} \alpha_k B_j B_k\right)\underline{t} = B_j\underline{P},$$

$$\left(\sum_{k=0}^{m} \alpha_k \sum_{i=0}^{m} p^i_{jk} B_i\right)\underline{t} = B_j\underline{P},$$

or

$$\left[\sum_{i=0}^{m}\left(\sum_{k=0}^{m} \alpha_k p^i_{jk}\right)B_i\right]\underline{t} = B_j\underline{P}. \tag{5.3.21}$$

The coefficient of B_0 in (5.3.21) is

$$\sum_{k=0}^{m} \alpha_k p_{jk}^{o} = \alpha_j n_j. \tag{5.3.22}$$

We take the constraint

$$t_1 + t_2 + \cdots + t_v = 0.$$

This is equivalent to

$$(B_0 + B_1 + \cdots + B_m)\underline{t} = 0. \tag{5.3.23}$$

Hence

$$(a_{j1}B_1 + a_{j2}B_2 + \cdots + a_{jm}B_m)\underline{t} = B_j\underline{P}, \tag{5.3.24}$$

where

$$a_{jl} = -\alpha_j n_j + \sum_{k=0}^{m} p_{jk}^{l} \alpha_k \tag{5.3.25}$$

$$= \alpha_0 \delta_{jl} + \alpha_1 p_{j1}^{l} + \alpha_2 p_{j2}^{l} + \cdots + \alpha_j (p_{3j}^{l} - n_j)$$

$$+ \cdots + \alpha_m p_{jm}^{l} \qquad (j = 1,2,\ldots,m).$$

The ith component of the vector $B_j\underline{P}$ is the sum of the treatment effects of the jth associates of the ith treatment. Let us denote it by $t_j(i)$. Then

$$B_j\underline{t} = \begin{bmatrix} t_j(1) \\ t_j(2) \\ \vdots \\ t_j(v) \end{bmatrix},$$

and hence the ith component of the vector equation (5.3.24) is

$$a_{j1}t_1(i) + a_{j2}t_2(i) + \cdots + a_{jm}t_m(i) = P_j(i) \quad (j = 1,\ldots,m), \tag{5.3.26}$$

where $P_j(i)$ is the sum of the P's for all the treatments that are the jth associates of i, and

$$P_i = \omega Q_i + \tilde{\omega} Q_i,$$

$$P_j(i) = \omega Q_j(i) + \tilde{\omega}\tilde{Q}_j(i).$$

In practical use, $m = 2$ or 3. Solving these equations, we get

(5.3.27) $$t_j(i) = a^*_{j1} P_1(i) + \cdots + a^*_{jm} P_m(i).$$

Also the ith component of (5.3.20) is

(5.3.28) $$\alpha_0 t_i + \alpha_1 t_1(i) + \cdots + \alpha_m t_m(i) = P_i.$$

Substituting the values of $t_j(i)$ expressed in terms of $P_i(i)$, we get

(5.3.29) $$\begin{aligned}\alpha_0 \hat{t}_i = P_i &- (\alpha_1 a^*_{11} + \alpha_2 a^*_{21} + \cdots + \alpha_m a^*_m) P_1(i)\\ &- (\alpha_1 a^*_{12} + \alpha_2 a^*_{22} + \cdots + \alpha_m a^*_m) P_2(i)\\ &\cdots\cdots\cdots\cdots\cdots\cdots\\ &- (\alpha_1 a^*_{1m} + \alpha_2 a^*_{2m} + \cdots + \alpha_m a^*_{mm}) P_m(i)\end{aligned}$$

$$= P_i - (\alpha_1 \cdots \alpha_m) \begin{Vmatrix} a^*_{11} & a^*_{12} \cdots & a^*_{1m} \\ a^*_{21} & a^*_{22} \cdots & a^*_{2m} \\ \cdot & \cdots & \cdot \\ a^*_m & a^*_m \cdots & a^*_{mm} \end{Vmatrix} \begin{bmatrix} P_1(i) \\ P_2(i) \\ \vdots \\ P_m(i) \end{bmatrix},$$

(5.3.30) $$\begin{aligned}Var(t_i - t_\alpha) &= f^*_{ii} + f^*_{\alpha\alpha} - f^*_{i\alpha} - f^*_{\alpha i}\\ &= \frac{2}{\alpha_0}(1 + \alpha_1 a^*_{1j} + \alpha_2 a^*_{2j} + \cdots + \alpha_m a^*_{mj})\end{aligned}$$

if i and α are the jth associates.

5.4. Analysis of Variance

In the interblock analysis the underlying model is

$$\underline{y} = \underline{e} + \gamma\underline{j}_n + \Phi\underline{\tau} + \Psi\underline{x},$$

$$Var(\underline{y}) = \sigma^2 I + \Psi\Psi'\sigma_1^2.$$

Since

$$\underline{T} = \Phi'\underline{y} \quad \text{and} \quad \underline{B} = \Phi'\underline{y},$$

we have

$$E(T_{i\cdot}) = r_{i\cdot}(\tau_{i\cdot} + \gamma), \qquad Var(T_{i\cdot}) = r_{i\cdot}(\sigma^2 + \sigma_1^2),$$

$$E(B_{\cdot j}) = k\gamma + n_{1j\cdot}\tau_1 + n_{2j\cdot}\tau_2 + \cdots + n_{vj\cdot}\tau_v,$$

$$Var(B_{\cdot j}) = k(\sigma^2 + k\sigma_1^2),$$

as already calculated.

The correction for the mean is

$$E\left(\frac{G^2}{n}\right) = \frac{1}{n}\left[[E(G)]^2 + Var(G)\right]$$

$$= \frac{1}{n}[(n\gamma + r_1\tau_1 + \cdots + r_v\tau_v)^2 + n(\sigma^2 + k\sigma_1^2)]$$

$$= n(\gamma + \bar{\tau})^2 + (\sigma^2 + k\sigma_1^2),$$

where

$$\bar{\tau} = \frac{r_1\tau_1 + r_2\tau_2 + \cdots + r_v\tau_v}{n}.$$

The sum of squares due to the treatment totals is

$$E\left(\frac{T_{i\cdot}^2}{r_{i\cdot}}\right) = \frac{1}{r_{i\cdot}}[r_{i\cdot}^2(\tau_{i\cdot} + \gamma)^2 + r_{i\cdot}(\sigma^2 + \sigma_1^2) = r_{i\cdot}(\tau_{i\cdot} + \gamma)^2 + (\sigma^2 + \sigma_1^2).$$

The sum of squares due to treatment unadjusted is defined by

$$S_t'^2 = \sum_{i=1}^{v} \frac{T_{i.}^2}{r_i} - \frac{G^2}{n},$$

has the expectation

$$E(S_t'^2) = \sum_{i=1}^{v} r_i(\tau_i + \gamma)^2 + v(\sigma^2 + \sigma_1^2) - n(\gamma + \bar{\tau})^2 - (\sigma^2 + k\sigma_1^2) \tag{5.4.1}$$

$$= \sum_{i=1}^{v} r_i(\tau_i - \bar{\tau})^2 + (v - 1)\sigma^2 + (v - k)\sigma_1^2.$$

Since

$$E\left(\frac{B_{.j}^2}{k}\right) = \frac{1}{k}\left\{[E(B_{.j})]^2 + Var(B_{.j})\right\} - \frac{1}{k}(k\gamma + \tau_1 n_{1j} + \tau_2 n_{2j} + \cdots + \tau_v n_{vj})^2 + \sigma^2 + k\sigma_1^2,$$

the sum of squares due to block unadjusted

$$S_b'^2 = \sum_{j=1}^{b} \frac{B_{.j}^2}{k} - \frac{G^2}{n},$$

has the expectation

$$E(S_b'^2) = \frac{1}{k}\sum_{j=1}^{b} (k\gamma + \tau_1 n_{1j} + \cdots + \tau_v n_{vj})^2 + b\sigma^2 + n\sigma_1^2$$

$$-n(\gamma + \bar{\tau})^2 - (\sigma^2 + k\sigma_1^2)$$

$$= n\gamma^2 + 2\gamma(r_1\tau_1 + \cdots + r_v\tau_v) + \frac{1}{k}\sum_{i,\alpha} \lambda_{i\alpha}\tau_i\tau_\alpha$$

$$- n(\gamma + \bar{\tau})^2 + (b - 1)(\sigma^2 + k\sigma_1^2)$$

$$= \sum_{i=1}^{v} r_i(\tau_i - \bar{\tau})^2 - \left\{\sum_{i=1}^{v} r_i\tau_i^2 - \frac{1}{k}\sum_{i,\alpha} \lambda_{i\alpha}\tau_i\tau_\alpha\right\}$$

$$+ (b - 1)(\sigma^2 + k\sigma_1^2)$$

$$(5.4.2) \qquad = \sum_{i=1}^{v} r_i(\tau_i - \bar{\tau})^2 - \sum_{i,\alpha} c_{i\alpha}\tau_i\tau_\alpha + (b - 1)(\sigma^2 + k\sigma_1^2).$$

Let $\underline{c}'\underline{y}$ be any linear function belonging to error in the intrablock model. The sum of squares corresponding to $\underline{c}'\underline{y}$ is defined as

$$(5.4.3) \qquad \frac{(\underline{c}'\underline{y})^2}{\underline{c}'\underline{c}}$$

irrespective of the model. We want to find the expectation of this quantity in the new model.

The intrablock model corresponding to our design would be

$$E(\underline{y}) = (\underline{j}, \Phi')\begin{bmatrix}\gamma \\ \\ \underline{\tau}\end{bmatrix} + \Psi\underline{b},$$

where

$$\underline{b} = \begin{bmatrix} b_1 \\ \vdots \\ b_b \end{bmatrix}.$$

If a linear form $\underline{c}'\underline{y}$ belongs to error, then $E(\underline{c}'\underline{y}) = 0$ independently of the parameters. This gives us

$$\underline{c}'\underline{j} = 0, \quad \underline{c}'\Phi = 0, \quad \underline{c}'\Psi = 0.$$

This ensures that $E(\underline{c}'\underline{y}) = 0$ in the interblock model as well. Since in the interblock model

$$\underline{y} = \underline{e} + \gamma\underline{j}_n + \Phi\underline{\tau} + \Psi\underline{x},$$

$$E(\underline{y}) = \gamma\underline{j}_n + \Phi\underline{\tau},$$

$$(5.4.3) \qquad E\left[\frac{(\underline{c}'\underline{y})^2}{\underline{c}'\underline{c}}\right] = \frac{1}{\underline{c}'\underline{c}}[(E\underline{c}'\underline{y})^2 + Var(\underline{c}'\underline{y})] = \frac{Var(\underline{c}'\underline{y})}{\underline{c}'\underline{c}} = \frac{\underline{c}'\Sigma\underline{c}}{\underline{c}'\underline{c}},$$

where

$$\Sigma = \begin{Vmatrix} K & & & \\ & K & & \\ & & \ddots & \\ & & & K \end{Vmatrix} \quad \text{and} \quad K = \begin{Vmatrix} \sigma^2 + \sigma_1^2 & \sigma_1^2 & \cdots & \sigma_1^2 \\ \sigma_1^2 & \sigma^2 + \sigma_1^2 & \cdots & \sigma_1^2 \\ \vdots & \vdots & & \vdots \\ \sigma_1^2 & \sigma_1^2 & \cdots & \sigma^2 + \sigma_1^2 \end{Vmatrix}.$$

The condition $\underline{c}'\Psi = 0$ shows that the sum of the first k components of $\underline{c}$ is zero, the sum of the second k components of $\underline{c}$ is zero, and so on. Thus

$$\underline{c}'\Sigma = \sigma^2 \cdot \underline{c}',$$

and therefore

$$\underline{c}'\Sigma\underline{c} = \sigma^2 \cdot \underline{c}'\underline{c}.$$

Hence we get

$$E\left(\frac{(\underline{c}'\underline{y})^2}{\underline{c}'\underline{c}}\right) = \frac{\underline{c}'\Sigma\underline{c}}{\underline{c}'\underline{c}} = \sigma^2. \tag{5.4.4}$$

Hence, even in the new model, the expectation of the sum of squares corresponding to a linear function belonging to error is σ^2. Thus

$$E(S_e{}^2) = n_e\sigma^2, \tag{5.4.5}$$

where n_e is the number of degrees of freedom belonging to error.

Indeed,

$$S_t^2 + S_b'^2 = S_t'^2 + S_b^2,$$

$$\begin{aligned} E(S_b^2) &= E(S_t^2 + S_b'^2 - S_t'^2) = E(S_t^2) + E(S_b'^2) - E(S_t'^2) \\ &= \sum_{i=1}^{v} r_i(\tau_i - \bar{\tau})^2 + (b + v - 2)\sigma^2 + k(b - 1)\sigma_1^2 \\ &\quad - \sum_{i=1}^{v} r_i(\tau_i - \bar{\tau})^2 - (v - 1)\sigma^2 - (v - k)\sigma_1^2 \end{aligned}$$

$$= (b - 1)\sigma^2 + (n - v)\sigma_1^2.$$

$$s_e^{\ 2} = \frac{S_e^{\ 2}}{n - b - v + 1}, \qquad E(s_e^{\ 2}) = \sigma^2,$$

$$s_b^{\ 2} = \frac{S_b^{\ 2}}{b - 1}, \qquad E(s_b^{\ 2}) = \sigma^2 + \frac{n - v}{b - 1}\sigma_1^2.$$

Therefore we get

$$\sigma_1^2 = E\left[\frac{b - 1}{n - v}(s_b^{\ 2} - s_e^{\ 2})\right] \quad \text{and} \quad \sigma^2 = E(s_e^{\ 2}).$$

As estimators we have

$$\hat{\omega} = \frac{1}{s_e^{\ 2}} \quad \text{for} \quad \omega = \frac{1}{\sigma^2},$$

$$\hat{\tilde{\omega}} = \frac{n - v}{(n - k)s_b^{\ 2} - (v - k)s_e^{\ 2}} \quad \text{for} \quad \tilde{\omega} = \frac{1}{\sigma^2 + k\sigma_1^2}.$$

These are not unbiased estimators. Interblock Analysis of Variance Table I may be used for the purpose of the significance test of the treatment effects and Interblock Analysis of Variance Table II may be used for the estimation of the variance component σ_1^2 which is estimated by

$$\frac{b - 1}{n - v}(s_b^{\ 2} - s_e^{\ 2})$$

and consequently

$$\hat{\omega} = 1/s_e^{\ 2}, \qquad \hat{\tilde{\omega}} = (n - v)/[(n - k)s_b^{\ 2} - (v - k)s_e^{\ 2}]$$

as mentioned above.

TABLE 5.1

Interblock Analysis of Variance TABLE I

Sources of variation	Degrees of freedom	Sum of squares	Estimated sum of squares
Treatments adjusted	$v - 1$	$\sum_{i=1}^{v} t_i Q_i$	$\sum c_{i\alpha}\tau_i\tau_\alpha + (v - 1)\sigma^2$
Blocks unadjusted	$b - 1$	$\sum \frac{B_j^2}{k} - \frac{G^2}{n}$	$\sum r_i(\tau_i - \bar{\tau})^2 - \sum c_{i\alpha}\tau_i\tau_\alpha + (b - 1)(\sigma^2 + k\sigma_1^2)$
Error	$n - v - b + 1$	S_e^2	$(n - v - b + 1)\sigma^2$
Total	$n - 1$	$\sum_{f=1}^{n} y_f^2 - \frac{G^2}{n}$	$\sum r_i(\tau_i - \bar{\tau})^2 + (n - 1)\sigma^2 + k(b - 1)\sigma_1^2$

TABLE 5.2

Interblock Analysis of Variance TABLE II

Sources of variation	Degrees of freedom	Sum of squares	Estimated sum of square
Treatments unadjusted	$v - 1$	$S_t'^2 = \sum \frac{T_i^2}{r_i} - \frac{G^2}{n}$	$\sum r_i(\tau_i - \bar{\tau})^2 + (v-1)\sigma^2 + (v-k)\sigma_1^2$
Blocks adjusted	$b - 1$	S_b^2 (by subtraction)	$(b-1)\sigma^2 + (b-v)\sigma_1^2$
Error	$n - v - b + 1$	S_e^2	$(n-v-b+1)\sigma^2$
Total	$n - 1$	$\sum_{f=1}^{n} y_f^2 - \frac{G^2}{n}$	$\sum r_i(\tau_i - \bar{\tau})^2 + (n-1)\sigma^2 + k(b-1)\sigma_1^2$

Chapter 6
RANDOMIZATION OF PARTIALLY BALANCED INCOMPLETE BLOCK DESIGNS

6.1. The Null Distribution of the F-Statistic for Testing a Partial Null Hypothesis in a PBIB Design with m Associate Classes under the Neyman Model before the Randomization

We have treated the randomization procedure starting with the Neyman model in a randomized block design (Section 2.3), a Latin-square design (Sections 2.5 and 2.6), and a balanced incomplete block design (Section 4.5) in a rather heuristic way, that is, by the moment method. We are going to present a mathematically rigorous treatment of the randomization in a PBIB design, so that all former cases follow automatically as its special cases.

We shall be concerned with a PBIB design that has m associate classes, v treatments with an association being defined among them, b blocks of size k each, and r replications of each treatment, the number of incidence of any pair of treatments being λ_u if they are uth associates, $u = 1,2,\ldots,m$.

Let the association matrices be $A_0 = I_v, A_1, \ldots, A_m$ and let their regular representations be $P_0 = I_m, P_1, \ldots, P_m$, respectively, where

$$P_u = \left\| \begin{matrix} p_{0u}^0 & p_{0u}^1 & \cdots & p_{0u}^m \\ p_{1u}^0 & p_{1u}^1 & \cdots & p_{1u}^m \\ \cdot & \cdot & \cdots & \cdot \\ p_{mu}^0 & p_{mu}^1 & \cdots & p_{mu}^m \end{matrix} \right\| \qquad (u = 0,1,2,\ldots,m).$$

Let the characteristic roots of P_u be $z_{0u} = n_u, z_{1u}, \ldots, z_{mu}$. Then there exists a non-singular matrix

$$C = \left\| \begin{matrix} c_{00} & c_{01} & \cdots & c_{0m} \\ c_{10} & c_{11} & \cdots & c_{1m} \\ \cdot & \cdot & \cdots & \cdot \\ c_{m0} & c_{m1} & \cdots & c_{mm} \end{matrix} \right\|, \tag{6.1.2}$$

such that

$$CP_uC^{-1} = \left\| \begin{matrix} z_{0u} & & & 0 \\ & z_{1u} & & \\ & & \ddots & \\ 0 & & & z_{mu} \end{matrix} \right\| \qquad (u = 0,1,2,\ldots,m) \tag{6.1.2}$$

simultaneously. It should be remarked here that

$$n_u > z_{iu} \qquad (i = 1,2,\ldots,m;\ u = 1,2,\ldots,m), \tag{6.1.3}$$

and by the relation

$$P_uP_{u'} = \sum_{k=0}^{m} p_{uu'}^k P_k \tag{6.1.4}$$

we get $\sum_{k=0}^{m} p_{uu'}^{k} z_{ik} = z_{iu} z_{iu'}$ and this can be written in the form

$$\sum_{k=0}^{m} \frac{z_{ik}}{n_k} p_{ku'}^{u} = \frac{z_{iu}}{n_u} z_{iu'}, \tag{6.1.5}$$

because of the relation $n_k p_{uu'}^{k} = n_u p_{ku'}^{u}$. Hence we may put

$$c_{ui} = \frac{z_{ui}}{n_i} \qquad (i,u = 0,1,2,\ldots,m). \tag{6.1.6}$$

The $m + 1$ orthogonal idempotents of the association algebra $\mathfrak{A}$ generated by the association matrices over the field of all real numbers are given by

$$A_u^{\#} = \left(\sum_{i=0}^{m} c_{ui} z_{ui} \right)^{-1} \sum_{i=0}^{m} c_{ui} A_i \quad (u = 0,1,2,\ldots,m), \tag{6.1.7}$$

with the respective ranks $\alpha_0 = 1, \alpha_1, \alpha_2, \ldots, \alpha_m$. By taking the trace of the matrix $A_u^{\#} A_{u'}^{\#}$, we get the relation

$$\sum_{i=0}^{m} \frac{z_{ui} z_{u'i}}{n_i} = \delta_{uu'} \frac{v}{\alpha_u} \qquad (u,u' = 0,1,2,\ldots,m), \tag{6.1.8}$$

where $\delta_{uu'}$ denotes the Kronecker delta. It is also noted that

$$A_0^{\#} = \frac{1}{v} G_v \quad \text{and} \quad \sum_{u=0}^{m} A_u^{\#} = I_v. \tag{6.1.9}$$

If we denote the incidence matrix of the PBIB design under consideration by N, then the spectral decomposition of the matrix NN' is given by

$$NN' = \rho_0 A_0^{\#} + \rho_1 A_1^{\#} + \rho_2 A_2^{\#} + \cdots + \rho_m A_m^{\#}, \tag{6.1.10}$$

where $\rho_0 = rk, \rho_1, \rho_2, \ldots, \rho_m$ are the eigenvalues of NN' with the respective multiplicities $\alpha_0, \alpha_1, \alpha_2, \ldots, \alpha_m$ and

(6.1.11) $$\rho_u = \sum_{i=0}^{m} \lambda_i z_{ui} \qquad (u = 0,1,2,\ldots,m),$$

where we have put $\lambda_0 = r$.

Let the incidence matrices of treatments and blocks be Φ and Ψ respectively. Then the Neyman model assuming no interaction between the treatments and the blocks is given by

(6.1.12) $$\underline{x} = \gamma\underline{1} + \Phi\underline{\tau} + \Psi\underline{\beta} + \underline{\pi} + \underline{e},$$

where $\underline{x}' = (x_1,x_2,\ldots,x_n)$ is the observation vector, $\underline{\tau}' = (\tau_1,\ldots,\tau_v)$ and $\underline{\beta}' = (\beta_1,\ldots,\beta_b)$ are the treatment effects and block effects being subjected to the restrictions

(6.1.13) $$\tau_1 + \cdots + \tau_v = 0 \quad \text{and} \quad \beta_1 + \cdots + \beta_b = 0,$$

respectively, and $\underline{\pi}' = (\pi_1,\pi_2,\ldots,\pi_n)$ stands for the unit effect subject to the restriction

(6.1.14) $$\underline{\Psi}'\pi = \underline{0}.$$

Finally, $\underline{e}' = (e_1,e_2,\ldots,e_n)$ is the technical errors being distributed as $N(\underline{0}',\sigma^2 I)$.

The sum of squares due to treatments adjusted is given by

(6.1.15) $$S_t^2 = \underline{x}'(V_1^{\#} + \cdots + V_m^{\#})\underline{x},$$

where

(6.1.16) $$V_u^{\#} = c_u\left(I - \frac{1}{k}B\right)T_u^{\#}\left(I - \frac{1}{k}B\right)$$

with

$$u = 1,2,\ldots,m.$$

(6.1.17) $$T_u^{\#} = \Phi A_u^{\#}\Phi', \quad B = \Psi\Psi', \quad c_u = \frac{k}{rk - \rho_u}.$$

The sum of squares due to error is

(6.1.18) $$S_e^2 = \underline{x}'(I - V_1^{\#} - \cdots - V_m^{\#})\underline{x}.$$

Now, one is interested in testing a "partial null hypothesis" that some of the hypotheses $A_u^{\#}\underline{\tau} = 0$, $u = 1,2,\ldots,m$ are true. One can take, without any loss of generality, the null hypothesis as

(6.1.19) $$H_{0(h)} : A_u^{\#}\underline{\tau} = 0 \qquad (u = 1,2,\ldots,h;\ h \leqq m).$$

Clearly this hypothesis is equivalent to

(6.1.20) $$\sum_{u=1}^{h} A_u^{\#}\underline{\tau} = \underline{0},$$

and when $h = m$ this reduces to the total null hypothesis

(6.1.21) $$H_0 : \underline{\tau} = \underline{0}.$$

Indeed, since $A_0^{\#} = \frac{1}{v}G_0$, $A_0^{\#}\underline{\tau} = \underline{0}$ due to the restriction imposed on $\underline{\tau}$ and

(6.1.22) $$\sum_{i=0}^{m} A_u^{\#} = I_v,$$

To test the null hypothesis $H_{0(h)}$, we consider the partial sum of squares

(6.1.23) $$S_{t(h)}^2 = \underline{x}'(V_1^{\#} + \cdots + V_h^{\#})\underline{x}$$

instead of S_t^2. Then it follows under the Neyman model that

(6.1.24) $$\begin{aligned} S_{t(h)}^2 &= \underline{\pi}'\left[I - \frac{1}{k}B\right]\Phi(c_1A_1^{\#} + \cdots + c_hA_h^{\#})\Phi'\left[I - \frac{1}{k}B\right]\underline{\pi} \\ &+ 2\underline{\pi}'\left[I - \frac{1}{k}B\right]\Phi(c_1A_1^{\#} + \cdots + c_hA_h^{\#})\Phi'\left[I - \frac{1}{k}B\right]\underline{e} \\ &+ \underline{e}'\left[I - \frac{1}{k}B\right]\Phi(c_1A_1^{\#} + \cdots + c_hA_h^{\#})\Phi'\left[I - \frac{1}{k}B\right]\underline{e} \end{aligned}$$

provided that the null hypothesis $H_{0(h)}$ is true.

Hence the null distribution of the variate

$$\chi_1^2 = \frac{S_{t(h)}^2}{\sigma^2}$$

before randomization is the non-central χ^2-distribution with

$$\bar{\alpha} = \alpha_1 + \alpha_2 + \cdots + \alpha_h \tag{6.1.25}$$

degrees of freedom and with the non-centrality parameter

$$\bar{K}_1 = \underline{\pi}'\Phi(c_1A_1^{\#} + \cdots + c_hA_h^{\#})\Phi'\underline{\pi}/\sigma^2. \tag{6.1.26}$$

Hence the probability element is given by

$$exp\left(-\frac{\bar{K}_1}{2}\right) \sum_{\mu=0}^{\infty} \frac{(\bar{K}_1/2)^{\mu}}{\mu!} \frac{(\chi_1^2/2)^{(\bar{\alpha}/2)+\mu-1}}{\Gamma[(\bar{\alpha}/2) + \mu]} exp\left(-\frac{\chi_1^2}{2}\right) d\left(\frac{\chi_1^2}{2}\right). \tag{6.1.27}$$

The sum of squares due to error, S_e^2, becomes

$$\begin{aligned} S_e^2 = {} & \underline{\pi}'\left(I - \frac{1}{k}B\right)\left[I - \Phi(c_1A_1^{\#} + \cdots + c_mA_m^{\#})\Phi'\right]\left(I - \frac{1}{k}B\right)\underline{\pi} \\ & + 2\underline{\pi}'\left(I - \frac{1}{k}B\right)\left[I - \Phi(c_1A_1^{\#} + \cdots + c_mA_m^{\#})\Phi'\right]\left(I - \frac{1}{k}B\right)\underline{e} \\ & + \underline{e}'\left(I - \frac{1}{k}B\right)\left[I - \Phi(c_1A_1^{\#} + \cdots + c_mA_m^{\#})\Phi'\right]\left(I - \frac{1}{k}B\right)\underline{e} \end{aligned} \tag{6.1.28}$$

independently of $H_{0(h)}$. The distribution of the variate

$$\chi_2^2 = \frac{S_e^2}{\sigma^2}$$

before randomization is the non-central χ^2-distribution with $n - b - v + 1$ degrees of freedom and with the non-centrality parameter

$$(6.1.29) \qquad K_2 = \underline{\pi}'\left[I - \Phi(c_1A_1^{\#} + \cdots + c_mA_m^{\#})\Phi'\right]\underline{\pi}/\sigma^2$$

$$= \Delta/\sigma^2 - \bar{K}_1 - \bar{\bar{K}}_1,$$

where

$$(6.1.30) \qquad \Delta = \underline{\pi}'\underline{\pi}$$

and

$$(6.1.31) \qquad \bar{\bar{K}}_1 = \underline{\pi}'\Phi(c_{h+1}A_{h+1}^{\#} + \cdots + c_mA_m^{\#})\Phi'\underline{\pi}/\sigma^2.$$

The probability element of χ_2^2 is given by

$$(6.1.32) \qquad exp\left(-\frac{K_2}{2}\right)\sum_{\nu=0}^{\infty}\frac{(K_2/2)^{\nu}}{\nu!}\frac{(\chi_2^2/2)^{[(n-b-v+1)/2]+\nu-1}}{\Gamma\{[(n-v-v+1)/2]+\nu\}}exp\left(-\frac{\chi_2^2}{2}\right)d\left(\frac{\chi_2^2}{2}\right)$$

The variates χ_1^2 and χ_2^2 are mutually independent for the following reason : since

$$(6.1.33) \qquad \left(I - \frac{1}{k}B\right)\Phi A_u^{\#}\Phi'\left(I - \frac{1}{k}B\right)\left(I - \frac{1}{k}B\right)\Phi A_{u'}^{\#}\Phi'\left(I - \frac{1}{k}B\right)$$

$$= \left(I - \frac{1}{k}B\right)\Phi A_u^{\#}\Phi'\left(I - \frac{1}{k}\Psi\Psi'\right)\Phi A_{u'}^{\#}\Phi'\left(I - \frac{1}{k}B\right)$$

$$= \left(I - \frac{1}{k}B\right)\Phi A_u^{\#}\left(rI_v - \frac{1}{k}NN'\right)A_{u'}^{\#}\Phi'\left(I - \frac{1}{k}B\right)$$

$$= \left(I - \frac{1}{k}B\right)\Phi A_u\left(\sum_{i=1}^{m}\frac{rk - \rho_i}{k}A_i^{\#}\right)A_{u'}^{\#}\Phi'\left(I - \frac{1}{k}B\right)$$

$$= \frac{rk - \rho_u}{k}\delta_{uu'}\left(I - \frac{1}{k}B\right)\Phi A_u^{\#}\Phi'\left(I - \frac{1}{k}B\right),$$

one can easily confirm the orthogonality of the coefficient matrices. Hence the null distribution of the F-statistic

$$F = \frac{n - b - v + 1}{\bar{\alpha}} \frac{S_{t(h)}^2}{S_e^2} \tag{6.1.34}$$

before randomization is the noncentral F-distribution, whose probability element is given by

$$\begin{aligned} & exp\left(-\frac{\bar{K}_1 + K_2}{2}\right) \sum_{\mu,\nu=0}^{\infty} \frac{(\bar{K}_1/2)^{\mu}(K_2/2)^{\nu}}{\mu!\nu!} \\ & \times \frac{\Gamma\{[(n - b - \bar{\bar{\alpha}})/2] + \mu + \nu\}}{\Gamma[(\bar{\alpha}/2) + \mu]\Gamma\{[(n - b - v + 1)/2] + \nu\}} \\ & \times \left(\frac{\bar{\alpha}}{n - b - v + 1}F\right)^{(\bar{\alpha}/2)+\mu-1} \\ & \times \left(1 + \frac{\bar{\alpha}}{n - b - v + 1}F\right)^{-[(n-b-\bar{\bar{\alpha}})/2]} d\left(\frac{\bar{\alpha}}{n - b - v + 1}F\right). \end{aligned} \tag{6.1.35}$$

where

$$\bar{\bar{\alpha}} = \alpha_{h+1} + \cdots + \alpha_m. \tag{6.1.36}$$

If we put

$$\bar{\theta} = \Delta^{-1} \cdot \underline{\pi}'\Phi(c_1A_1^{\#} + \cdots + c_hA_h^{\#})\Phi'\underline{\pi} \tag{6.1.37}$$

and

$$\bar{\bar{\theta}} = \Delta^{-1} \cdot \underline{\pi}'\Phi(c_{h+1}A_{h+1}^{\#} + \cdots + c_mA_m^{\#})\Phi'\underline{\pi}, \tag{6.1.38}$$

then the probability element of F can be written as

(6.1.39)

$$exp\left(-\frac{\Delta}{2\sigma^2}\right) \sum_{\ell=0}^{\infty} \frac{[\Delta/2\sigma^2]}{\ell!} \sum_{\mu+\nu+\gamma=\ell} \frac{\ell!}{\mu!\nu!\gamma!} \bar{\theta}^{\mu}\bar{\bar{\theta}}^{\gamma}(1 - \bar{\theta} - \bar{\bar{\theta}})^{\nu}$$

$$\times \frac{\Gamma\{[(n - b - \bar{\bar{\alpha}})/2] + \mu + \nu\}}{\Gamma[(\bar{\alpha}/2) + \mu]\Gamma\{[(n - b - v + 1)/2] + \nu\}}$$

$$\times \left(\frac{\bar{\alpha}}{n - b - v + 1}F\right)^{(\bar{\alpha}/2)+\mu-1}$$

$$\times \left(1 + \frac{\bar{\alpha}}{n - b - v + 1}F\right)^{-[(n-b-\bar{\bar{\alpha}})/2]-\mu-\nu} d\left(\frac{\bar{\alpha}}{n - b - v + 1}F\right).$$

The null distribution of the F-statistic after randomization should be obtained by taking the average of the above expression with respect to the permutation distribution of $(\bar{\theta},\bar{\bar{\theta}})$ due to randomization, that is,

(6.1.40)
$$\frac{(n - b - \bar{\bar{\alpha}})/2}{\Gamma(\bar{\alpha}/2)\Gamma[(n - b - v + 1)/2]} \left(\frac{\bar{\alpha}}{n - b - v + 1}F\right)^{(\bar{\alpha}/2)-1}$$

$$\times \left(1 + \frac{\bar{\alpha}}{n - b - v + 1}F\right)^{(n-b-\bar{\bar{\alpha}})/2} d\left(\frac{\bar{\alpha}}{n - b - v + 1}F\right)$$

$$\times\ exp\left(- \frac{\Delta}{2\sigma^2}\right) \sum_{\ell=0}^{\infty} \frac{(\Delta/2\sigma^2)^{\ell}}{\ell!} \sum_{\mu+\nu+\gamma=\ell} \frac{\ell!}{\mu!\nu!\gamma!}$$

$$\times\ E\left[\bar{\theta}^{\mu}\bar{\bar{\theta}}^{\gamma}(1 - \bar{\theta} - \bar{\bar{\theta}})^{\nu}\right]\left(1 + \frac{\bar{\alpha}}{n - b - v + 1}F\right)^{-(\mu+\nu)}$$

$$\times \left(\frac{\bar{\alpha}}{n - b - v + 1}F\right)^{\mu}$$

$$\times \frac{\Gamma\{[(n - b - \bar{\bar{\alpha}})/2] + \mu + \nu\}\Gamma(\bar{\alpha}/2)\Gamma[(n - b - v + 1)/2]}{\Gamma[(\bar{\alpha}/2) + \mu]\Gamma\{[(n - b - v + 1)/2] + \nu\}\Gamma[(n - b - \bar{\bar{\alpha}})/2]}$$

where the operator E stands for the expectation with respect to the permutation distribution of $(\bar{\theta},\bar{\bar{\theta}})$ due to randomization.

6.2. Calculation of the Exact Moments of $\bar{\theta}$ and $\bar{\bar{\theta}}$ with respect to the Permutation Distribution Due to Randomization

Necessary Notations. Let us put

$$
\begin{aligned}
\bar{T}^{\#} &= c_1 T_1^{\#} + \cdots + c_h T_h^{\#} = \Phi(c_1 A_1^{\#} + \cdots + c_h A_h^{\#})\Phi' \\
&= \bar{\mu}_0 T_0 + \bar{\mu}_1 T_1 + \cdots + \bar{\mu}_m T_m \\
\bar{\bar{T}}^{\#} &= c_{h+1} T_{h+1}^{\#} + \cdots + c_m T_m^{\#} = \Phi(c_{h+1} A_{h+1}^{\#} + \cdots + c_m A_m^{\#})\Phi' \\
&= \bar{\bar{\mu}}_0 T_0 + \bar{\bar{\mu}}_1 T_1 + \cdots + \bar{\bar{\mu}}_m T_m,
\end{aligned}
\tag{6.2.1}
$$

where

$$
\bar{\mu}_\alpha = \sum_{i=1}^{h} \mu_{i\alpha} \quad \text{and} \quad \bar{\bar{\mu}}_\alpha = \sum_{i=h+1}^{m} \mu_{i\alpha}
\tag{6.2.2}
$$

with

$$
\mu_{i\alpha} = \frac{\alpha_i c_i z_{i\alpha}}{v n_\alpha} \qquad (i = 1,2,\ldots,m; \quad \alpha = 0,1,\ldots,m).
\tag{6.2.3}
$$

Numbering all the experimental units from 1 through n in such a way that the ith unit in the pth block bears the number

$$f = (p - 1)k + i,$$

let us put

$$
T_\alpha = \left\| T_{pq}^{(\alpha)} \right\|_{(p,q=1,\ldots,b)}, \quad T_{pq}^{(\alpha)} = \left\| t_{ij}^{(\alpha)pq} \right\|_{(i,j=1,\ldots,k)},
\tag{6.2.4}
$$

where

$$
t_{ij}^{(\alpha)pq} = \begin{cases} 1, \text{if the } f = [(p - 1)k + i]\text{th and the } f' = [(q - 1)k + j]\text{th units receive the treatments that are } \alpha\text{th associates,} \\ 0, \text{otherwise.} \end{cases}
\tag{6.2.5}
$$

Clearly

$$t_{ij}^{(\alpha)pq} = t_{ji}^{(\alpha)qp}, \quad t_{ij}^{(0)pp} = \delta_{ij},$$

$$t_{ij}^{(0)pq} + t_{ij}^{(1)pq} + \cdots + t_{ij}^{(m)pq} = 1. \tag{6.2.6}$$

If we put

$$\bar{T}^{\#} = \left\|\bar{T}_{pq}^{\#}\right\|_{(p,q=1,\ldots,b)}, \quad \bar{T}_{pq}^{\#} = \left\|\bar{t}_{ij}^{\#pq}\right\|_{(i,j=1,\ldots,k)}, \tag{6.2.7}$$

and

$$\bar{\bar{T}}^{\#} = \left\|\bar{\bar{T}}_{pq}^{\#}\right\|_{(p,q=1,\ldots,b)}, \quad \bar{\bar{T}}_{pq}^{\#} = \left\|\bar{\bar{t}}_{ij}^{\#pq}\right\|_{(i,j=1,\ldots,k)}, \tag{6.2.8}$$

then

$$\begin{aligned} \bar{t}_{ij}^{\#pq} &= \bar{\mu}_0 t_{ij}^{(0)pq} + \bar{\mu}_1 t_{ij}^{(1)pq} + \cdots + \bar{\mu}_m t_{ij}^{(m)pq}, \\ \bar{\bar{t}}_{ij}^{\#pq} &= \bar{\bar{\mu}}_0 t_{ij}^{(0)pq} + \bar{\bar{\mu}}_1 t_{ij}^{(1)pq} + \cdots + \bar{\bar{\mu}}_m t_{ij}^{(m)pq}. \end{aligned} \tag{6.2.9}$$

Further let us put

$$\pi_f = \pi_i^{(p)} \quad \text{if} \quad f = (p - 1)k + i \tag{6.2.10}$$

and

$$\underline{\pi}^{(p)\prime} = \left[\pi_1^{(p)}, \ldots, \pi_k^{(p)}\right],$$

$$\Delta_p = \underline{\pi}^{(p)\prime}\,\underline{\pi}^{(p)}, \quad \Gamma_p = \sum_{i=1}^{k} \pi_i^{(p)^4} \qquad (p = 1,\ldots,b).$$

Other quantities that will appear in calculations in this section are as follows for the case when $k \geqq 4$:

$$\lambda_{pp}^{(1)\alpha\alpha} = \sum_{i\neq j} t_{ij}^{(\alpha)pp} \qquad (\alpha = 0,1,\ldots,m), \tag{6.2.12}$$

$$\lambda_{pp}^{(2)\alpha\beta} = \sum_{i\neq j\neq l} t_{ij}^{(\alpha)pp} t_{il}^{(\beta)pp} \qquad (\alpha,\beta = 0,1,\ldots,m), \tag{6.2.13}$$

(6.2.14) $$\lambda_{pp}^{(3)\alpha\beta} = \sum_{i\neq j\neq l\neq s} t_{ij}^{(\alpha)pp} t_{ls}^{(\beta)pp} \qquad (\alpha,\beta = 0,1,\ldots,m),$$

(6.2.15) $$\lambda_{pp}^{(4)\alpha\alpha} = \sum_{i,j} t_{ij}^{(\alpha)pq} = \sum_{i} t_{ii}^{(\alpha)pq} + \sum_{i\neq j} t_{ij}^{(\alpha)pq} \quad (\alpha,\beta = 0,1,\ldots,m),$$

(6.2.16) $$\lambda_{pq}^{(5)\alpha\beta} = \sum_{i\neq j} t_{ii}^{(\alpha)pq} t_{jj}^{(\beta)pq} + \sum_{i\neq j} t_{ij}^{(\alpha)pq} t_{ji}^{(\beta)pq}$$
$$+ \sum_{i\neq j\neq l} t_{ii}^{(\alpha)pq} t_{jl}^{(\beta)pq} + \sum_{i\neq j\neq l} t_{jl}^{(\alpha)pq} t_{ii}^{(\beta)pq}$$
$$+ \sum_{i\neq j\neq l} t_{ij}^{(\alpha)pq} t_{li}^{(\beta)pq} + \sum_{i\neq j\neq l} t_{li}^{(\alpha)pq} t_{ij}^{(\beta)pq}$$
$$+ \sum_{i\neq j\neq l\neq s} t_{ij}^{(\alpha)pq} t_{ls}^{(\beta)pq} \qquad (\alpha,\beta = 0,1,\ldots,m),$$

(6.2.17) $$2\lambda_{pq}^{(6)\alpha\beta} = \sum_{i\neq j} t_{ii}^{(\alpha)pq} t_{ij}^{(\beta)pq} + \sum_{i\neq j} t_{ii}^{(\alpha)pq} t_{ji}^{(\beta)pq}$$
$$+ \sum_{i\neq j\neq l} t_{ij}^{(\alpha)pq} t_{il}^{(\beta)pq} + \sum_{i\neq j} t_{ij}^{(\alpha)pq} t_{ii}^{(\beta)pq}$$
$$+ \sum_{i\neq j} t_{ji}^{(\alpha)pq} t_{ii}^{(\beta)pq} + \sum_{i\neq j\neq l} t_{ij}^{(\alpha)pq} t_{lj}^{(\beta)pq}.$$

<u>The Means of $\bar{\theta}$ and $\bar{\bar{\theta}}$.</u> Since

$$\Delta\bar{\theta} = \underline{\pi}'\bar{T}^{\#}\underline{\pi} = \sum_{\alpha=0}^{m} \bar{\mu}_{\alpha}\underline{\pi}'T_{\alpha}\underline{\pi},$$

(6.2.18) $$\underline{\pi}'T_{\alpha}\underline{\pi} = \sum_{p,q=1}^{b} \underline{\pi}^{(p)'}T_{pq}^{(\alpha)}\underline{\pi}^{(q)},$$

$$\underline{\pi}^{(p)'}T_{pq}^{(\alpha)}\underline{\pi}^{(q)} = \sum_{i,j=1}^{k} t_{ij}^{(\alpha)pq}\pi_i^{(p)}\pi_j^{(q)},$$

and

$$E\left[\underline{\pi}^{(p)'}T_{pq}^{(\alpha)}\underline{\pi}^{(q)}\right] = \frac{1}{(k!)^2}\sum_{i,j=1}^{k} t_{ij}^{(\alpha)pq}$$

$$\times \sum_{\sigma}\sum_{\tau}\pi_{\sigma(i)}^{(p)}\pi_{\tau(j)}^{(q)} = 0 \quad \text{if} \quad p \neq q,$$

(6.2.19)

$$E\left[\underline{\pi}^{(p)'}T_{pp}^{(\alpha)}\underline{\pi}^{(p)}\right] = \frac{1}{k!}\sum_{i,j} t_{ij}^{(\alpha)pp}\sum_{\sigma}\pi_{\sigma(i)}^{(p)}\pi_{\sigma(j)}^{(p)}$$

$$= \frac{\Delta_p}{k}\sum_{i=1}^{k} t_{ii}^{(\alpha)pp} - \frac{\Delta_p}{k(k-1)}\sum_{i\neq j} t_{ij}^{(\alpha)pp},$$

we obtain

$$\Delta E(\bar{\theta}) = \sum_{\alpha=0}^{m}\bar{\mu}_\alpha E(\underline{\pi}'T_\alpha\underline{\pi}) = \sum_{\alpha=0}^{m}\bar{\mu}_\alpha\sum_{p,q} E\left[\underline{\pi}^{(p)'}T_{pq}^{(\alpha)}\underline{\pi}^{(q)}\right]$$

(6.2.20)

$$= \sum_{\alpha=0}^{m}\bar{\mu}_\alpha\sum_{p=1}^{b} E\left[\underline{\pi}^{(p)'}T_{pp}^{(\alpha)}\underline{\pi}^{(p)}\right]$$

$$= \sum_{\alpha=0}^{m}\bar{\mu}_\alpha\sum_{p=1}^{b}\left[\frac{\Delta_p}{k}\sum_{i=1}^{k} t_{ii}^{(\alpha)pp} - \frac{\Delta_p}{k(k-1)}\sum_{i\neq j} t_{ij}^{(\alpha)pp}\right].$$

Taking into account the facts that

(6.2.21) $$t_{ii}^{(0)pp} = 1 \quad \text{and} \quad t_{ii}^{(\alpha)pp} = 0 \qquad (\alpha = 1,2,\ldots,m),$$

we get

(6.2.22) $$\Delta E(\bar{\theta}) = \sum_{p=1}^{b}\Delta_p\left[\bar{\mu}_0 - \frac{1}{k(k-1)}\sum_{\alpha=0}^{m}\bar{\mu}_\alpha\lambda_{pp}^{(1)\alpha\alpha}\right].$$

Since

$$\sum_{\alpha=0}^{m} t_{ij}^{(\alpha)pp} = 1 \quad \text{for all } i,j$$

and

$$\sum_{\alpha=0}^{m}\lambda_{pp}^{(1)\alpha\alpha} = k(k-1),$$

we have

$$(6.2.23)\qquad E(\bar{\theta}) = \frac{1}{\Delta k(k-1)} \sum_{p=1}^{b} {}_p\Delta \sum_{\alpha=1}^{m} (\bar{\mu}_0 - \bar{\mu}_\alpha)\lambda_{pp}^{(1)\alpha\alpha}.$$

In a completely similar manner we obtain

$$(6.2.24)\qquad E(\bar{\bar{\theta}}) = \frac{1}{\Delta k(k-1)} \sum_{p=1}^{b} \Delta_p \sum_{\alpha=1}^{m} (\bar{\bar{\mu}}_0 - \bar{\bar{\mu}}_\alpha)_{pp}^{(1)\alpha\alpha}.$$

Variances of $\bar{\theta}$ and $\bar{\bar{\theta}}$. Now

$$(6.2.25)\qquad \Delta^2 E(\bar{\theta}^2) = E(\underline{\pi}'\bar{T}^{\#}\underline{\pi})^2$$

$$= \frac{1}{(k!)^b} \sum_{\sigma_1} \cdots \sum_{\sigma_b} \left[\sum_{p,q} \underline{\pi}^{(p)'} S_p' \bar{T}_{pq}^{\#} S_q \underline{\pi}^{(q)}\right]^2$$

$$= \frac{1}{(k!)^b} \sum_{\sigma_1 \cdots \sigma_b} \left[\sum_p \underline{\pi}^{(p)'} S_{\sigma_p}' \bar{T}_{pp}^{\#} S_{\sigma_p} \underline{\pi}^{(p)} + \sum_{p \neq q} \underline{\pi}^{(p)'} S_{\sigma_p}' \bar{T}_{pq}^{\#} S_{\sigma_q} \underline{\pi}^{(q)}\right]^2.$$

After the expansion, one puts the right-hand side of (6.2.25) in the following form:

$$(6.2.26)\qquad \Delta^2 E(\bar{\theta}^2) = E(\bar{A}) + E(\bar{B}) + 2E(\bar{C}),$$

where

$$\bar{A} = \sum_{p=1}^{b} \left[\underline{\pi}^{(p)'} S_{\sigma_p}' \bar{T}_{pp}^{\#} S_{\sigma_p} \underline{\pi}^{(p)}\right]^2,$$

$$(6.2.27)\qquad \bar{B} = \sum_{p \neq q} \underline{\pi}^{(p)'} S_{\sigma_p}' \bar{T}_{pp}^{\#} S_{\sigma_p} \underline{\pi}^{(p)} \underline{\pi}^{(q)'} S_{\sigma_q}' \bar{T}_{qq}^{\#} S_{\sigma_q} \underline{\pi}^{(q)},$$

$$\bar{C} = \sum_{p \neq q} \left[\underline{\pi}^{(p)'} S_{\sigma_p}' \bar{T}_{pq}^{\#} S_{\sigma_q} \underline{\pi}^{(q)}\right]^2.$$

Because all such other terms as

$$\underline{\pi}^{(p)\prime} S'_{\sigma_p} \bar{T}^{\#}_{pp} S_{\sigma_p} \underline{\pi}^{(p)} \underline{\pi}^{(p)\prime} S'_{\sigma_p} \bar{T}^{\#}_{pq} S_{\sigma_r} \underline{\pi}^{(r)},$$

$$\underline{\pi}^{(p)\prime} S'_{\sigma_p} \bar{T}^{\#}_{pq} S_{\sigma_q} \underline{\pi}^{(q)} \underline{\pi}^{(p)\prime} S'_{\sigma_p} \bar{T}^{\#}_{pr} S_{\sigma_r} \underline{\pi}^{(r)},$$

$$\underline{\pi}^{(p)\prime} S'_{\sigma_p} \bar{T}^{\#}_{pq} S_{\sigma_q} \underline{\pi}^{(q)} \underline{\pi}^{(r)\prime} S'_{\sigma_r} \bar{T}^{\#}_{rs} S_{\sigma_s} \underline{\pi}^{(s)},$$

vanish after taking their expectations. Only square terms such as

$$\left[\underline{\pi}^{(p)\prime} S'_{\sigma_p} \bar{T}^{\#}_{pp} S_{\sigma_p} \underline{\pi}^{(p)}\right]^2 = \left[\sum_i \bar{t}^{\#pp}_{ii} \pi^{(p)2}_{\sigma(i)} + \sum_{i \neq j} \bar{t}^{\#pp}_{ij} \pi^{(p)}_{\sigma(i)} \pi^{(p)}_{\sigma(j)}\right]^2$$

$$= \sum_i \bar{t}^{\#pp\,2}_{ii} \pi^{(p)4}_{\sigma(i)} + \sum_{i \neq j} \bar{t}^{\#pp\,2}_{ij} \pi^{(p)2}_{\sigma(i)} \pi^{(p)2}_{\sigma(j)} + \sum_{i \neq j} \bar{t}^{\#pp}_{ii} \bar{t}^{\#pp}_{jj} \pi^{(p)2}_{\sigma(i)} \pi^{(p)2}_{\sigma(j)}$$

$$+ 2 \sum_{i \neq j} \left(\bar{t}^{\#pp}_{ii} \bar{t}^{\#pp}_{ij} + \bar{t}^{\#pp}_{ii} \bar{t}^{\#pp}_{ji}\right) \pi^{(p)3}_{\sigma(i)} \pi^{(p)}_{\sigma(j)}$$

(6.2.28)
$$+ 2 \sum_{i \neq j \neq l} \bar{t}^{\#pp}_{ii} \bar{t}^{\#pp}_{jl} \pi^{(p)2}_{\sigma(i)} \pi^{(p)}_{\sigma(j)} \pi^{(p)}_{\sigma(l)} + \sum_{i \neq j} \bar{t}^{\#pp}_{ij} \bar{t}^{\#pp}_{ji} \pi^{(p)2}_{\sigma(i)} \pi^{(p)2}_{\sigma(j)}$$

$$+ \sum_{i \neq j \neq l} \left(\bar{t}^{\#pp}_{ij} \bar{t}^{\#pp}_{il} + \bar{t}^{\#pp}_{ij} \bar{t}^{\#pp}_{li}\right) \pi^{(p)2}_{\sigma(i)} \pi^{(p)}_{\sigma(j)} \pi^{(p)}_{\sigma(l)}$$

$$+ \sum_{i \neq j \neq l} \left(\bar{t}^{\#pp}_{ij} \bar{t}^{\#pp}_{li} + \bar{t}^{\#pp}_{ij} \bar{t}^{\#pp}_{jl}\right) \pi^{(p)}_{\sigma(i)} \pi^{(p)2}_{\sigma(j)} \pi^{(p)}_{\sigma(l)}$$

$$+ \sum_{i \neq j \neq l \neq s} \bar{t}^{\#pp}_{ij} \bar{t}^{\#pp}_{ls} \pi^{(p)}_{\sigma(i)} \pi^{(p)}_{\sigma(j)} \pi^{(p)}_{\sigma(l)} \pi^{(p)}_{\sigma(s)}.$$

should be calculated. Since

$$E\left[\pi^{(p)4}_{\sigma(i)}\right] = \frac{1}{k} \Gamma_p,$$

$$E\left[\pi_{\sigma(i)}^{(p)^3}\pi_{\sigma(j)}^{(p)}\right] = \frac{-1}{k(k-1)}\Gamma_p,$$

$$E\left[\pi_{(i)}^{(p)^2}\pi_{(j)}^{(p)}\pi_{(l)}^{(p)}\right] = E\left[\pi_{\sigma(i)}^{(p)}\pi_{\sigma(j)}^{(p)^2}\pi_{\sigma(l)}^{(p)}\right]$$

$$= \frac{1}{k(k-1)(k-2)}\left(2\Gamma_p - \Delta_p^2\right),$$

$$E\left[\pi_{\sigma(i)}^{(p)^2}\pi_{\sigma(j)}^{(p)^2}\right] = \frac{1}{k(k-1)}\left(\Delta_p^2 - \Gamma_p\right),$$

$$E\left[\pi_{\sigma(i)}^{(p)}\pi_{\sigma(j)}^{(p)}\pi_{\sigma(l)}^{(p)}\pi_{\sigma(s)}^{(p)}\right] = \frac{-3}{k(k-1)(k-2)(k-3)}\left(2\Gamma_p - \Delta_p^2\right),$$

and

$$\sum_i \bar{t}_{ii}^{\#pp^2} = \sum_{\alpha,\beta}\sum_i t_{ii}^{(\alpha)pp}t_{ii}^{(\beta)pp}\bar{\mu}_\alpha\bar{\mu}_\beta = k\bar{\mu}_0^2,$$

$$\sum_{i\neq j} \bar{t}_{ij}^{\#pp^2} = \sum_{\alpha,\beta}\sum_{i\neq j} t_{ij}^{(\alpha)pp}t_{ij}^{(\beta)pp}\bar{\mu}_\alpha\bar{\mu}_\beta = \sum_{\alpha=1}^{m}\lambda_{pp}^{(1)\alpha\alpha}\bar{\mu}_\alpha^2,$$

$$\sum_{i\neq j} \bar{t}_{ii}^{\#pp}\bar{t}_{jj}^{\#pp} = \sum_{\alpha,\beta}\sum_{i\neq j} t_{ii}^{(\alpha)pp}t_{jj}^{(\beta)pp}\bar{\mu}_\alpha\bar{\mu}_\beta$$

$$= \sum_{i\neq j} t_{ii}^{(0)pp}t_{jj}^{(0)pp}\bar{\mu}_0^2 = k(k-1)\bar{\mu}_0^2,$$

$$\sum_{i\neq j} \bar{t}_{ii}^{\#pp}\bar{t}_{ij}^{\#pp} = \sum_{i\neq j} \bar{t}_{ii}^{\#pp}\bar{t}_{ji}^{\#pp} = \sum_{\alpha,\beta}\sum_{i\neq j} t_{ii}^{(\alpha)pp}t_{ij}^{(\beta)pp}\bar{\mu}_\alpha\bar{\mu}_\beta$$

$$= \bar{\mu}_0\sum_{\beta=1}^{m}\lambda_{pp}^{(1)\beta\beta}\bar{\mu}_\beta^2,$$

$$\sum_{i\neq j} \bar{t}_{ij}^{\#pp}\bar{t}_{ji}^{\#pp} = \sum_{\alpha=1}^{m}\lambda_{pp}^{(1)\alpha\alpha}\bar{\mu}_\alpha^2,$$

$$\sum_{i\neq j\neq l} \bar{t}_{ii}^{\#pp}\bar{t}_{jl}^{\#pp} = (k-2)\sum_{\beta=1}^{m} \lambda_{pp}^{(1)\beta\beta}\bar{\mu}_0\bar{\mu}_\beta,$$

$$\sum_{i\neq j\neq l} \bar{t}_{ij}^{\#pp}\bar{t}_{il}^{\#pp} = \sum_{i\neq j\neq l} \bar{t}_{ij}^{\#pp}\bar{t}_{li}^{\#pp} = \sum_{\alpha,\beta=1}^{m} \lambda_{pp}^{(2)\alpha\beta}\bar{\mu}_\alpha\bar{\mu}_\beta,$$

$$\sum_{i\neq j\neq l} \bar{t}_{ij}^{\#pp}\bar{t}_{lj}^{\#pp} = \sum_{i\neq j\neq l} \bar{t}_{ij}^{\#pp}\bar{t}_{jl}^{\#pp} = \sum_{\alpha,\beta=1}^{m} \lambda_{pp}^{(2)\alpha\beta}\bar{\mu}_\alpha\bar{\mu}_\beta,$$

$$\sum_{i\neq j\neq l\neq s} \bar{t}_{ij}^{\#pp}\bar{t}_{ls}^{\#pp} = \sum_{\alpha,\beta=1}^{m} \lambda_{pp}^{(3)\alpha\beta}\bar{\mu}_\alpha\bar{\mu}_\beta,$$

it follows that

$$E\left[\underline{\pi}^{(p)'}S'_{\sigma_p}\bar{T}^{\#}_{pp}S_{\sigma_p}\underline{\pi}^{(p)}\right]^2 = \Gamma_p\bar{\mu}_0^2 + \frac{2\left(\Delta_p^2-\Gamma_p\right)}{k(k-1)}\sum_{\alpha=1}^{m}\lambda_{pp}^{(1)\alpha\alpha}\bar{\mu}_\alpha^2 + \left(\Delta_p^2-\Gamma_p\right)\bar{\mu}_0^2$$

$$- \frac{4\Gamma_p}{k(k-1)}\sum_{\alpha=1}^{m}\lambda_{pp}^{(1)\alpha\alpha}\bar{\mu}_0\bar{\mu}_\alpha + \frac{2\left(2\Gamma_p-\Delta_p^2\right)}{k(k-1)}\sum_{\beta=1}^{m}\lambda_{pp}^{(1)\beta\beta}\bar{\mu}_0\bar{\mu}_\beta$$

$$+ \frac{4\left(2\Gamma_p-\Delta_p^2\right)}{k(k-1)(k-2)}\sum_{\alpha,\beta=1}^{m}\lambda_{pp}^{(2)\alpha\beta}\bar{\mu}_\alpha\bar{\mu}_\beta - \frac{3\left(2\Gamma_p-\Delta_p^2\right)}{k(k-1)(k-2)(k-3)}$$

$$\times \sum_{\alpha,\beta=1}^{m}\lambda_{pp}^{(3)\alpha\beta}\bar{\mu}_\alpha\bar{\mu}_\beta = \frac{\Delta_p^2}{k(k-1)}\sum_{\alpha=1}^{m}\left(\bar{\mu}_0-\bar{\mu}_\alpha\right)^2\lambda_{pp}^{(1)\alpha\alpha} + \left(\Delta_p^2 - 2\Gamma_p\right)$$

$$\times \left[\frac{1}{k(k-1)}\sum_{\alpha=1}^{m}\lambda_{pp}^{(1)\alpha\alpha}\bar{\mu}_\alpha^2 - \frac{4}{k(k-1)(k-2)}\sum_{\alpha,\beta=1}^{m}\lambda_{pp}^{(2)\alpha\beta}\bar{\mu}_\alpha\bar{\mu}_\beta\right.$$

$$\left. + \frac{3}{k(k-1)(k-2)(k-3)}\sum_{\alpha,\beta=1}^{m}\lambda_{pp}^{(3)\alpha\beta}\bar{\mu}_\alpha\bar{\mu}_\beta\right].$$

Hence one obtains

$$\text{(6.2.29)}\quad E(\bar{A}) = \sum_{p=1}^{b}\left\{\frac{\Delta_p^2}{k(k-1)}\sum_{\alpha}\left(\bar{\mu}_0-\bar{\mu}_\alpha\right)^2\lambda_{pp}^{(1)\alpha\alpha} + \left(\Delta_p^2 - 2\Gamma_p\right)\right.$$

$$\times \left[\frac{1}{k(k-1)} \sum_{\alpha} \lambda_{pp}^{(1)\alpha\alpha} \bar{\mu}_{\alpha}^2 - \frac{4}{k(k-1)(k-2)} \sum_{\alpha,\beta} \lambda_{pp}^{(2)\alpha\beta} \bar{\mu}_{\alpha}\bar{\mu}_{\beta} + \frac{3}{k(k-1)(k-2)(k-3)} \sum_{\alpha,\beta} \lambda_{pp}^{(3)\alpha\beta} \bar{\mu}_{\alpha}\bar{\mu}_{\beta}\right]\Bigg\}.$$

Next we calculate $E(\bar{B})$ as follows: Since

(6.2.30)
$$\sum_{i} \bar{t}_{ii}^{\#pp} = \sum_{\alpha} \sum_{i} t_{ii}^{(\alpha)pp} \bar{\mu}_{\alpha} = k\bar{\mu}_0,$$
$$\sum_{i \neq j} \bar{t}_{ij}^{\#pp} = \sum_{\alpha} \sum_{i \neq j} t_{ij}^{(\alpha)pp} \bar{\mu}_{\alpha} = \sum_{\alpha} \lambda_{pp}^{(1)\alpha\alpha} \bar{\mu}_{\alpha}$$

it follows that

(6.2.31) $$E\left[\underline{\pi}^{(p)'} S_{\sigma}' \bar{T}_{pp}^{\#} S_{\sigma} \underline{\pi}^{(p)} \underline{\pi}^{(q)'} S_{\sigma}' \bar{T}_{qq}^{\#} S_{\sigma} \underline{\pi}^{(q)}\right]$$

$$= \left\{\sum_{i} \bar{t}_{ii}^{\#pp} E\left[\pi_{\sigma(i)}^{(p)2}\right] + \sum_{i \neq j} \bar{t}_{ij}^{\#pp} E\left[\pi_{\sigma(i)}^{(p)} \pi_{\sigma(j)}^{(p)}\right]\right\}$$
$$\times \left\{\sum_{i} \bar{t}_{ii}^{\#qq} E\left[\pi_{\tau(i)}^{(q)2}\right] + \sum_{i \neq j} \bar{t}_{ij}^{\#qq} E\left[\pi_{\tau(i)}^{(q)} \pi_{\tau(j)}^{(q)}\right]\right\}.$$

$$= \left[\bar{\mu}_0 \Delta_p - \frac{\Delta_p}{k(k-1)} \sum_{\alpha=1}^{m} \lambda_{pp}^{(1)\alpha\alpha} \bar{\mu}_{\alpha}\right]\left[\bar{\mu}_0 \Delta_q - \frac{\Delta_q}{k(k-1)} \sum_{\beta=1}^{m} \lambda_{qq}^{(1)\beta\beta} \bar{\mu}_{\beta}\right]$$

$$= \Delta_p \Delta_q \left[\bar{\mu}_0^2 - \frac{\bar{\mu}_0}{k(k-1)} \sum_{\alpha=1}^{m} \lambda_{pp}^{(1)\alpha\alpha} \bar{\mu}_{\alpha} - \frac{\bar{\mu}_0}{k(k-1)} \sum_{\beta=1}^{m} \lambda_{qq}^{(1)\beta\beta} \bar{\mu}_{\beta} + \frac{1}{k^2(k-1)^2} \sum_{\alpha,\beta=1}^{m} \lambda_{pp}^{(1)\alpha\alpha} \lambda_{qq}^{(1)\beta\beta} \bar{\mu}_{\alpha}\bar{\mu}_{\beta}\right],$$

we have

$$E(\bar{B}) = \sum_{p \neq q} \Delta_p \Delta_q \left[\bar{\mu}_0^2 - \frac{\bar{\mu}_0}{k(k-1)} \sum_{\alpha=1}^{m} \lambda_{pp}^{(1)\alpha\alpha} \bar{\mu}_\alpha - \frac{\bar{\mu}_0}{k(k-1)} \right. \tag{6.2.32}$$

$$\left. \times \sum_{\beta=1}^{m} \lambda_{qq}^{(1)\beta\beta} \bar{\mu}_\beta + \frac{1}{k^2(k-1)^2} \sum_{\alpha,\beta=1}^{m} \lambda_{pp}^{(1)\alpha\alpha} \lambda_{qq}^{(1)\beta\beta} \bar{\mu}_\alpha \bar{\mu}_\beta \right].$$

Similarly we get

$$E(\bar{C}) = \sum_{p \neq q} \Delta_p \Delta_q \left[\frac{1}{k^2} \sum_{\alpha=0}^{m} \lambda_{pq}^{(4)\alpha\alpha} \bar{\mu}_\alpha^2 + \frac{1}{k^2(k-1)^2} \sum_{\alpha,\beta=0}^{m} \right. \tag{6.2.33}$$

$$\left. \times \lambda_{pq}^{(5)\alpha\beta} \bar{\mu}_\alpha \bar{\mu}_\beta - \frac{4}{k^2(k-1)} \sum_{\alpha,\beta=0}^{m} \lambda_{pq}^{(6)\alpha\beta} \bar{\mu}_\alpha \bar{\mu}_\beta \right].$$

Consequently we obtain

$$Var(\bar{\theta}) = \Delta^{-2} \left[E(\bar{\theta}^2) - E^2(\bar{\theta}) \right] \tag{6.2.34}$$

$$= \frac{1}{\Delta^2} \sum_{p=1}^{b} \left[\frac{\Delta_p^2}{k(k-1)} \sum_{\alpha=1}^{m} \left(\bar{\mu}_0 - \bar{\mu}_\alpha \right)^2 \lambda_{pp}^{(1)\alpha\alpha} + \left(\Delta_p^2 - 2\Gamma_p \right) \right.$$

$$\left. \times \left[\frac{1}{k(k-1)} \sum_{\alpha=1}^{m} \lambda_{pp}^{(1)\alpha\alpha} \bar{\mu}_\alpha^2 - \frac{4}{k(k-1)(k-2)} \sum_{\alpha,\beta=0}^{m} \lambda_{pp}^{(3)\alpha\beta} \bar{\mu}_\alpha \bar{\mu}_\beta \right] \right]$$

$$+ \frac{1}{\Delta^2} \sum_{p \neq q} \Delta_p \Delta_q \left[\bar{\mu}_0^2 - \frac{\bar{\mu}_0}{k(k-1)} \sum_{\alpha=1}^{m} \lambda_{pp}^{(1)\alpha\alpha} \bar{\mu}_\alpha - \frac{\bar{\mu}_0}{k(k-1)} \sum_{\beta=1}^{m} \right.$$

$$\times \lambda_{qq}^{(1)\beta\beta} \bar{\mu}_\beta + \frac{1}{k^2(k-1)^2} \sum_{\alpha,\beta=0}^{m} \lambda_{pp}^{(1)\alpha\alpha} \lambda_{qq}^{(1)\beta\beta} \bar{\mu}_\alpha \bar{\mu}_\beta + \frac{2}{k^2} \sum_{\alpha=0}^{m}$$

$$\times \lambda_{pq}^{(4)\alpha\alpha} \bar{\mu}_\alpha^2 + \frac{2}{k^2(k-1)^2} \sum_{\alpha,\beta=0}^{m} \lambda_{pq}^{(5)\alpha\beta} \bar{\mu}_\alpha \bar{\mu}_\beta - \frac{4}{k^2(k-1)}$$

$$\times \sum_{\alpha,\beta=0}^{m} \lambda_{pq}^{(6)\alpha\beta}\bar{\mu}_\alpha\bar{\mu}_\beta\Bigg] - \frac{1}{\Delta^2}\sum_{p=1}^{b}\Delta_p^2\Bigg[\bar{\mu}_0^2 - \frac{\bar{\mu}_0}{k(k-1)}\sum_{\alpha=1}^{m}\lambda_{pp}^{(1)\alpha\alpha}\bar{\mu}_\alpha$$

$$- \frac{\bar{\mu}_0}{k(k-1)}\sum_{\beta=1}^{m}\lambda_{pp}^{(1)\beta\beta}\bar{\mu}_\beta^2 + \frac{1}{k^2(k-1)^2}\sum_{\alpha,\beta=1}^{m}\lambda_{pp}^{(1)\alpha\alpha}\lambda_{pp}^{(1)\beta\beta}\bar{\mu}_\alpha\bar{\mu}_\beta\Bigg]$$

$$- \frac{1}{\Delta^2}\sum_{p\neq q}^{b}\Delta_p\Delta_q\Bigg[\bar{\mu}_0^2 - \frac{\bar{\mu}_0}{k(k-1)}\sum_{\alpha=1}^{m}\lambda_{pp}^{(1)\alpha\alpha}\bar{\mu}_\alpha - \frac{\bar{\mu}_0}{k(k-1)}$$

$$\times \sum_{\beta=1}^{m}\lambda_{qq}^{(1)\beta\beta}\bar{\mu}_\beta + \frac{1}{k^2(k-1)^2}\sum_{\alpha,\beta=1}^{m}\lambda_{pp}^{(1)\alpha\alpha}\lambda_{qq}^{(1)\beta\beta}\bar{\mu}_\alpha\bar{\mu}_\beta\Bigg]$$

$$= \frac{1}{\Delta^2}\sum_{p}\left(\Delta_p^2 - 2\Gamma_p\right)\Bigg[\frac{1}{k(k-1)}\sum_{\alpha=1}^{m}\lambda_{pp}^{(1)\alpha\alpha}\bar{\mu}_\alpha^2 - \frac{4}{k(k-1)(k-2)}$$

$$\times \sum_{\alpha,\beta=0}^{m}\lambda_{pp}^{(2)\alpha\beta}\bar{\mu}_\alpha\bar{\mu}_\beta + \frac{3}{k(k-1)(k-2)(k-3)}\sum_{\alpha,\beta=0}^{m}\lambda_{pp}^{(3)\alpha\beta}\bar{\mu}_\alpha\bar{\mu}_\beta\Bigg]$$

$$+ \frac{1}{\Delta^2}\sum_{p}\Delta_p^2\Bigg[\frac{1}{k(k-1)}\sum_{\alpha=0}^{m}\lambda_{pp}^{(1)\alpha\alpha}\bar{\mu}_\alpha^2 - \frac{1}{k^2(k-1)^2}$$

$$\times \sum_{\alpha,\beta=1}^{m}\lambda_{pp}^{(1)\alpha\alpha}\lambda_{pp}^{(1)\beta\beta}\bar{\mu}_\alpha\bar{\mu}_\beta\Bigg] + \frac{1}{\Delta^2}\sum_{p\neq q}\Delta_p\Delta_q\Bigg[\frac{2}{k^2}\sum_{\alpha=1}^{m}\lambda_{pq}^{(4)\alpha\alpha}\bar{\mu}_\alpha^2$$

$$+ \frac{2}{k^2(k-1)^2}\sum_{\alpha,\beta=0}^{m}\lambda_{pq}^{(5)\alpha\beta}\bar{\mu}_\alpha\bar{\mu}_\beta - \frac{4}{k^2(k-1)}\sum_{\alpha,\beta=0}^{m}\lambda_{pq}^{(6)\alpha\beta}\bar{\mu}_\alpha\bar{\mu}_\beta\Bigg].$$

On replacing $\bar{\mu}_\alpha$ by $\bar{\bar{\mu}}_\alpha$ we get

(6.2.35)

$$Var(\bar{\bar{\theta}}) = \frac{1}{\Delta^2}\sum_{p}\left(\Delta_p^2 - 2\Gamma_p\right)\Bigg[\frac{1}{k(k-1)}\sum_{\alpha}\lambda_{pp}^{(1)\alpha\alpha}\bar{\bar{\mu}}_\alpha^2 - \frac{4}{k(k-1)(k-2)}$$

$$\times \sum_{\alpha,\beta}\lambda_{pp}^{(2)\alpha\beta}\bar{\bar{\mu}}_\alpha\bar{\bar{\mu}}_\beta + \frac{3}{k(k-1)(k-2)(k-3)}\sum_{\alpha,\beta}\lambda_{pp}^{(3)\alpha\beta}\bar{\bar{\mu}}_\alpha\bar{\bar{\mu}}_\beta\Bigg]$$

$$+ \frac{1}{\Delta^2}\sum_p \Delta_p^2 \left[\frac{1}{k(k-1)}\sum_\alpha \lambda_{pp}^{(1)\alpha\alpha}\bar{\bar{\mu}}_\alpha^2 - \frac{1}{k^2(k-1)^2}\sum_{\alpha,\beta}\lambda_{pp}^{(1)\alpha\alpha}\right.$$

$$\left.\times\, \lambda_{pp}^{(1)\beta\beta}\bar{\bar{\mu}}_\alpha\bar{\bar{\mu}}_\beta\right] + \frac{1}{\Delta^2}\sum_{p\neq q}\Delta_p\Delta_q\left[\frac{2}{k^2}\sum_\alpha \lambda_{pq}^{(4)\alpha\alpha}\bar{\bar{\mu}}_\alpha^2 + \frac{2}{k^2(k-1)^2}\right.$$

$$\left.\times \sum_{\alpha,\beta}\lambda_{pq}^{(5)\alpha\beta}\bar{\bar{\mu}}_\alpha\bar{\bar{\mu}}_\beta - \frac{4}{k^2(k-1)}\sum_{\alpha,\beta}\lambda_{pq}^{(6)\alpha\beta}\bar{\bar{\mu}}_\alpha\bar{\bar{\mu}}_\beta\right].$$

<u>Calculation of $Cov\ (\bar{\theta},\bar{\bar{\theta}})$.</u> Since

$$\theta = \bar{\theta} + \bar{\bar{\theta}} \quad \text{and} \quad \mu_\alpha = \bar{\mu}_\alpha + \bar{\bar{\mu}}_\alpha,$$

we get

(6.2.36)

$$Var(\theta) = \frac{1}{\Delta^2}\sum_p \left(\Delta_p^2 - 2\Gamma_p\right)\left[\frac{1}{k(k-1)}\sum_\alpha \lambda_{pp}^{(1)\alpha\alpha}\mu_\alpha^2 - \frac{4}{k(k-1)(k-2)}\right.$$

$$\left.\times \sum_{\alpha,\beta}\lambda_{pp}^{(2)\alpha\beta}\mu_\alpha\mu_\beta + \frac{3}{k(k-1)(k-2)(k-3)}\sum_{\alpha,\beta}\lambda_{pp}^{(3)\alpha\beta}\mu_\alpha\mu_\beta\right]$$

$$+ \frac{1}{\Delta^2}\sum_p \Delta_p^2 \left[\frac{1}{k(k-1)}\sum_\alpha \lambda_{pp}^{(1)\alpha\alpha}\mu_\alpha^2 - \frac{1}{k^2(k-1)^2}\sum_{\alpha,\beta}\lambda_{pp}^{(1)\alpha\alpha}\right.$$

$$\left.\times\, \lambda_{pp}^{(1)\beta\beta}\mu_\alpha\mu_\beta\right] + \frac{1}{\Delta^2}\sum_{p\neq q}\Delta_p\Delta_q\left[\frac{2}{k^2}\sum_\alpha \lambda_{pq}^{(4)\alpha\alpha}\mu_\alpha^2 + \frac{2}{k^2(k-1)^2}\right.$$

$$\left.\times \sum_{\alpha,\beta}\lambda_{pq}^{(5)\alpha\beta}\mu_\alpha\mu_\beta - \frac{4}{k^2(k-1)}\sum_{\alpha,\beta}\lambda_{pq}^{(6)\alpha\beta}\mu_\alpha\mu_\beta\right].$$

The covariance $Cov(\bar{\theta},\bar{\bar{\theta}})$ can be calculated as

(6.2.37) $$2Cov(\bar{\theta},\bar{\bar{\theta}}) = Var(\bar{\theta}+\bar{\bar{\theta}}) - Var(\bar{\theta}) - Var(\bar{\bar{\theta}}).$$

Hence we obtain

(6.2.38) $$Cov(\bar{\theta},\bar{\bar{\theta}}) = \frac{1}{\Delta^2}\sum_p \left(\Delta_p^2 - 2\Gamma_p\right)\left[\frac{1}{k(k-1)}\sum_\alpha \lambda_{pp}^{(1)\alpha\alpha}\bar{\mu}_\alpha\bar{\bar{\mu}}_\alpha\right.$$

$$- \frac{4}{k(k-1)(k-2)}\sum_{\alpha,\beta}\lambda_{pp}^{(2)\alpha\beta}\bar{\mu}_\alpha\bar{\bar{\mu}}_\beta + \frac{3}{k(k-1)(k-2)(k-3)}$$

$$\left.\times \sum_{\alpha,\beta}\lambda_{pp}^{(3)\alpha\beta}\bar{\mu}_\alpha\bar{\bar{\mu}}_\beta\right] + \frac{1}{\Delta^2}\sum_p \Delta_p^2\left[\frac{1}{k(k-1)}\sum_{\alpha,\beta}\lambda^{(2)\alpha\beta}\bar{\mu}_\alpha\bar{\bar{\mu}}_\beta\right.$$

$$\left.- \frac{1}{k\,(k-1)^2}\sum_{\alpha,\beta}\lambda_{pp}^{(1)\alpha\alpha}\lambda_{pp}^{(1)\beta\beta}\bar{\mu}_\alpha\bar{\bar{\mu}}_\beta\right] + \frac{1}{\Delta^2}\sum_{p\neq q}\Delta_p\Delta_q\left[\frac{2}{k^2}\sum_\alpha \lambda_{pq}^{(4)\alpha\alpha}\bar{\mu}_\alpha\bar{\bar{\mu}}_\alpha\right.$$

$$\left.+ \frac{2}{k^2(k-1)^2}\sum_{\alpha,\beta}\lambda_{pq}^{(5)\alpha\beta}\bar{\mu}_\alpha\bar{\bar{\mu}}_\beta - \frac{4}{k^2(k-1)}\sum_{\alpha,\beta}\lambda_{pq}^{(6)\alpha\beta}\bar{\mu}_\alpha\bar{\bar{\mu}}_\beta\right].$$

6.3. Heuristic Derivation of the Asymptotic Null Distribution of the *F*-Statistic after Randomization

We are going to find an asymptotic continuous distribution of $(\bar{\theta},\bar{\bar{\theta}})$ in a rather heuristic way when $b \to \infty$ whereas v, k, n_α, $p_{\alpha\beta}^\gamma$ are fixed. For this purpose we are impcsing rather strong uniformity conditions on the unit error; *i.e.*

(6.3.1) $$\Delta_p = \Delta_0 = \frac{1}{b}\,\Delta, \qquad \Gamma_p = \Gamma_0 \qquad (p = 1,2,\ldots,b).$$

Under these uniformity conditions we obtain

$$E(\bar{\theta}) = \frac{1}{bk(k-1)}\sum_{\alpha=1}^m \left(\bar{\mu}_0 - \bar{\mu}_\alpha\right)\sum_{p=1}^b \lambda_{pp}^{(1)\alpha\alpha},$$

(6.3.2) $$E(\bar{\bar{\theta}}) = \frac{1}{bk(k-1)}\sum_{\alpha=1}^m \left(\bar{\bar{\mu}}_0 - \bar{\bar{\mu}}_\alpha\right)\sum_{p=1}^b \lambda_{pp}^{(1)\alpha\alpha},$$

$$Var(\bar{\theta}) = \left(1 - \frac{2\Gamma_0}{\Delta_0^2}\right)\left[\frac{1}{b^2k(k-1)} \sum_{\alpha=1}^{m} \bar{\mu}_\alpha^2 \sum_p \lambda_{pp}^{(1)\alpha\alpha} - \frac{4}{b^2k(k-1)(k-2)}\right.$$

$$\times \sum_{\alpha,\beta=0}^{m} \bar{\mu}_\alpha \bar{\mu}_\beta \sum_p \lambda_{pp}^{(2)\alpha\beta} + \frac{3}{b^2k(k-1)(k-2)(k-3)}$$

$$\left.\times \sum_{\alpha,\beta=0}^{m} \bar{\mu}_\alpha \bar{\mu}_\beta \sum_p \lambda_{pp}^{(3)\alpha\beta}\right] + \frac{1}{b^2k(k-1)} \sum_{\alpha=1}^{m} \bar{\mu}_\alpha^2 \sum_p \lambda_{pp}^{(1)}$$

(6.3.3)

$$- \frac{1}{b^2k^2(k-1)^2} \sum_{\alpha,\beta=1}^{m} \bar{\mu}_\alpha \bar{\mu}_\beta \sum_p \lambda_{pp}^{(1)\alpha\alpha} \lambda_{pp}^{(1)\beta\beta}$$

$$+ \frac{2}{b^2k^2} \sum_{\alpha=0}^{m} \bar{\mu}_\alpha^2 \sum_{p \neq q} \lambda_{pq}^{(4)\alpha\alpha} + \frac{2}{b^2k^2(k-1)^2} \sum_{\alpha,\beta=0}^{m} \bar{\mu}_\alpha \bar{\mu}_\beta \sum_{p \neq q} \lambda_{pq}^{(5)\alpha\beta}$$

$$- \frac{4}{b^2k^2(k-1)} \sum_{\alpha,\beta=0}^{m} \bar{\mu}_\alpha \bar{\mu}_\beta \sum_{p \neq q} \lambda_{pq}^{(6)\alpha\beta},$$

$$Var(\bar{\bar{\theta}}) = \left(1 - \frac{2\Gamma_0}{\Delta_0^2}\right)\left[\frac{1}{b^2k(k-1)} \sum_{\alpha} \bar{\bar{\mu}}_\alpha^2 \sum_p \lambda_{pp}^{(1)\alpha\alpha} - \frac{4}{b^2k(k-1)(k-2)}\right.$$

$$\times \sum_{\alpha,\beta} \bar{\bar{\mu}}_\alpha \bar{\bar{\mu}}_\beta \sum_p \lambda_{pp}^{(2)\alpha\beta} + \frac{3}{b^2k(k-1)(k-2)(k-3)} \sum_{\alpha,\beta} \bar{\bar{\mu}}_\alpha \bar{\bar{\mu}}_\beta$$

(6.3.4)

$$\left.\times \sum_p \lambda_{pp}^{(3)\alpha\beta}\right] + \frac{1}{b^2k(k-1)} \sum_{\alpha} \bar{\bar{\mu}}_\alpha^2 \sum_p \lambda_{pp}^{(1)\alpha\alpha} - \frac{1}{b^2k^2(k-1)^2}$$

$$\times \sum_{\alpha,\beta=1}^{m} \bar{\bar{\mu}}_\alpha \bar{\bar{\mu}}_\beta \sum_p \lambda_{pp}^{(1)\alpha\alpha} \beta_{pp}^{(1)\beta\beta} + \frac{2}{b^2k^2} \sum_{\alpha=1}^{m} \bar{\bar{\mu}}_\alpha^2 \sum_{p \neq q} \lambda_{pq}^{(4)\alpha\alpha}$$

$$+ \frac{2}{b^2k^2(k-1)^2} \sum_{\alpha,\beta} \bar{\bar{\mu}}_\alpha \bar{\bar{\mu}}_\beta \left[\sum_{p \neq q} \lambda_{pq}^{(5)\alpha\beta} - 2(k-1) \sum_{p \neq q} \lambda_{pq}^{(6)\alpha\beta}\right],$$

and

$$
\begin{aligned}
Cov(\bar{\theta},\bar{\bar{\theta}}) = {} & \left(1 - \frac{2\Gamma_0}{\Delta_0^2}\right)\left[\frac{1}{b^2k(k-1)} \sum_\alpha \bar{\mu}_\alpha \bar{\bar{\mu}}_\alpha \sum_p \lambda_{pp}^{(1)\alpha\alpha} - \frac{4}{b^2k(k-1)(k-2)}\right. \\
& \times \sum_{\alpha,\beta} \bar{\mu}_\alpha \bar{\bar{\mu}}_\beta \sum_p \lambda_{pp}^{(2)\alpha\beta} + \frac{3}{b^2k(k-1)(k-2)(k-3)} \\
& \left.\times \sum_{\alpha,\beta} \bar{\mu}_\alpha \bar{\bar{\mu}}_\beta \sum_p \lambda_{pp}^{(3)\alpha\beta}\right] + \frac{1}{b^2k(k-1)} \sum_\alpha \bar{\mu}_\alpha \bar{\bar{\mu}}_\alpha \sum_p \lambda_{pp}^{(1)\alpha\alpha} \\
& - \frac{1}{b^2k^2(k-1)^2} \sum_{\alpha,\beta} \bar{\mu}_\alpha \bar{\bar{\mu}}_\beta \sum_p \lambda_{pp}^{(1)\alpha\alpha} \lambda_{pp}^{(1)\beta\beta} + \frac{2}{b^2k^2} \\
& \times \sum_\alpha \bar{\mu}_\alpha \bar{\bar{\mu}}_\alpha \sum_{p\neq q} \lambda_{pq}^{(4)\alpha\alpha} + \frac{2}{b^2k^2(k-1)^2} \sum_{\alpha,\beta} \bar{\mu}_\alpha \bar{\bar{\mu}}_\beta \sum_{p\neq q} \lambda_{pq}^{(5)\alpha\beta} \\
& - \frac{4}{b^2k^2(k-1)} \sum_{\alpha,\beta} \bar{\mu}_\alpha \bar{\bar{\mu}}_\beta \sum_{p\neq q} \lambda_{pq}^{(6)\alpha\beta}.
\end{aligned}
\tag{6.3.5}
$$

In order to evaluate the asymptotic values of $Var(\bar{\theta})$, $Var(\bar{\bar{\theta}})$, and $Cov(\bar{\theta},\bar{\bar{\theta}})$ as $b \to \infty$ while v, k, n_α, and $p_{\beta\gamma}^{\alpha}$ remain fixed we need the following lemmas:

<u>Lemma 6.3.1</u>

(i) $\lambda_{pp}^{(1)\alpha\alpha}\lambda_{pp}^{(1)\beta\beta} = 2\delta_{\alpha\beta}\lambda_{pp}^{(1)\alpha\alpha} + 4\lambda_{pp}^{(2)\alpha\beta} + \lambda_{pp}^{(3)\alpha\beta}$,

(ii) $\lambda_{pq}^{(4)\alpha\alpha}\lambda_{pq}^{(4)\beta\beta} = \delta_{\alpha\beta}\lambda_{pq}^{(4)\alpha\alpha} + \lambda_{pq}^{(5)\alpha\beta} + 2\lambda_{pq}^{(6)\alpha\beta}$

<u>Lemma 6.3.2</u>

(i) $\sum_p \lambda_{pp}^{(1)\alpha\alpha} = v(1 - \delta_{0\alpha})\lambda_\alpha n_\alpha$,

(ii) $$\sum_{p\neq q} \lambda_{pq}^{(4)\alpha\alpha} = v(r^2 - \lambda_\alpha)n_\alpha ,$$

(iii) $$\sum_{p\neq q} \lambda_{pq}^{(4)\alpha\alpha}\lambda_{pq}^{(4)\beta\beta} = v \sum_{\sigma,\tau,\gamma\geqq 0} \lambda_\sigma\lambda_\tau p_{\sigma\tau}^\gamma p_{\alpha\beta}^\gamma n_\gamma - \delta_{\alpha 0}\delta_{\beta 0} vrk$$

$$- (1 - \delta_{\alpha 0})(\delta_{\beta 0}k + (1 - \delta_{\beta 0})\delta_{\alpha\beta})v\lambda_\alpha n_\alpha - (1 - \delta_{\beta 0})(\delta_{\alpha 0}k$$

$$+ (1 - \delta_{\alpha 0})\delta_{\alpha\beta})v\lambda_\beta n_\beta - (1 - \delta_{\alpha 0})(1 - \delta_{\beta 0})(4\Lambda_{\alpha\beta}^{(1)} + \Lambda_{\alpha\beta}^{(2)})$$

where

$$\Lambda_{\alpha\beta}^{(1)} = \sum_{p\neq q} \lambda_{pq}^{(2)\alpha\beta} \quad \text{and} \quad \Lambda_{\alpha\beta}^{(2)} = \sum_{p\neq q} \lambda_{pq}^{(3)\alpha\beta} ,$$

(iv) $$\sum_{p\neq q} \lambda_{pq}^{(6)\alpha\beta} = (1 - \delta_{\alpha 0})(1 - \delta_{\beta 0})(vr \sum_{\gamma\geqq 1} \lambda_\gamma p_{\alpha\beta}^\gamma n_\gamma - \Lambda_{\alpha\beta}^{(1)}) ,$$

(v) $$\sum_{p\neq q} \lambda_{pq}^{(5)\alpha\beta} = v \sum_{\sigma,\tau,\gamma\geqq 0} \lambda_\sigma\lambda_\tau p_{\sigma\tau}^\gamma p_{\alpha\beta}^\gamma n_\gamma - \delta_{\alpha 0}\delta_{\beta 0} vrk$$

$$- \delta_{\alpha\beta}v(r^2 - \lambda_\alpha)n_\alpha - (1 - \delta_{\alpha 0})(\delta_{\beta 0}k + (1 - \delta_{\beta 0})\delta_{\alpha\beta})vn_\alpha\lambda_\alpha$$

$$- (1 - \delta_{\beta 0})(\delta_{\alpha 0}k + (1 - \delta_{\alpha 0})\delta_{\alpha\beta})vn_\beta\lambda_\beta - (1 - \delta_{\alpha 0})(1 - \delta_{\beta 0})$$

$$\times (2\Lambda_{\alpha\beta}^{(1)} + \Lambda_{\alpha\beta}^{(2)}) - (1 - \delta_{\alpha 0})(1 - \delta_{\beta 0})2vr \sum_{\gamma\geqq 1} \lambda_\gamma p_{\alpha\beta}^\gamma n_\gamma .$$

Under the uniformity conditions

(6.3.6) $$\Delta_p = \Delta_0, \quad \Gamma_p = \Gamma_0 \qquad (p = 1,\ldots,b),$$

one obtains

$$(6.3.7)\quad Var(\bar{\theta}) = \left[1 - \frac{2\Gamma_0}{\Delta_0^2}\right]\left[\frac{v}{b^2k(k-1)} \sum_{\alpha=1}^{m} \bar{\mu}_\alpha^2 \lambda_\alpha n_\alpha - \frac{4}{b^2k(k-1)(k-2)}\right.$$

$$\times \sum_{\alpha,\beta=0}^{m} \bar{\mu}_\alpha \bar{\mu}_\beta \Lambda_{\alpha\beta}^{(1)} + \left.\frac{3}{b^2k(k-1)(k-2)(k-3)} \sum_{\alpha,\beta=0}^{m} \bar{\mu}_\alpha \bar{\mu}_\beta \Lambda_{\alpha\beta}^{(2)}\right] + \frac{v}{b^2k(k-1)}$$

$$\times \sum_{\alpha=1}^{m} \bar{\mu}_\alpha^2 \lambda_\alpha n_\alpha - \frac{1}{b^2k^2(k-1)^2} \sum_{\alpha=1}^{m} \bar{\mu}_\alpha \bar{\mu}_\beta \left[2\delta_{\alpha\beta} v \lambda_\alpha n_\alpha + 4\Lambda_{\alpha\beta}^{(1)} + \Lambda_{\alpha\beta}^{(2)}\right] + \frac{2v}{b^2k^2}$$

$$\times \sum_{\alpha=0}^{m} \bar{\mu}_\alpha^2 (r^2 - \lambda_\alpha) n_\alpha + \frac{2}{b^2k^2(k-1)^2} \sum_{\alpha,\beta=0}^{m} \bar{\mu}_\alpha \bar{\mu}_\beta \left[v \sum_{\sigma,\tau,\gamma=0}^{m} \lambda_\sigma \lambda_\tau p_{\sigma\tau}^{\gamma} p_{\alpha\beta}^{\gamma} n_\gamma\right.$$

$$- 2vr \sum_{\gamma=0}^{m} \lambda_\gamma p_{\alpha\beta}^{\gamma} n_\gamma - \delta_{\alpha\beta} v (r^2 - \lambda_\alpha) n_\alpha - \delta_{\alpha 0}\delta_{\beta 0} vrk - (1-\delta_{\alpha 0})\{\delta_{\beta 0}(k-2)$$

$$+ (1-\delta_{\beta 0})\delta_{\alpha\beta}\} v \lambda_\alpha n_\alpha - (1-\delta_{\beta 0})\{\delta_{\alpha 0}(k-2) + (1-\delta_{\alpha 0})\delta_{\alpha\beta}\} v \lambda_\alpha n_\alpha - (1-\delta_{\alpha 0})$$

$$\times \left.(1-\delta_{\beta 0})(2\Lambda_{\alpha\beta}^{(1)} + \Lambda_{\alpha\beta}^{(2)})\right] - \frac{4}{b^2k^2(k-1)} \sum_{\alpha,\beta=0}^{m} \bar{\mu}_\alpha \bar{\mu}_\beta \left[vr \sum_{\gamma=1}^{m} \lambda_\gamma p_{\alpha\beta}^{\gamma} n_\gamma\right.$$

$$\left. - \delta_{\alpha 0}(1-\delta_{\beta 0}) v \lambda_\beta n_\beta - \delta_{\beta 0}(1-\delta_{\alpha 0}) v \lambda_\alpha n_\alpha - (1-\delta_{\alpha 0})(1-\delta_{\beta 0})\Lambda_{\alpha\beta}^{(1)}\right].$$

Now, since

$$(6.3.8)\qquad \bar{\mu}_\alpha = O\left[\frac{1}{b}\right], \qquad \lambda_\alpha = O(b),$$

$$(6.3.9)\qquad \begin{cases} \Lambda_{\alpha\beta}^{(1)} \leqq bk(k-1)(k-2), \\ \Lambda_{\alpha\beta}^{(2)} \leqq bk(k-1)(k-2)(k-3), \end{cases}$$

(6.3.10) $$\sum_{\alpha=0}^{m} \bar{\mu}_\alpha^2 n_\alpha = \sum_{i=1}^{h} \frac{c_i \alpha_i}{v}$$

(6.3.11) $$\sum_{\alpha,\beta \geqq 0} \sum_{\gamma \geqq 1} \bar{\mu}_\alpha \bar{\mu}_\beta \lambda_\gamma p_{\alpha\beta}^\gamma n_\gamma = \sum_{i=1}^{h} \frac{c_i^2 \alpha_i (\rho_i - r)}{v},$$

(6.3.12) $$\sum_{\alpha,\beta \geqq 0} \sum_{\sigma,\tau,\gamma \geqq 0} \bar{\mu}_\alpha \bar{\mu}_\beta \lambda_\sigma \lambda_\tau p_{\sigma\tau}^\gamma p_{\alpha\beta}^\gamma n_\gamma = \sum_{i=1}^{h} \frac{c_i^2 \alpha_i \rho_i^2}{v},$$

the terms of the lowest order $O(1/b^2)$ of $Var(\theta)$ become

(6.3.13) $$\frac{2v}{b^2k^2(k-1)^2}\left[r^2k(k-2)\sum_{\alpha \geqq 0} \bar{\mu}_\alpha^2 n_\alpha + \sum_{\substack{\alpha,\beta \geqq 0 \\ \sigma,\tau,\gamma \geqq 0}} \bar{\mu}_\alpha \bar{\mu}_\beta \lambda_\sigma \lambda_\tau p_{\sigma\tau}^\gamma p_{\alpha\beta}^\gamma n_\gamma - 2rk \sum_{\substack{\alpha,\beta \geqq 0 \\ \gamma \geqq 1}} \bar{\mu}_\alpha \bar{\mu}_\beta \lambda_\gamma p_{\alpha\beta}^\gamma n_\gamma\right] = \frac{2}{b^2k^2(k-1)^2} \sum_{i=1}^{h} \alpha_i c_i^2 (rk - \rho_i)^2$$

$$= \frac{2\bar{\alpha}}{b^2(k-1)^2}$$

Thus

(6.3.14) $$Var(\bar{\theta}) = \frac{2\bar{\alpha}}{b^2(k-1)^2}\left[1 + O\left(\frac{1}{b}\right)\right].$$

In a similar manner we can show that

(6.3.15) $$Var(\bar{\bar{\theta}}) = \frac{2\bar{\bar{\alpha}}}{b^2(k-1)^2}\left[1 + O\left(\frac{1}{b}\right)\right]$$

and

(6.3.16) $$Cov(\bar{\theta}, \bar{\bar{\theta}}) = O\left(\frac{1}{b^3}\right).$$

For the Dirichlet distribution,

$$\frac{\Gamma\left(\frac{n-b}{2}\right)}{\Gamma\left(\frac{\bar{\alpha}}{2}\right)\Gamma\left(\frac{\bar{\bar{\alpha}}}{2}\right)\Gamma\left(\frac{n-b-v+1}{2}\right)} \bar{\theta}^{(\bar{\alpha}/2)-1} \bar{\bar{\theta}}^{(\bar{\bar{\alpha}}/2)-1}$$

$$\times (1 - \bar{\theta} - \bar{\bar{\theta}})^{[(n-b-v+1)/2]-1}\, d\bar{\theta}\, d\bar{\bar{\theta}}$$

$$(\bar{\theta} \geqq 0,\ \bar{\bar{\theta}} \geqq 0,\ \bar{\theta} + \bar{\bar{\theta}} \leqq 1),$$

we get

$$E^*(\bar{\theta}) = \frac{\bar{\alpha}}{n - b} = \frac{\bar{\alpha}}{b(k - 1)},\quad E^*(\bar{\bar{\theta}}) = \frac{\bar{\alpha}}{n - b} = \frac{\bar{\bar{\alpha}}}{b(k - 1)}, \tag{6.3.17}$$

and

$$Var^*(\bar{\theta}) = \frac{2\bar{\alpha}(n - b - \bar{\alpha})}{(n - b)^2(n - b + 2)} = \frac{2\bar{\alpha}}{b^2(k - 1)^2}\left[1 + O\left(\frac{1}{b}\right)\right],$$

$$Var^*(\bar{\bar{\theta}}) = \frac{2\bar{\bar{\alpha}}}{b^2(k - 1)^2}\left[1 + O\left(\frac{1}{b}\right)\right], \tag{6.3.18}$$

$$Cov^*(\bar{\theta},\bar{\bar{\theta}}) = O\left(\frac{1}{b^3}\right) .$$

Thus one can take the above Dirichlet distribution (6.3.16) as an approximation to the permutation distribution of $(\bar{\theta},\bar{\bar{\theta}})$ when $b \to \infty$. Hence

$$E^*\left(\bar{\theta}^{\mu}\bar{\bar{\theta}}^{\gamma}(1 - \bar{\theta} - \bar{\bar{\theta}})^{\nu}\right) \tag{6.3.19}$$

$$= \frac{\Gamma\left(\frac{n - b}{2}\right)\Gamma\left(\frac{\bar{\alpha}}{2} + \mu\right)\Gamma\left(\frac{\bar{\bar{\alpha}}}{2} + \gamma\right)\Gamma\left(\frac{n - b - v + 1}{2} + \nu\right)}{\Gamma\left(\frac{\bar{\alpha}}{2}\right)\Gamma\left(\frac{\bar{\bar{\alpha}}}{2}\right)\Gamma\left(\frac{n - b - v + 1}{2}\right)\Gamma\left(\frac{n - b - \bar{\bar{\alpha}}}{2} + \mu + \nu + \gamma\right)},$$

and consequently one obtains the null distribution of the F-statistic after randomization:

$$\frac{\Gamma\left(\frac{n - b - \bar{\bar{\alpha}}}{2}\right)}{\Gamma\left(\frac{\bar{\alpha}}{2}\right)\Gamma\left(\frac{n - b - v + 1}{2}\right)}\left(\frac{\bar{\alpha}}{n - b - v + 1}F\right)^{\frac{\bar{\alpha}}{2} - 1}$$

$$\times \left(1 + \frac{\bar{\alpha}}{n - b - v + 1}F\right)^{-(n-b-\bar{\bar{\alpha}})/2} d\left(\frac{\bar{\alpha}}{n - b - v + 1}F\right) .$$

where we have made use of the identity

$$(6.3.20) \qquad \sum_{\mu+\nu=\ell} \frac{\ell!}{\mu!\nu!} \frac{\Gamma\left(\frac{n-b-\bar{\bar{\alpha}}}{2}+\mu\right)\Gamma\left(\frac{\bar{\bar{\alpha}}}{2}+\nu\right)}{\Gamma\left(\frac{n-b}{2}+\ell\right)} = \frac{\Gamma\left(\frac{n-b-\bar{\bar{\alpha}}}{2}\right)\Gamma\left(\frac{\bar{\bar{\alpha}}}{2}\right)}{\Gamma\left(\frac{n-b}{2}\right)}.$$

6.4. Asymptotic Equivalence of Probability Distributions

For any given positive integer n, let R_n be the n-dimensional Euclidean space and let $\mathcal{B}_n$ be the usual Borel field of subsets of R_n. Let us denote the family of all probability distributions or of all random variables defined over the measurable space $(R_n,\mathcal{B}_n)$ by $F(R_n,\mathcal{B}_n)$, the members of which will be designated by random variables $X^{(n)}, Y^{(n)},\ldots$, with the corresponding probability measures $P^{X^{(n)}}, P^{Y^{(n)}},\ldots$, respectively. Let $\nu^{(n)}$ be a given σ-finite measure over $(R_n,\mathcal{B}_n)$ and let $P\left[R_n,\mathcal{B}_n,\nu^{(n)}\right]$ be the family of all probability distributions over $(R_n,\mathcal{B}_n)$ that are absolutely continuous with respect to $\nu^{(n)}$. Throughout the present section $\nu^{(n)}$ stands for the Lebesgue measure over $(R_n,\mathcal{B}_n)$.

Let $\mathcal{C}_n$ be any given non-empty subclass of $\mathcal{B}_n$ and let us define

$$(6.4.1) \qquad d\left[X^{(n)},\ Y^{(n)}\ :\ \mathcal{C}_n\right] = \sup_{E\in\mathcal{C}_n}\left[P^{X^{(n)}}(E) - P^{Y^{(n)}}(E)\right].$$

This quantity defines a distance over the family $F(R_n,\mathcal{B}_n)$ if we we identify those random variables that have the same probability measure over $\mathcal{C}_n$. The random variables $X^{(n)}$ and $Y^{(n)}$ are said to have the same probability measures over $\mathcal{C}_n$ if $P^{X^{(n)}}(E) = P^{Y^{(n)}}(E)$ for all $E \in \mathcal{C}_n$. It should also be noted that, in the case when both $X^{(n)}$ and $Y^{(n)}$ belong to the family $P\left[R_n,\mathcal{B}_n,\nu^{(n)}\right]$ for some σ-finite measure $\nu^{(n)}$, we have

(6.4.2) $$2d\left[X^{(n)},Y^{(n)}:\mathcal{B}_n\right] = \int_{R_n} |f - g|d\nu^{(n)},$$

where f and g are the generalized probability density functions of $X^{(n)}$ and $Y^{(n)}$ with respect to $\nu^{(n)}$ respectively.

It is immediately obvious that if $m \leq n$ and $\mathcal{C}_n$ contains all the subsets of the form $E_m \times E_{n-m}$, $E_m \in \mathcal{C}_m$, then

(6.4.3) $$d\left[X^{(m)},Y^{(m)}:\mathcal{C}_m\right] \leq d\left[X^{(n)},Y^{(n)}:\mathcal{C}_n\right],$$

where $X^{(n)} = \left[X^{(m)},X^{(n-m)}\right]$ and $Y^{(n)} = \left[Y^{(m)},Y^{(n-m)}\right]$ are the decompositions of random variables corresponding to the decomposition of the space $R_n = R_m \times R_{n-m}$.

Now we consider some of the familiar subclasses of $\mathcal{B}_n$:

1. Let $\mathcal{M}_n$ be the class of all subsets of R_n such that

(6.4.4) $$M = \{\underline{x}' = (x_1,\ldots,x_n) : -\infty < x_i < a_i,\ i = 1,\ldots,n\}$$

for all extended real vectors $\underline{a}' = (a_1,\ldots,a_n)$, a_i being admitted to assume the values $\pm\infty$. This is a multiplicative class and contains the empty set and the whole space as its element.

2. The class $\mathcal{S}_n$ is defined to be the class of all subsets of the form

(6.4.5) $$S = \{\underline{x}' = (x_1,\ldots,x_n) : b_i \leq x_i < a_i,\ i = 1,\ldots,n\},$$

where $\underline{a}' = (a_1,\ldots,a_n)$ and $\underline{b}' = (b_1,\ldots,b_n)$ are any extended real vectors.

3. Let $\mathcal{A}_n$ be a finitely additive class over $\mathcal{M}_n$, and finally

4. Let $\mathcal{G}_n$ be the class of all open subsets of R_n with respect to the Euclidean distance.

It is well known that (1) the class $\mathcal{A}_n$ consists of all finite unions of members of $\mathcal{S}_n$ that may or may not be disjoint, and (2) for any subset $E \in \mathcal{S}_n$, there exists a set of finite numbers of $\mathcal{M}_n$, say $\{F_1,\ldots,F_N\}$, and a set of constants c_i, taking the values ±1,

$$\nu^{(n)}(E) = \sum_{i=1}^{N} c_i \, \nu^{(n)}(F_i) \tag{6.4.6}$$

for any σ-finite measure $\nu^{(n)}$ over $(R_n, \mathcal{B}_n)$ and $N \leq 2^n$.

<u>Lemma 6.4.1</u>

$$d\left[X^{(n)}, Y^{(n)} : \mathcal{M}_n\right] \leq d\left[X^{(n)}, Y^{(n)} : \mathcal{S}_n\right]$$

$$\leq 2^n d\left[X^{(n)}, Y^{(n)} : \mathcal{M}_n\right]$$

<u>Proof</u>. The first part is evident because $\mathcal{M}_n \subseteq \mathcal{S}_n$. For the second part, for $E \in \mathcal{S}_n$,

$$\left|P^{X^{(n)}}(E) - P^{Y^{(n)}}(E)\right| \leq \sum_{i=1}^{N} \left|P^{X^{(n)}}(F_i) - P^{Y^{(n)}}(F_i)\right| \tag{6.4.7}$$

$$\leq 2^N \, d\left[X^{(n)}, Y^{(n)} : \mathcal{M}_n\right]$$

implies that

$$d\left[X^{(n)}, Y^{(n)} : \mathcal{S}_n\right] \leq 2^n d\left[X^{(n)}, Y^{(n)} : \mathcal{M}_n\right]. \tag{6.4.8}$$

<u>Lemma 6.4.2</u>

$$d\left[X^{(n)}, Y^{(n)} : \mathcal{A}_n\right] = d\left[X^{(n)}, Y^{(n)} : \mathcal{B}_n\right].$$

Indeed, for any σ-finite measure $\nu^{(n)}$ over $(R_n, \boldsymbol{B}_n)$ and for any subset $E \in \boldsymbol{B}_n$,

$$\nu^{(n)}(E) = \inf \left[\sum_i \nu^{(n)}(F_i);\ E \subseteq \bigcup_i F_i,\ \ F_i \in \boldsymbol{A}_n\right]. \tag{6.4.9}$$

Hence, for any given $\varepsilon > 0$, there are coverings of E by the disjoint element of $\boldsymbol{A}_n$

$$E \subseteq \bigcup_i F_i, \quad E \subseteq \bigcup_i G_i, \quad F_i, G_i \in \boldsymbol{A}_n, \tag{6.4.10}$$

such that

$$P^{X^{(n)}}(E) < \sum_i P^{X^{(n)}}(F_i) < P^{X^{(n)}}(E) + \varepsilon \tag{6.4.11}$$

and

$$P^{Y^{(n)}}(E) < \sum_i P^{Y^{(n)}}(F_i) < P^{Y^{(n)}}(E) + \varepsilon. \tag{6.4.12}$$

Therefore

$$\begin{aligned} P^{X^{(n)}}(E) - P^{Y^{(n)}}(E) - \varepsilon &< \sum_i P^{X^{(n)}}(F_i) - \sum_i P^{Y^{(n)}}(F_i) \\ &< P^{X^{(n)}}(E) - P^{Y^{(n)}}(E) + \varepsilon, \end{aligned} \tag{6.4.13}$$

which implies the equality

$$d\left[X^{(n)}, Y^{(n)} : \boldsymbol{B}_n\right] = d\left[X^{(n)}, Y^{(n)} : \boldsymbol{A}_n\right]. \tag{6.4.14}$$

Let $\left\{X_s^{(n_s)}\right\}$ and $\left\{Y_s^{(n_s)}\right\}$, $s = 1,2,\ldots$, be two sequences of random variables belonging to the family $F(R_{n_s}, \boldsymbol{B}_{n_s})$ for each s. The sequence of the underlying spaces $\left\{R_{n_s}\right\}$ is called *the sequence of basic spaces*. The case where the dimensions of the basic space n_s are independent of s is called *the case of equal basic spaces*;

otherwise it is called *the case of unequal basic spaces.* An important case of unequal basic spaces is when n_s tends to infinity as $s \to \infty$.

Definition 6.4.1. Two sequences of random variables $\left\{X_s^{(n_s)}\right\}$, $\left\{Y_s^{(n_s)}\right\}$, $s = 1,2,\ldots$ are said to be *asymptotically equivalent in the sense of the type*

$$d\left(X_s^{(n_s)}, Y_s^{(n_s)} : \mathcal{C}_{n_s}\right) \to 0 \quad \text{as} \qquad s \to \infty.$$

and are denoted by

$$X_s^{(n_s)} \sim Y_s^{(n_s)}, (\mathcal{C}_{n_s})_d \qquad (s \to \infty).$$

This is the uniform asymptotic equivalence.

In the case of equal basic spaces where $n_s = k$, if we fix $Y_s^{n_s}$ independently of s, that is, $Y_s^{(n_s)} = Y^{(k)}$, then Definition 6.4.1 gives us the notion of uniform convergence.

In the case of equal basic spaces there is a weaker definition of asymptotic equivalence.

Definition 6.4.2. Two sequences of random variables $\left\{X_s^{(n)}\right\}$, $\left\{Y_s^{(n)}\right\}$, $s = 1,2,\ldots$ are said to be *asymptotically equivalent in the sense of the type* $((\mathcal{C}_{n_s}))_d$ if

$$\left| P^{X_s^{(n)}}(E) - P^{Y_s^{(n)}}(E) \right| \to 0 \quad \text{as} \quad s \to \infty$$

for every subset $E \in \mathcal{C}_{n_s}$, and this is denoted by

$$X_s^{(n)} \sim Y_s^{(n)}, ((\mathcal{C}_{n_s}))_d \qquad (s \to \infty).$$

Lemma 6.4.3. In order for the two sequences of random variables $\left\{X_s^{(n_s)}\right\}, \left\{Y_s^{(n_s)}\right\}$, $s = 1,2,\ldots,$ to be asymptotically equivalent in the sense of the type $((\mathcal{C}_{n_s}))_d$ as $s \to \infty$, it is necessary and sufficient that

$$\left| P^{X_s^{(n_s)}}\left[E_s^{(n_s)}\right] - P^{Y_s^{(n_s)}}\left[E_s^{(n_s)}\right] \right| \to 0 \quad \text{as} \quad s \to \infty$$

for any given sequence of the subset $\left\{E_s^{(n_s)}\right\}$, $s = 1,2,\ldots,$ with $E_s^{(n_s)}$ belonging to $\mathcal{C}_{n_s}$ for each s.

The following inclusion relations are known for the case of *unequal basic spaces* :

(6.4.15) $$(\boldsymbol{G})_d \rightleftarrows (\boldsymbol{B})_d \rightleftarrows (\boldsymbol{A})_d \to (\boldsymbol{S})_d \to (\boldsymbol{M})_d,$$

and for the case of *equal basic spaces* we have

(6.4.16) $$(\boldsymbol{G})_d \rightleftarrows (\boldsymbol{B})_d \rightleftarrows (\boldsymbol{A})_d \to (\boldsymbol{S})_d \rightleftarrows (\boldsymbol{M})_d$$

and

(6.4.17) $$\begin{array}{ccccccc} & & & ((\boldsymbol{G}))_d & & & \\ & \nearrow & & & \searrow & & \\ ((\boldsymbol{B}))_d & \to & ((\boldsymbol{A}))_d & \rightleftarrows & ((\boldsymbol{S}))_d & \rightleftarrows & ((\boldsymbol{M}))_d \end{array}$$

Let $\left\{X_s^{(n_s)}\right\}$, $s = 1,2,\ldots,$ be a sequence of random variables belonging to $F(R_n, \boldsymbol{B}_n)$ and let $\mathcal{C}_n$ be a subclass of $\boldsymbol{B}_n$ for which the following definitions are meaningful.

Definition 6.4.3. We say that the sequence $\left\{X_s^{(n)}\right\}$, $s = 1,2,\ldots,$ has the property $B(\mathcal{C}_n)$ if, for any $\varepsilon > 0$, there exist a bounded subset, say B belonging to $\mathcal{C}_n$ and a positive integer s_0 such that

$$P^{X_s^{(n)}}(B) > 1 - \varepsilon \quad \text{for all} \quad s \geq s_0.$$

Note that $B(\boldsymbol{B})$, $B(\boldsymbol{G})$, $B(\boldsymbol{A})$, and $B(\boldsymbol{S})$ are mutually equivalent properties. We therefore use the property $B(\boldsymbol{S})$ henceforth.

Definition 6.4.4. The sequence $\left\{X_s^{(n)}\right\}$, $s = 1,2,\ldots,$ is said to have the property $C(\mathcal{C}_n)$ if, for any $\varepsilon > 0$, there exist a positive number δ and a positive integer s_0 such that the conditions

$$E \in \mathcal{C}_n \quad \text{and} \quad \nu^{(n)}(E) < \delta$$

imply that

$$P^{X_s^{(n)}}(E) < \varepsilon \quad \text{for all} \quad s \geq s_0,$$

where $\nu^{(n)}$ is the Lebesgue measure over R_n.

Lemma 6.4.4.

(i) If $X_s^{(n)} \sim Y_s^{(n)}, ((\boldsymbol{M}))_d$ $(s \to \infty)$, and either one of the sequences has the property $B(\boldsymbol{S})$, then the other has the same property.

(ii) If $X_s^{(n)} \sim Y_s^{(n)}, ((\boldsymbol{M}))_d$ $(s \to \infty)$, then the sequence $\left\{X_s^{(n)}\right\}$ has the property $B(\boldsymbol{S})$.

(iii) A necessary and sufficient condition for $\left\{X_s^{(n)}\right\}$, $(s = 1,2,\ldots)$, to have the property $B(\boldsymbol{S})$ is that for some decompositions $X_s^{(n)} = \left[X_s^{1(m_1)},\ldots,X_s^{k(m_k)}\right]$, k and $m_1,\ldots,m_k$ being independent of s, every sequence of marginals $\left\{X_s^{i(m_i)}\right\}$, $s = 1,2,\ldots$, has the property $B(\boldsymbol{S})$, $i = 1,\ldots,k$.

(iv) If $\left\{X_s^{(n)}\right\}$, $s = 1,2,\ldots$, where $\bar{X}_s^{(n)} = \left(X_s^1,\ldots,X_s^n\right)$, has the property $B(\boldsymbol{S})$, then $\left\{\bar{X}_s^{(m)}\right\}$, $s = 1,2,\ldots$, where $\bar{X}_s^{(m)} = \left(X_s^{i_1},\ldots,X_s^{i_m}\right)$ has the property $B(\boldsymbol{S})$. Although m is independent of s, the choice of $(i_1,\ldots,i_m)$ out of $(1,2,\ldots,n)$ may depend on s.

Lemma 6.4.5

(i) $C(\mathcal{C}_n) \to C(\mathcal{C}'_n)$ if $\mathcal{C}'_n \subseteq \mathcal{C}_n$.

(ii) The property of $C(\mathcal{C}_n)$ is carried over by the asymptotic equivalence in the sense of the type $((\mathcal{C}_n))_d$.

(iii) If $\left\{X_s^{(n)}\right\}$, $s = 1,2,\ldots$, $X_s^{(n)} = \left(X_s^1 \ldots X_s^n\right)$, has the properties $C(\boldsymbol{S})$ and $B(\boldsymbol{S})$, then the sequence of any marginals $\left\{\bar{X}_s^{(m)}\right\}$, $s = 1,2,\ldots$, has the same properties when m is fixed independently of s while the choice of $(i_1,\ldots,i_n)$ may depend on s.

(iv) If $\mathcal{C}_n^*$ is the class of all finite unions of the members of $\mathcal{C}_n$, then $C(\mathcal{C}_n) \rightleftarrows C(\mathcal{C}_n^*)$. Thus $C(\boldsymbol{S}) \rightleftarrows C(\boldsymbol{A})$.

(v) If $X_s^{(n)} \sim Y^{(n)}((\mathcal{C}_n))_d$ $(s \to \infty)$ for some $Y^{(n)} \in P\left[R_n, \boldsymbol{B}_n, \mu^{(n)}\right]$, then $\left\{X_s^{(n)}\right\}$, $s = 1,2,\ldots$, has the property $C(\mathcal{C}_n)$.

Theorem 6.4.1

(i) If $X_s^{(n)} \sim Y_s^{(n)}((\boldsymbol{M}))_d$, $(s \to \infty)$ and both sequences have the properties $C(\boldsymbol{S})$ and $B(\boldsymbol{S})$, then $X_s^{(n)} \sim Y_s^{(n)}((\boldsymbol{G}))_d$, $(s \to \infty)$.

(ii) If both the sequences $\left\{X_s^{(n)}\right\}$, $\left\{Y_s^{(n)}\right\}$ have the properties $C(\boldsymbol{S})$ and $B(\boldsymbol{S})$, then $X_s^{(n)} \sim Y_s^{(n)}, ((\boldsymbol{G}))_d$, $(s \to \infty)$, implies that $X_s^{(n)} \sim Y_s^{(n)}, ((\boldsymbol{B}))_d$, $(s \to \infty)$.

<u>Theorem 6.4.2.</u> If $X_s^{(n)} \sim Y_s^{(n)}, ((\boldsymbol{M}))_d$, $(s \to \infty)$, and the sequences have the properties $C(\boldsymbol{S})$ and $B(\boldsymbol{S})$, then

$$X_s^{(n)} \sim Y_s^{(n)}, (\boldsymbol{M})_d, \quad (s \to \infty).$$

<u>Corollary 6.4.1.</u> If $X_s^{(n)} \sim Y^{(n)}, ((\boldsymbol{M}))_d$, $(s \to \infty)$, and $Y^{(n)} \in P(R_n, \boldsymbol{B}_n, \nu^{(n)})$, then the convergence is of the type $(\boldsymbol{M})_d$.

Note that the type $((\boldsymbol{M}))_d$ convergence in the above case is equivalent to the *convergence in law*.

<u>Lemma 6.4.6.</u> If $X_s^{(n)} \sim Y_s^{(n)}, ((\boldsymbol{M}))_d$, $(s \to \infty)$, and $Y_s^{(n)} \to c_{(n)}$ in probability as $s \to \infty$, then $X_s^{(n)} \to c_{(n)}$ in probability as $s \to \infty$.

<u>Theorem 6.4.3.</u> Suppose the following conditions are satisfied :

(i) $\left\{X_s^{(n)}\right\}$, $s = 1,2,\ldots$, $X_s^{(n)} = \left(X_s^1 \ldots X_s^n\right)$, has the properties $C(\boldsymbol{S})$ and $B(\boldsymbol{S})$.

(ii) $\left\{c_s^{(n)}\right\}$, $s = 1,2,\ldots$, $c_s^{(n)} = \left(c_s^1,\ldots,c_s^n\right)$, is any given sequence of points of R_n that converges to the point $1^{(n)} = (1,\ldots,1)$.

Then

$$X_s^{(n)} \sim Y_s^{(n)}, ((\boldsymbol{M}))_d \qquad (s \to \infty),$$

where

$$Y_s^{(n)} = \left(c_s^1 X_s^1, \ldots, c_s^n X_s^n\right).$$

Next we consider the $(\mathbf{M})_d$ asymptotic equivalence of marginal random variables.

Suppose that a sequence of random variables $\left\{\left[X_s^{(n_s)}, Z_s^{(m_s)}\right]\right\}$ with $\left\{\left[X_s^{(n_s)}, Z_s^{(m_s)}\right]\right\} \in F(R_{n_s+m_s}, \mathcal{B}_{n_s+m_s})$ for each s is given. Let the cumulative distribution function of $X_s^{(n_s)}$ be $F_s\left[x^{(n_s)}\right]$ and the conditional cumulative distribution function of $Z_s^{(m_s)}$, given $X_s^{(n_s)} = x^{(n_s)}$, be $P_s\left[z^{(m_s)} \middle| x^{(n_s)}\right]$. The cumulative distribution function of $Z_s^{(m_s)}$ is then given by

$$(6.4.18) \qquad H_s\left[z^{(m_s)}\right] = \int_{R_{n_s}} P_s\left[z^{(m_s)} \middle| x^{(n_s)}\right] dF_s\left[x^{(n_s)}\right].$$

Let $\left\{\bar{X}_s^{(n_s)}\right\}$, $s = 1,2,\ldots$, be a sequence of random variables with cumulative distribution function $\bar{F}_s\left[x^{(n_s)}\right]$ and let $\left\{\bar{Z}_s^{(m_s)}\right\}$ be a sequence of random variables whose cumulative distribution is given by

$$(6.4.19) \qquad \bar{H}_s\left[z^{(m_s)}\right] = \int_{R_{n_s}} P_s\left[z^{(m_s)} \middle| x^{(n_s)}\right] d\bar{F}_s\left[x^{(n_s)}\right] \qquad (s = 1,2,\ldots)$$

Our problem is to establish the conditions under which the asymptotic equivalence

$$X_s^{(n_s)} \sim \bar{X}_s^{(n_s)}, (\boldsymbol{M})_d \qquad (s \to \infty)$$

implies that

$$Z_s^{(m_s)} \sim \bar{Z}_s^{(m_s)}, (\boldsymbol{M})_d \qquad (s \to \infty)$$

The same problem arises in the case when $Z_s^{(m_s)}$ has the probability density function $p_s\left[z^{(m_s)} \middle| x^{(n_s)}\right]$ given $X_s^{(n_s)} = x^{(n_s)}$, for each s. In this case $\bar{Z}_s^{(m_s)}$ has the probability density function

$$\bar{h}_s\left[z^{(m_s)}\right] = \int_{R_{n_s}} p_s\left[z^{(m_s)} \middle| x^{(n_s)}\right] dF_s\left[x^{(n_s)}\right]. \tag{6.4.20}$$

Theorem 6.4.4.

(i) Let $\bar{Z}_s^{(m_s)}$ be a random variable with the cumulative distribution function given by (6.4.19) for each s. Then the condition

$$X_s^{(n_s)} \sim \bar{X}_s^{(n_s)}, (\boldsymbol{B})_d \qquad (s \to \infty)$$

implies that

$$Z_s^{(m_s)} \sim \bar{Z}_s^{(m_s)}, (\boldsymbol{M})_d \qquad (s \to \infty).$$

(ii) In the case when $Z_s^{(m_s)}$ has the conditional probability density function and $\bar{Z}_s^{(m_s)}$ is a random variable with the probability density function given by (6.4.20), then

$$X_s^{(n_s)} \sim \bar{X}_s^{(n_s)}, (\boldsymbol{B})_d \qquad (s \to \infty)$$

implies that

$$Z_s^{(m_s)} \sim \bar{X}_s^{(m_s)}, (\mathcal{B})_d \qquad (s \to \infty)$$

Theorem 6.4.5. Let $\bar{Z}_s^{(m_s)}$ be a random variable with the probability density function given by (6.4.20) and assume the following conditions

(i) For some real $c_s^{(n)} = \left(c_s^1 \ldots c_s^n\right)$ and $d_s^{(n)} = \left(d_s^1 \ldots d_s^n\right)$ the sequence of random variables $\left\{\bar{Y}_s^{(n)}\right\}$, $s = 1,2,\ldots$, has the property $B(S)$, where

$$\bar{Y}_s^{(n)} = \left(\frac{\bar{X}_s^1 - d_s^1}{c_s^1}, \ldots, \frac{\bar{X}_s^n - d_s^n}{c_s^n}\right)$$

for each s.

(ii) For the constants given in (i), put

$$x_s^{(n)} = (c_s^1 x^1 + d_s^1, \ldots, c_s^n x^n + d_s^n)$$

for $x^{(n)} = (x^1, \ldots, x^n)$ and

$$Q_s\left[z^{(m)} \mid x^{(n)}\right] = P_s\left[z^{(m)} \mid x_s^{(n)}\right]$$

for each s. Then, for any given $\varepsilon > 0$ and any given bounded subset $B \in \mathcal{S}_n$, there exist a positive number $\delta = \delta(\varepsilon, B)$ and a positive integer $s_0 = s_0(\varepsilon, B)$ such that $\left|x^{(n)} - y^{(n)}\right| < \delta$, with $x^{(n)}, y^{(n)} \in B$, implies that

$$\sup_{z^{(m)} \in R_m} \left| Q_s\left[z^{(m)} \mid x^{(n)}\right] - Q_s\left[z^{(m)} \mid y^{(n)}\right] \right| < \varepsilon$$

for all $s \geq s_0$.

Under these conditions $X_s^{(n)} \sim \bar{X}_s^{(n)}, (\boldsymbol{M})_d$, $(s \to \infty)$, implies that $Z_s^{(m)} \sim \bar{Z}_s^{(m)}, (\boldsymbol{M})_d$, $(s \to \infty)$. If $X_s^{(n)} \to \bar{X}^{(n)}, (\boldsymbol{M})_d$, then $Z_s^{(m)} \to \bar{Z}^{(m)}$, $(\boldsymbol{M})_d$, $s \to \infty$. $X^{(n)}$ being a member of $\mathcal{F}(R_n, B_n)$ the first condition of Theorem 5 is automatically satisfied.

Corollary 6.4.2. Suppose that, for any given $\varepsilon > 0$ and any given bounded subset $B \in \boldsymbol{S}_n$, there exist a positive number $\delta = \delta(\varepsilon, B)$ and a positive integer $s_0 = s_0(\varepsilon, B)$ such that $\left|x^{(n)} - y^{(n)}\right| < \delta$, with $x^{(n)}, y^{(n)} \in B$, implies that

$$\sup_{z^{(m)} \in R_m} \left| P_s\left[z^{(m)} \middle| x^{(n)}\right] - P_s\left[z^{(m)} \middle| y^{(n)}\right] \right| < \varepsilon$$

for all $s \geq s_0$.

Then $X_s^{(n)} \to \bar{X}^{(n)}, (\boldsymbol{M})_d$, $(s \to \infty)$, implies that $Z_s^{(n)} \sim \bar{Z}_s^{(m)}, (\boldsymbol{M})_d$, $(s \to \infty)$. Here $\bar{Z}^{(m)}$ stands for the random variable whose cumulative distribution function is given by

$$\bar{H}_s\left(z^{(m)}\right) = \int_{R_n} P_s\left[z^{(m)} \middle| x^{(n)}\right] d\bar{F}\left[x^{(n)}\right],$$

with $\bar{F}\left[x^{(n)}\right]$ being the cumulative distribution function of $\bar{X}^{(n)}$.

Lemma 6.4.7. For the condition (ii) of Theorem 6.4.5 to hold, it is sufficient that

(i) For any given $z^{(m)}$ in R_m and for each s, the function $q_s\left[z^{(m)} \middle| x^{(n)}\right] = p_s\left[z^{(m)} \middle| x_s^{(n)}\right]$ is totally differentiable with respect to $x^{(n)}$ in R_n.

(ii) For any given bounded subset $B \in \boldsymbol{S}_n$, there exist a positive number $\eta = \eta(B)$ and a positive integer $s_0 = s_0(B)$ such that

$$\sup_{x^{(n)} \in B} \int_{R_m} \left| \phi_{si}\left[z^{(m)} \middle| x^{(n)}\right] d\nu^{(m)}\left[z^{(m)}\right] \right| \leq \eta \qquad (i = 1,\ldots,n)$$

for all $s \geq s_0$, where

$$\phi_{si}\left[z^{(m)} \middle| x^{(n)}\right] = \frac{\partial}{\partial x_i} q_s\left[z^{(m)} \middle| x^{(n)}\right] \qquad (i = 1,\ldots,n).$$

Now we consider the case of equal basic spaces. Let $\left\{\left(X_s^{(n)}, Z_s^{(m)}\right)\right\}$, $s = 1,2,\ldots$ be a sequence of random variables belonging to $F(R_{n+m}, B_{n+m})$, and let $F_s\left(x^{(n)}\right)$ and $P_s\left(z^{(m)} \middle| x^{(n)}\right)$ be cumulative distribution function of $X_s^{(n)}$ and conditional cumulative distribution function of $Z_s^{(m)}$ given $X_s^{(n)} = x^{(n)}$, respectively. When $Z_s^{(m)}$ has the conditional probability density function, then let it be $p_s\left(z^{(m)} \middle| x^{(n)}\right)$. Suppose, for some real c_s^i and d_s^i, $i = 1,\ldots, n$ and $s = 1,2,\ldots,$ the random variables

$$Y_s^{(n)} = \left(\frac{X_s^1 - d_s^1}{c_s^1}, \ldots, \frac{X_s^n - d_s^n}{c_s^n}\right)$$

converges in probability to the point $\underline{1}^{(n)} = (1,\ldots,1)$ as $s \to \infty$. Under this situation, a question is asked whether $\left\{Z_s^{(m)}\right\}$ and $\left\{\bar{Z}_s^{(m)}\right\}$ are asymptotically equivalent in the sense of the type $(\boldsymbol{M})_d$, where

$\bar{Z}_s^{(n)}$ stands for a random variable whose cumulative distribution function is given by $P_s\left(z^{(m)} \middle| c_s^{(n)} + d_s^{(n)}\right)$ for each s, provided that this is a member of $\mathcal{F}(R_m, B_m)$.

Theorem 6.4.6. Under the situation stated above suppose that

$$Y_s^{(n)} \to \underline{1}^{(n)} \quad \text{in probability as } s \to \infty,$$

and $P_s\left(z^{(m)} \middle| c_s^{(n)} + d_s^{(n)}\right)$ is the cumulative function of $\bar{Z}^{(m)} \in \mathcal{F}(R_m, B_m)$ for each s. Then in order that $Z_s^{(m)} \sim \bar{Z}_s^{(m)}, (\mathcal{M})_d$, $s \to \infty$, it is sufficient that

(i) For any given $\varepsilon > 0$, there exist a positive number $\delta = \delta(\varepsilon)$ and a positive integer $s_0 = s_0(\varepsilon)$ such that

$$\left| x^{(n)} - \underline{1}^{(n)} \right| < \delta$$

implies

$$\sup_{z^{(m)} \in R_m} \left| Q_s\left(z^{(m)} \middle| x^{(n)}\right) - Q_s\left(z^{(m)} \middle| \underline{1}^{(n)}\right) \right| < \varepsilon$$

for all $s \geqq s_0$.

6.5. Rigorous Derivation of the Asymptotic Null Distribution of the F-Statistic After Randomization

We have been concerned with the distribution of the F-statistic occurring in the analysis of variance of a partially balanced incomplete block design and m associate classes, v treatments with an association of m associate classes, b blocks of size k each and

r replications of each treatment, the number of incidence in blocks of any pair of treatments being λ_u if they are uth associates. The randomization procedure has been applied in the allocation of k treatments of k experimental units in each block independently from block to block.

Let us fix a special way of numbering all the $n(= vr = bk)$ experimental units in such a way that the ith unit of the pth block bears the number $f = (p - 1)k + i$. Let the incidence matrices of treatments and blocks be Φ and Ψ, respectively. Then the Neyman model assuming no interaction between the treatments and the units is given by

(6.5.1) $$\underline{x} = \gamma\underline{1} + \Phi\underline{\tau} + \Psi\underline{\beta} + \underline{\pi} + \underline{e} ,$$

where $\underline{x} = (x_1,\ldots,x_n)'$ is the observation vector; γ is the general mean; $\underline{1} = (1,\ldots,1)'$, $\underline{\tau} = (\tau_1,\ldots,\tau_v)'$, and $\underline{\beta} = (\beta_1,\ldots,\beta_b)'$ are treatment effect and block effect vectors subject to the restrictions $\tau_1 + \cdots + \tau_v = 0$ and $\beta_1 + \cdots + \beta_b = 0$, respectively; $\underline{\pi} = (\pi_1,\ldots,\pi_n)'$ stands for the unit-error vector subject to the restriction $\Psi'\underline{\pi} = \mathbf{0}$; and finally $\underline{e} = (e_1,\ldots,e_n)'$ is the technical-error vector, which is supposed to be distributed as the normal distribution $N(0,\sigma^2 I)$, with σ^2 being unknown.

Let $A_0^{\#} = G_v/v, A_1^{\#},\ldots,A_m^{\#}$ be $m + 1$ mutually orthogonal idempotent matrices in the association algebra and let h be any given integer such that $1 \leqq h \leqq m$. We have been interested in testing a "partial null hypothesis"

(6.5.2) $$H_{0(h)} : A_u^{\#}\underline{\tau} = 0 \qquad (u = 1,\ldots,h),$$

which evidently reduces to the total null hypothesis $H_0 : \underline{\tau} = 0$ if $h = m$.

To test the null hypothesis $H_{0(h)}$, we considered the partial sum of squares due to treatments adjusted by blocks,

$$(6.5.3) \qquad S^2_{t(h)} = \underline{x}'\left(\sum_{u=1}^{h} V^{\#}_u\right)\underline{x},$$

and the sum of squares due to error

$$(6.5.4) \qquad S^2_e = \underline{x}'\left(I_n - \frac{B}{k} - \sum_{u=1}^{h} V^{\#}_u\right)\underline{x},$$

where

$$(6.5.5) \qquad V^{\#}_u = c_u(I_n - \frac{B}{k})T^{\#}_u(I_n - \frac{B}{k}) \qquad (u = 1,\ldots,m),$$

with

$$(6.5.6, \qquad B = \Psi\Psi', \quad T^{\#}_u = \Phi A^{\#}_u \Phi', \quad \text{and} \quad c_u = \frac{k}{(rk - \rho_u)} \qquad (u = 1,\ldots,m).$$

Here $\rho_0 = rk, \rho_1, \ldots, \rho_m$ are the characteristic roots of the symmetric matrix NN', where $N = \Phi'\Psi$ is the incidence matrix of the design under consideration with the respective multiplicities $\alpha_0 = 1, \alpha_1, \ldots, \alpha_m$ for which $\alpha_0 + \alpha_1 + \cdots + \alpha_m = v$.

Here some words on the statistical meaning of the above statistics would be in order. Let T_α be the sum of the yields obtained from the units to which the treatment α is allocated and let B_a be the sum or the yields from the units in block a. Then

$$\begin{bmatrix} T_1 \\ \vdots \\ T_v \end{bmatrix} = \Phi'\underline{x} \quad \text{and} \quad \begin{bmatrix} B_1 \\ \vdots \\ B_b \end{bmatrix} = \Psi'\underline{x}.$$

Hence

$$\Phi'(I_n - \frac{B}{k})\underline{x} = \Phi'\underline{x} - \frac{N}{k}\Psi'\underline{x} = \begin{bmatrix} Q_1 \\ \vdots \\ Q_v \end{bmatrix}, \text{ say}$$

One obtains the treatment sum adjusted by block :

$$Q_\alpha = T_\alpha - \frac{1}{k}(n_{\alpha 1}B_1 + \cdots + n_{\alpha b}B_b) \qquad \alpha = 1,\ldots,v.$$

Thus one gets

$$\underline{x}'V_u^{\#}\underline{x} = \frac{k}{rk - \rho_u}\underline{Q}'A_u^{\#}\underline{Q}.$$

Let us put $\bar{\alpha} = \alpha_1 + \cdots + \alpha_h$ and $\bar{\bar{\alpha}} = \alpha_{h+1} + \cdots + \alpha_m$. Then it was shown that the null distribution of the F-statistic

$$F = \frac{n - b - v + 1}{\bar{\alpha}} \frac{S_{t(b)}^2}{S_e^2} \tag{6.5.7}$$

before randomization is the non-central F-distribution whose probability element is given by

(6.5.8)

$$exp\left(-\frac{\Delta}{2\sigma^2}\right) \sum_{l=0}^{\infty} \frac{(\Delta/2\sigma^2)^l}{l!} \sum_{\mu+\nu+k=l} \frac{l!}{\mu!\nu!k!} \bar{\theta}^{\mu}\bar{\bar{\theta}}^{\kappa}$$

$$\times (1 - \bar{\theta} - \bar{\bar{\theta}})^{\nu} \frac{\Gamma\{(n - b - \bar{\bar{\alpha}})/2] + \mu + \nu\}}{\Gamma\left[(\bar{\alpha}/2) + \mu\right]\Gamma\{[(n - b - v + 1)/2] + \nu\}}$$

$$\times \left(\frac{\bar{\alpha}}{n - b - v + 1}F\right)^{(\bar{\alpha}/2)+\mu-1}$$

$$\times \left(1 + \frac{\bar{\alpha}}{n - b - v + 1}F\right)^{-[(n-b-\bar{\bar{\alpha}})/2+\mu+\nu]} d\left(\frac{\bar{\alpha}}{n - b - v + 1}F\right),$$

where

$$\Delta = \underline{\pi}'\underline{\pi}, \quad \bar{\theta} = \underline{\pi}'\left(\sum_{u=1}^{h} V_u^{\#}\right)\underline{\pi}/\Delta, \quad \text{and} \quad \bar{\bar{\theta}} = \underline{\pi}'\left(\sum_{u=h+1}^{m} V_u^{\#}\right)\underline{\pi}/\Delta. \tag{6.5.9}$$

In Section 6.3 we showed by comparing the means, variances, and covariances of both distributions in a quite tedious way, that the permutation distribution of $(\bar{\theta},\bar{\bar{\theta}})$ due to randomization can be approximated asymptotically when $b \to \infty$ by the Dirichlet distribution, whose probability element is given by

(6.5.10)
$$\frac{\Gamma[(n-b)/2]}{\Gamma(\bar{\alpha}/2)\Gamma(\bar{\bar{\alpha}}/2)\Gamma[(n-b-v-1)/2]} \bar{\theta}^{(\bar{\alpha}/2)-1}\,\bar{\bar{\theta}}^{(\bar{\bar{\alpha}}/2)-1}$$
$$\times (1 - \bar{\theta} - \bar{\bar{\theta}})^{[(n-b-v+1)/2]-1}\, d\bar{\theta}\, d\bar{\bar{\theta}}$$
$$(0 < \bar{\theta},\ \bar{\bar{\theta}} < 1,\ \bar{\theta} + \bar{\bar{\theta}} < 1),$$

provided the rather rigid uniformity conditions

(6.5.11) $$\Delta_p = \sum_{i=1}^{k} \pi_i^{(p)^2} = \Delta_0 \quad \text{and} \quad \Gamma_p = \sum_{i=1}^{k} \pi_i^{(p)^4} = \Gamma_0 \quad (p = 1,\ldots,b),$$

where we have put $\pi_f = \pi_i^{(p)}$ for $f = (p-1)k + i$, are satisfied. Then we integrated out the factors $\bar{\theta}^{\mu}\bar{\bar{\theta}}^{\kappa}(1 - \bar{\theta} - \bar{\bar{\theta}})^{\nu}$ in 6.5.8. with respect to the Dirichlet distribution (6.5.10) to obtain the asymptotically approximate distribution of the F-statistic (6.5.7) after the randomization. This turned out to be the central F-distribution with $(\bar{\alpha},\ n - b - v + 1)$ degrees of freedom. When $h = m$ i.e., for testing the null hypothesis $H_0 : \tau = 0$, this reduces to the familiar central F-distribution with $(v - 1,\ n - b - v + 1)$ degrees of freedom.

However up to Section 6.3, all the arguments concerning the asymptotic approximation were quite heuristic, and from the rigorous mathematical point of view there are two questions to be settled : (1) in what sense or in terms of what kind of distance is the permutation distribution of $(\bar{\theta},\bar{\bar{\theta}})$ due to randomization asymptotically approximated by the Dirichlet distribution (6.5.10) as $b \to \infty$? and (2) if the permutation distribution of $(\bar{\theta},\bar{\bar{\theta}})$ is asymptotically equivalent to the Dirichlet distribution (6.5.10) in terms of a certain distance, is it true that the null distribution of the F-statistic is asymptotically equivalent to the probability distribution whose probability element is obtained by integrating out the factors $\bar{\theta}^{\mu}\bar{\bar{\theta}}^{\kappa}(1 - \bar{\theta} - \bar{\bar{\theta}})^{\nu}$ with respect to the Dirichlet distribution (6.5.10), that is, the central F-distribution with $(\bar{\alpha}, n - b - v + 1)$ in terms of the same distance?

Based on the theory of the asymptotic equivalence of probability distributions developed in Section 6.4, we treat those two problems and give satisfactory answers in the affirmative. The calculations are much reduced, with the uniformity conditions (6.5.11) being somewhat relaxed.

In this section it is shown that the permutation distribution of $(\bar{\theta},\bar{\bar{\theta}})$ due to randomization is asymptotically equivalent $\left((\boldsymbol{M})_d\right)$, under rather mild uniformity conditions, to the joint distribution of the two mutually independent χ^2 variates $\left[\chi^2_{\bar{\alpha}}/(n-b),\ \chi^2_{\bar{\bar{\alpha}}}/(n-b)\right]$ as $b \to \infty$. This can be proved by making use of the central limit theorem due to W. Feller.

Then, it is first shown that the joint distribution of mutually independent χ^2 variates above is asymptotically equivalent to the Dirichlet distribution (6.5.10) in the sense of the type $(\boldsymbol{M})_d$, which was briefly explained in Section 6.4. Then a result in the theory of asymptotic equivalence, that is, Corollary 6.4.2 is applied to show that the central F-distribution with $(\bar{\alpha}, n - b - v + 1)$ degrees of freedom is asymptotically equivalent in the sense of the type $(\boldsymbol{M})_d$ to the null distribution of the F-statistic (6.5.7) after randomization as $b \to \infty$.

Asymptotic Distribution of the Permutation Distribution of $(\bar{\theta},\bar{\bar{\theta}})$ Due to Randomization. We derive a probability distribution that is asymptotically equivalent in the sense of the type $(\boldsymbol{M})_d$ to the permutation distribution of $(\bar{\theta},\bar{\bar{\theta}})$ due to randomization as $b \to \infty$.

The limiting process, under which our discussions are made, is as follows :

(6.5.12) $b \to \infty$ while $v, k, n_1, \ldots, n_m$ and p^i_{jk}, $i, j, k = 0, 1, \ldots, m$, are kept fixed.

Thus r and at least one of the λ_u, $u = 1, \ldots, m$, should be of the same order of magnitude as b. Hereafter throughout this Section, we are concerned with this limiting process and denote it by $b \to \infty$. It should be noted that this limiting process cannot be realized by any symmetric PBIB design, but by an asymmetric PBIB design with considerably large b and r compared with v and k.

The randomization procedure under consideration is described as follows :

Let

$$\sigma_p = \begin{pmatrix} 1 & 2 & \cdots & k \\ \sigma_p(1) & \sigma_p(2) & \cdots & \sigma_p(k) \end{pmatrix} \qquad (p = 1, \ldots, b)$$

be the set of permutation associated with the randomization in the pth block and let

$$(6.5.13) \quad U_\sigma = \begin{Vmatrix} S\sigma_1 & & & 0 \\ & S\sigma_2 & & \\ & & \ddots & \\ 0 & & & S\sigma_b \end{Vmatrix}$$

be the permutation matrix corresponding to the randomization, where S_{σ_p} is the $k \times k$ permutation matrix corresponding to σ_p, $p = 1, \ldots, b$. For any fixed incidence matrix Φ of treatments under the special labeling of all the units mentioned in the preceding section, a realization of the incidence matrix as the result of the randomization is one of the $(k!)^b$ matrices $\{U_\sigma \Phi\}$ with equal probability $1/(k!)^b$.

It is also noted here that the matrix B defined by (6.5.6) is of the form

$$(6.5.14)\quad B = \begin{Vmatrix} G_k & & & 0 \\ & G_k & & \\ & & \ddots & \\ 0 & & & G_k \end{Vmatrix},$$

where G_k is the $k \times k$ matrix whose elements are all unity.

Now, since the $m + 3$ matrices $V_u^{\#}$, $u = 1,\ldots,m$,

$$(6.5.15)\quad \left(\frac{1}{k}\right)B - \left(\frac{1}{n}\right)G_n,\quad \left(\frac{1}{n}\right)G_n \text{ and } I - \left(\frac{1}{k}\right)B - \sum_{u=1}^{m} V_u^{\#}$$

with the respective ranks α_u, $u = 1,\ldots,m$, $b - 1$, 1, and $n - b - v + 1$ are mutually orthogonal idempotent matrices, there exists an orthogonal matrix of order n

$$(6.5.16)\quad P = \begin{Vmatrix} C \\ \cdots \\ D \end{Vmatrix},$$

with

$$(6.5.17)\quad C = \begin{Vmatrix} c_{11}^{(1)} & \cdots & c_{1k}^{(1)} & \cdots & c_{11}^{(p)} & \cdots & c_{1k}^{(p)} & \cdots & c_{11}^{(b)} & \cdots & c_{1k}^{(b)} \\ c_{21}^{(1)} & \cdots & c_{2k}^{(1)} & \cdots & c_{21}^{(p)} & \cdots & c_{2k}^{(p)} & \cdots & c_{21}^{(b)} & \cdots & c_{2k}^{(b)} \\ \vdots & & \vdots & & \vdots & & \vdots & & \vdots & & \vdots \\ c_{v-11}^{(1)} & \cdots & c_{v-1k}^{(1)} & \cdots & c_{v-11}^{(p)} & \cdots & c_{v-1k}^{(p)} & \cdots & c_{v-11}^{(b)} & \cdots & c_{v-1k}^{(b)} \end{Vmatrix}$$

such that

$$(6.5.18a)\quad PV_u^{\#}P' = \begin{Vmatrix} 0 & & & & 0 \\ & \ddots & & & \\ & & 0 \; I_{\alpha_u} \; 0 & & \\ & & & \ddots & \\ 0 & & & & 0 \end{Vmatrix} \qquad (u = 1,\ldots,m),$$

$$(6.5.18b)\quad P\left(I_n - \frac{1}{k}B - \sum_{u=}^{m} V_u^{\#}\right)P' = \left\| \begin{array}{cccccc} 0 & & & & & 0 \\ & \ddots & & & & \\ & & 0 & & & \\ & & & I_{n-b-v+1} & & \\ & & & & 0 & \\ & & & & & \ddots \\ 0 & & & & & 0 \end{array} \right\|,$$

$$(6.5.18c)\quad P\left(\frac{1}{k}B - \frac{1}{n}G_n\right)P' = \left\| \begin{array}{ccccccc} 0 & & & & & & 0 \\ & \ddots & & & & & \\ & & 0 & & & & \\ & & & 0 & & & \\ & & & & \ddots & & \\ & & & & & 0 & \\ & & & & & & I_{b-1} \\ 0 & & & & & & 0 \end{array} \right\|,$$

and

$$(6.5.18d)\quad P\left(\frac{1}{n}G\right)P' = \left\| \begin{array}{ccc} 0 & & 0 \\ & \ddots & \\ & & 0 \\ 0 & & 1 \end{array} \right\|$$

simultaneously. Partitioning the submatrix C of P into two parts

$$C = \left\| \begin{array}{c} C_1 \\ \cdots \\ C_2 \end{array} \right\|,$$

where C_1 and C_2 are matrices consisting of the first $\bar{\alpha}$ and the remaining $\bar{\bar{\alpha}}$ rows of C, respectively, we obtain, by (6.5.18a)

$$(6.5.19)\quad C_1\left(\sum_{u=1}^{m} V_u^{\#}\right)C_1' = I_{\bar{\alpha}} \quad \text{and} \quad C_2\left(\sum_{u=b+1}^{m} V_u^{\#}\right)C_2' = I_{\bar{\bar{\alpha}}}.$$

From (6.5.18c) and (6.5.18d) it follows that

$$(6.5.20)\quad \sum_{i=1}^{k} c_{si}^{(p)} = 0 \qquad (s = 1,\ldots, v-1;\ \ p = 1,\ldots,b).$$

Since the unit-error vector $\underline{\pi}$ is annihilated by B and hence by G_n, and the $m + 3$ matrices in (6.5.15) sum up to I_n, it follows from (6.5.18) that

$$\Delta = \underline{\pi}'\underline{\pi} = \underline{\pi}'U_\sigma'U_\sigma\underline{\pi}$$

$$= \underline{\pi}'U_\sigma'P' \begin{Vmatrix} I_{\bar{\alpha}} & 0 \\ 0 & 0 \end{Vmatrix} PU_\sigma\underline{\pi} + \underline{\pi}'U_\sigma'P' \begin{Vmatrix} 0 & 0 & 0 \\ 0 & I_{\bar{\bar{\alpha}}} & 0 \\ 0 & 0 & 0 \end{Vmatrix}$$

$$\times PU_\sigma\underline{\pi} + \underline{\pi}'U_\sigma'P' \begin{Vmatrix} 0 & 0 \\ 0 & I_{n-v+1} \end{Vmatrix} PU_\sigma\underline{\pi},$$

whence one may write

$$\bar{\theta} = \frac{(C_1U_\sigma\underline{\pi})'(C_1U_\sigma\underline{\pi})}{\Delta},$$

$$\bar{\bar{\theta}} = \frac{(C_2U_\sigma\underline{\pi})'(C_2U_\sigma\underline{\pi})}{\Delta}, \tag{6.5.21}$$

$$\theta = \bar{\theta} + \bar{\bar{\theta}} = \frac{(CU_\sigma\underline{\pi})'(CU_\sigma\underline{\pi})}{\Delta}.$$

Now we shall show the following :

Lemma 6.5.1. The elements of the matrix C given by (6.5.17) are uniformly of the order of magnitude $O(1/\sqrt{b})$ as $b \to \infty$; that is, there exists a positive constant κ that is independent of b such that

$$\max_{s,i,p} \left| c_{si}^{(p)} \right| \leqq \frac{\kappa}{\sqrt{b}} \tag{6.5.22}$$

for sufficiently large values of b.

Proof. First, it follows from (6.5.16) and (6.5.17) that there exists a positive number κ_1 independent of b such that

(6.5.23) $$\max_{i,j} \left| \left(\sum_{u=1}^{m} V_u^{\#} \right)_{(i,j)} \right| \leqq \kappa_1 \max_u |c_u|$$

for sufficiently large values of b, where $A_{(i,j)}$ denotes the (i,j)-element of the matrix A. But, since

$$rk = \sum_{i=0}^{m} n_i \lambda_i, \quad \rho_u = \sum_{i=0}^{m} z_{ui} \lambda_i \qquad (u = 1, \ldots, m),$$

it follows that

$$rk - \rho_u = \sum_{i=1}^{m} (n_i - z_{ui}) \lambda_i \qquad (u = 1, \ldots, m),$$

where z_{ui}, $u = 0, 1, \ldots, m$, are the characteristic roots of the matrix $\|p_{ji}^k\|$, $j,k = 0, 1, \ldots, m$. Here it is known that $z_{0i} = n_i$ and $|z_{ui}| < n_i$, $u = 1, \ldots, m$. Thus one sees that $rk - \rho_u = O(b)$ and hence

$$c_u = O\left(\frac{1}{b}\right) \quad \text{as} \quad b \to \infty \quad \text{for } u = 1, \ldots, m.$$

Hence it follows from (6.5.23) that there exists a positive constant being independent of b such that

(6.5.24) $$\max_{i,j} \left| \left(\sum_{u=1}^{m} V_u^{\#} \right)_{(i,j)} \right| \leqq \frac{\kappa_2}{b}$$

for sufficiently large values of b.

From (6.5.18) we can see that any given row, say $\underline{c}'$ of the matrix C is a linear combination of $v - 1$ linearly independent columns of the matrix $\sum_{u=1}^{m} V_u^{\#}$, that is,

(6.5.25) $$\underline{c} = \sum_{i=1}^{v-1} s_i \underline{v}_i,$$

where $\underline{v}_1, \underline{v}_2, \ldots, \underline{v}_{v-1}$ is a set of linearly independent column vectors of $\sum_{u=1}^{m} V_u^{\#}$ and $\underline{s}' = (s_1, \ldots, s_{v-1})$ is a real vector.

Let us put

$$W = r\left\|\underline{v}_i'\underline{v}_j\right\|.$$

Then, by the orthogonality of the matrix P, we get the relation

$$\underline{c}'\underline{c} = \frac{1}{r}\,\underline{s}'W\underline{s} = 1. \tag{6.5.26}$$

Since the matrix $\sum_{u=1}^{m} V_u^{\#}$ is idempotent, $\underline{v}_i\underline{v}_j$ must be an element of $\sum_{u=1}^{m} V_u^{\#}$ and hence it is seen from (6.5.24) that

$$\max_{i,j} \left|W_{(i,j)}\right| \leqq \kappa_2.$$

Furthermore, since W is positive definite, its characteristic roots $\delta_1, \ldots, \delta_{v-1}$, say, are all positive and of the order $O(1)$ as $b \to \infty$. Hence there exists an orthogonal matrix Q whose elements are of the order $O(1)$ as $b \to \infty$, such that

$$QWQ' = \begin{Vmatrix} \delta_1 & & & 0 \\ & \delta_2 & & \\ & & \ddots & \\ 0 & & & \delta_{v-1} \end{Vmatrix}.$$

Hence, putting $\underline{t} = Q\underline{s} = (t_1, \ldots, t_{v-1})'$, one can see by (6.5.26) that

$$\underline{t}'\underline{t} = \underline{s}'\underline{s} \quad \text{and} \quad \sum_{i=1}^{v-1} \delta_i t_i^2 = r,$$

from which is follows that

$$\max_i \left|t_i\right| < \kappa_3 \sqrt{b} \quad \text{and hence} \quad \max_i \left|s_i\right| < \kappa_3 \sqrt{b}$$

for sufficiently large values of b, where κ_3 is a positive constant independent of b. Hence one can conclude, from (6.5.25), that each element of the vector $\underline{c}$ is of the order $C(1/\sqrt{b})$ as $b \to \infty$, as was to be proved.

Suppose that the following uniformity conditions are satisfied :

$$\text{(6.5.27)} \quad \bar{\Delta} \equiv \frac{1}{b}\sum_{p=1}^{b} \Delta_p \to \Delta_0 \quad \text{and} \quad \frac{1}{b}\sum_{p=1}^{b}\left[\Delta_p - \bar{\Delta}\right]^{3/2+\delta} \to 0 \ , \text{ as } b \to \infty,$$

for some $\delta > 0$.

Lemma 6.5.2. The permutation distribution of $(C_1 U_{\underline{\sigma}}\underline{\pi}, C_2 U_{\underline{\sigma}}\underline{\pi})$ converges in law to the bivariate normal distribution

$$N\left[0, \frac{\Delta_0}{(k-1)} \cdot \left\| \begin{matrix} I_{\bar{\alpha}} & 0 \\ 0 & I_{\bar{\bar{\alpha}}} \end{matrix} \right\| \right] \quad \text{as } b \to \infty$$

under conditions (6.5.27).

Proof. It will be sufficient to show that $CU_{\underline{\sigma}}\underline{\pi}$ converges in law to $N\left[0, \Delta_0/(k-1) \cdot I_{v-1}\right]$ as $b \to \infty$. To prove this we make use of the following central limit theorem due to W. Feller.

Theorem 6.5.1. (W. Feller). Let $X_1^{(n)}, \ldots, X_n^{(n)}$ be a sequence of independent random variables of k dimensions such that

$$E\left[X_i^{(n)}\right] = 0 \quad \text{and} \quad D\left[X_i^{(n)}\right] = \Lambda_i^{(n)} \qquad (i = 1,\ldots,n; n = 1,2,\ldots)$$

Suppose that the conditions

$$\text{(6.5.28)} \qquad \frac{1}{n}\sum_{i=1}^{n} \Lambda_i^{(n)} \to \Lambda(\neq 0) \quad \text{as} \quad n \to \infty$$

and

$$(6.5.29)\quad \frac{1}{n}\sum_{i=1}^{n}\int_{\|\underline{x}\| > \varepsilon\sqrt{n}} \|\underline{x}\|^2 dF_i^{(n)} \to 0 \quad \text{as} \quad n \to \infty,$$

are satisfied, where $F_i^{(n)}$ denotes the distribution function of $X_i^{(n)}$ and $\|\underline{x}\|$ is the Euclidean norm of the vector $\underline{x}$. Then it holds that

$$L\left[\frac{1}{\sqrt{n}}\sum_{i=1}^{n} X_i^{(n)}\right] \to N(0,\Lambda) \quad \text{as} \quad n \to \infty.$$

Proof. Let us first put

$$(6.5.30)\quad CU_{\sigma}\underline{\pi} = \sum_{p=1}^{b} \underline{\xi}^{(p)}, \underline{\xi}^{(p)} = \begin{bmatrix} \sum_{i=1}^{k} c_{1i}^{(p)}\pi_{\sigma(i)}^{(p)} \\ \vdots \\ \sum_{i=1}^{k} c_{v-1i}^{(p)}\pi_{\sigma(i)}^{(p)} \end{bmatrix} \quad (p = 1,\ldots,b).$$

Since

$$\mathcal{E}\left[\pi_{\sigma(i)}^{(p)}\right] = 0, \quad \mathcal{E}\left[\pi_{\sigma(i)}^{(p)2}\right] = \Delta_p/k,$$

$$\mathcal{E}\left[\pi_{\sigma(i)}^{(p)}\pi_{\sigma(j)}^{(p)}\right] = -\Delta_p/k(k-1), \; i \neq j,$$

and $\{\underline{\xi}^{(1)},\ldots,\underline{\xi}^{(b)}\}$ is a system of independent $(v - 1)$-dimensional random vectors, we have

$$(6.5.31)\quad \mathcal{E}(CU_{\sigma}\underline{\pi}) = 0$$

and

$$(6.5.32)\quad \mathcal{E}[(CU_{\sigma}\underline{\pi})(CU_{\sigma}\underline{\pi})'] = \sum_{p=1}^{b} C\mathcal{E}\left[\underline{\pi}^{(p)}\underline{\pi}^{(p)'}\right]C'$$

$$= \sum_{p=1}^{b} \left\| \frac{\Delta_p}{k}\sum_{i=1}^{k} c_{si}^{(p)}c_{ti}^{(p)} - \frac{\Delta_p}{k(k-1)}\sum_{i\neq j} c_{si}^{(p)}c_{tj}^{(p)} \right\|,$$

$$= \sum_{p=1}^{b} \frac{\Delta_p}{k-1} \left\| \sum_{i=1}^{k} c_{si}^{(p)} c_{ti}^{(p)} \right\|$$

$$= \frac{\bar{\Delta}}{k-1} I_{v-1} + \frac{1}{k-1} \sum_{p=1}^{b} (\Delta_p - \bar{\Delta}) \left\| \sum_{i=1}^{k} c_{si}^{(p)} c_{ti}^{(p)} \right\|$$

$$(s, t = 1, \ldots, v-1).$$

By Hölder's inequality one gets

$$\left| \sum_{p=1}^{b} (\Delta_p - \bar{\Delta}) \sum_{i=1}^{k} c_{si}^{(p)} c_{ti}^{(p)} \right| \leqq \left(\sum_{p=1}^{b} |\Delta_p - \bar{\Delta}|^{\mu} \right)^{1/\mu} \tag{6.5.33}$$

$$\times \left| \sum_{p=1}^{b} \left| \sum_{i=1}^{k} c_{si}^{(p)} c_{ti}^{(p)} \right|^{\eta} \right|^{1/\eta} \qquad \left(\frac{1}{\mu} + \frac{1}{\eta} = 1, \quad \mu, \eta > 1 \right),$$

and from (6.5.22), it can be seen that

$$\left(\sum_{p=1}^{b} \left| \sum_{i=1}^{k} c_{si}^{(p)} c_{ti}^{(p)} \right|^{\eta} \right)^{1/\eta} \leqq \frac{M}{b^{1/\mu}} \tag{6.5.34}$$

for sufficiently large b, where M is some positive constant independent of b.

Hence one obtains from (6.5.33)

$$\left| \sum_{p=1}^{b} (\Delta_p - \bar{\Delta}) \sum_{i=1}^{k} c_{si}^{(p)} c_{ti}^{(p)} \right| \leqq M \left(\frac{1}{b} \sum_{p=1}^{b} |\Delta_p - \bar{\Delta}|^{\mu} \right)^{1/\mu}.$$

If we take $\mu = 3/2+\delta$, $\delta > 0$, then from the uniformity conditions (6.5.27) it follows that

$$\left| \sum_{p=1}^{b} (\Delta_p - \bar{\Delta}) \sum_{i=1}^{k} c_{si}^{(p)} c_{ti}^{(p)} \right| \to 0 \quad \text{as} \quad b \to \infty.$$

Hence

$$\mathcal{E}[(CU_{\sigma}\underline{\pi})(CU_{\sigma}\underline{\pi})'] \to \frac{\Delta_0}{k-1} I_{v-1} \text{ as } b \to \infty. \tag{6.5.35}$$

This shows that condition (6.5.28) is satisfied in our case.

One can show that condition (6.5.29) is also satisfied as follows : From (6.5.30) one can see that

$$\mathcal{E}\left[\underline{\xi}^{(p)'}\underline{\xi}^{(p)}\right] = \frac{\Delta_p}{k-1} \sum_{s=1}^{v-1} \sum_{i=1}^{k} c_{si}^{(p)2} \qquad (p = 1,\dots,b) \tag{6.5.36}$$

and

$$\underline{\xi}^{(p)'}\underline{\xi}^{(p)} = \sum_{s=1}^{v-1} \left[\sum_{i=1}^{k} c_{si}^{(p)} \pi_{\sigma(i)}^{(p)}\right]^2 \leqq \Delta_p \sum_{s=1}^{v-1} \sum_{i=1}^{k} c_{si}^{(p)2} \tag{6.5.37}$$

$$(p = 1,\dots,b).$$

Now the condition (6.5.29) turns out in the present case to be

$$\sum_{p=1}^{b} \int_{\underline{\xi}^{(p)'}\underline{\xi}^{(p)} > \varepsilon^2} \underline{\xi}^{(p)'}\underline{\xi}^{(p)} dF_p \to 0 \quad \text{as} \quad b \to \infty, \tag{6.5.38}$$

where F_p denotes the distribution function of the permutation distribution of $\underline{\xi}^{(p)}$ due to the randomization. From (6.5.36), (6.5.37) and the Markov inequality one gets

$$\sum_{p=1}^{b} \int_{\underline{\xi}^{(p)'}\underline{\xi}^{(p)} > \varepsilon^2} \underline{\xi}^{(p)'}\underline{\xi}^{(p)} \, dF_p \tag{6.5.39}$$

$$\leqq \sum_{p=1}^{b} \Delta_p \left[\sum_{s=1}^{v-1} \sum_{i=1}^{k} c_{si}^{(p)2}\right] P\left[\underline{\xi}^{(p)'}\underline{\xi}^{(p)} > \varepsilon^2\right]$$

$$\leqq \frac{1}{\varepsilon^2(k-1)} \sum_{p=1}^{b} \Delta_p^2 \left[\sum_{s=1}^{v-1} \sum_{i=1}^{k} c_{si}^{(p)2}\right]^2 .$$

Since

$$\frac{1}{b}\sum_{p=1}^{b} |\Delta_p - \bar{\Delta}| \leq \left(\frac{1}{b}\sum_{p=1}^{b} |\Delta_p - \bar{\Delta}|^{3/2+\delta}\right)^{2/(3+2\delta)} \to 0 \quad \text{as } b \to \infty,$$

$$\sum_{p=1}^{b} \int_{\underline{\xi}^{(p)'}\underline{\xi}^{(p)} > \varepsilon^2} \underline{\xi}^{(p)'}\underline{\xi}^{(p)}\, dF_p \to 0 \quad \text{as} \quad b \to \infty.$$

Therefore we see that (6.5.28) is satisfied.

Thus it follows from the central limit theorem that the permutation distribution of $CU_{\underline{\sigma}}^{\pi}$ due to randomization converges in law to the normal distribution $N\left(0, \Delta_0/(k - 1)\cdot I_{v-1}\right)$ as $b \to \infty$. Consequently

(6.5.40) $$\frac{\bar{\Delta}}{\Delta_0}(k - 1)\theta \to \chi^2_{v-1} \quad \text{in law as} \quad b \to \infty.$$

This may be put in the form

(6.5.41) $$\frac{\Delta}{\Delta_0}(k - 1)\begin{bmatrix} \bar{\theta} \\ \bar{\bar{\theta}} \end{bmatrix} \to \begin{bmatrix} \chi^2_{\bar{\alpha}} \\ \chi^2_{\bar{\bar{\alpha}}} \end{bmatrix} \quad \text{in law as} \quad b \to \infty,$$

where $\chi^2_{\bar{\alpha}}$ and $\chi^2_{\bar{\bar{\alpha}}}$ are mutually independent. Note that this convergence is of the type $(\boldsymbol{M})_d$ because the limiting distributions are of the continuous type.

The first condition of (6.5.26) implies that $\Delta/(b\Delta_0) \to 1$ as $b \to \infty$. Hence Theorem (6.4.2) implies that

(6.5.42) $$(n - b)\begin{bmatrix} \bar{\theta} \\ \bar{\bar{\theta}} \end{bmatrix} \to \begin{bmatrix} \chi^2_{\bar{\alpha}} \\ \chi^2_{\bar{\bar{\alpha}}} \end{bmatrix} \quad \text{in law as} \quad b \to \infty,$$

and the permutation distribution of $(\bar{\theta}, \bar{\bar{\theta}})$ due to randomization is asymptotically equivalent in the sense of the type $(\boldsymbol{M})_d$ to the distribution of mutually independent χ^2 variates $\{[1/(n-b)]\chi^2_{\bar{\alpha}}, [1/(n-b)]\chi^2_{\bar{\bar{\alpha}}}\}$ as $b \to \infty$.

The Dirichlet Distribution (6.5.10) and an Asymptotic Null Distribution of the F-Statistic. In this section it will be shown that the permutation distribution of $(\bar{\theta}, \bar{\bar{\theta}})$ due to randomization is asymptotically equivalent in the sense of the type $(\boldsymbol{M})_d$ to the Dirichlet distribution given by (6.5.10). This is done by showing that the Dirichlet distribution is asymptotically equivalent in the sense of the type $(\boldsymbol{B})_d$ to the distribution of the mutually independent χ^2 variates $\left(\frac{1}{n-b}\chi^2_{\bar{\alpha}}, \frac{1}{n-b}\chi^2_{\bar{\bar{\alpha}}}\right)$ mentioned in the last paragraph of the preceding section. Then a rigorous justification will be given to show that the null distribution of the F-statistic is asymptotically equivalent in the sense of the type $(\boldsymbol{M})_d$ to the central F-distribution with $(\bar{\alpha},\ n - b - v + 1)$ degrees of freedom, which is to be obtained by integrating the factors involving $\bar{\theta}$ and $\bar{\bar{\theta}}$ with respect to the Dirichlet distribution out of the probability element of the F-statistic before randomization.

Let $f(x,y)$ be the probability-density function of the Dirichlet distribution (6.5.10) :

$$f(x,y) = \frac{\Gamma\left(\frac{n-b}{2}\right)}{\Gamma\left(\frac{\bar{\alpha}}{2}\right)\Gamma\left(\frac{\bar{\bar{\alpha}}}{2}\right)\Gamma\left(\frac{n-b-v+1}{2}\right)} x^{(\bar{\alpha}/2)-1} y^{(\bar{\bar{\alpha}}/2)-1}$$

(6.5.43)

$$\times\ (1 - x - y)^{[(n-b-v+1)/2]-1} \quad (0 < x,\ y < 1,\ x + y < 1)$$

and let $g(x,y)$ be the probability-density function of the χ^2 variates $[\chi^2_{\bar{\alpha}}/(n - b),\ \chi^2_{\bar{\bar{\alpha}}}/(n - b)]$:

$$g(x,y) = \frac{\left(\frac{n-b}{2}\right)^{(v-1)/2}}{\Gamma\left(\frac{\bar{\alpha}}{2}\right)\Gamma\left(\frac{\bar{\bar{\alpha}}}{2}\right)} x^{(\bar{\alpha}/2)-1} y^{(\bar{\bar{\alpha}}/2)-1}$$

$$(6.5.44) \qquad \times\ exp\left(-\frac{(n-b)(x+y)}{2}\right) \qquad (0 < x,\ y < \infty).$$

Furthermore let us denote the random vector with probability element (6.5.43) by (X,Y). Then, by a theorem of asymptotic equivalence, it is sufficient for (X,Y) and $\left[\chi^2_{\bar{\alpha}}/(n-b),\ \chi^2_{\bar{\bar{\alpha}}}/(n-b)\right]$ to be asymptotically equivalent in the sense of the type $(B)_d$ that the Kullback-Leibler mean information, which is defined by

$$(6.5.45) \qquad f(x:y) = \iint_{\substack{0<x,y \\ x+y<1}} f(x,y)\ \log\frac{f(x,y)}{g(x,y)}\, dx\, dy,$$

tends to zero as $b \to \infty$.

From (6.5.42) and (6.5.43) one obtains

$$(6.5.46) \qquad I(f:g) = log\left\{\frac{\Gamma\left(\frac{n-b}{2}\right)}{\Gamma\left(\frac{n-b-v+1}{2}\right)\cdot\left(\frac{n-b}{2}\right)^{(v-1)/2}}\right\}$$

$$+ \left(\frac{n-b-v+1}{2} - 1\right) E_f\ log(1 - X - Y)$$

$$+ \frac{n-b}{2} E_f(X+Y),$$

where E_f denotes the mathematical expectation with respect to the distribution of (X,Y) whose probability density function is given by $f(x,y)$ of (6.5.43).

By applying the Stirling formula, the first term of the right-hand side of (6.5.46) is seen to tend to zero as $b \to \infty$.

Using the relations

$$\int_0^1 z^{p-1}(1 - z)^{q-1} \log z \, dz = \frac{\Gamma(p)\,\Gamma(q)}{(p + q)} \times \left[\frac{\Gamma'(p + q)}{(p + q)} - \frac{\Gamma'(p)}{\Gamma(p)}\right]$$

and

$$\frac{\Gamma'(p)}{\Gamma(p)} = \log p - \frac{1}{2p} + O\left(\frac{1}{p^2}\right) \quad \text{as} \quad p \to \infty,$$

and the fact that $X + Y$ is distributed according to the β-distribution $B[(v - 1)/2, (n - b - v + 1)/2]$, it is not difficult to confirm that the remaining terms on the right-hand side of (6.5.46) tend to zero as $b \to \infty$.

Thus $I(f{:}g) \to 0$ as $b \to \infty$, whence one concludes that (X,Y) and the variates $\left(\chi^2_{\bar{\alpha}}/(n - b),\ \chi^2_{\bar{\bar{\alpha}}}/(n - b)\right)$ are asymptotically equivalent in the sense of the type $(\boldsymbol{M})_d$. From this and the last statement of the preceding section, it follows that *the permutation distribution of* $(\bar{\theta},\bar{\bar{\theta}})$ *due to randomization is asymptotically equivalent in the sense of the type* $(\boldsymbol{M})_d$ *to the Dirichlet distribution whose probability density is given by (6.5.10) as* $b \to \infty$*, provided the uniformity conditions (6.5.27) are satisfied.*

Now the probability element (6.5.8) of the F-statistic before randomization is regarded as that of the conditional distribution of the F, given $\bar{\theta}$ and $\bar{\theta}$, and hence the conditional probability-density function of the F, given $\bar{\theta}$ and $\bar{\bar{\theta}}$, is written as

(6.5.47)
$$p(F|\bar{\theta},\bar{\bar{\theta}}) = exp\left(-\frac{\Delta}{2\Delta^2}\right) \sum_{l=0}^{\infty} \frac{(\Delta/2\sigma^2)^l}{l!} \sum_{\mu+\nu+\kappa=l} \frac{l!}{\mu!\nu!\kappa!} \times \bar{\theta}^{\mu}\bar{\bar{\theta}}^{\kappa}(1 - \bar{\theta} - \bar{\bar{\theta}})^{\nu} h_{\mu,\nu}(F),$$

where

(6.5.48)
$$h_{\mu,\nu}(F) = \frac{\Gamma\left(\frac{n-b-\bar{\bar{\alpha}}}{2}+\mu+\nu\right)}{\Gamma\left(\frac{\bar{\alpha}}{2}+\mu\right)\cdot\Gamma\left(\frac{n-b-v+1}{2}+\nu\right)}$$
$$\times\left(\frac{\bar{\alpha}}{n-b-v+1}F\right)^{(\alpha_1/2)+\mu-1}$$
$$\times\left(1+\frac{\bar{\alpha}}{n-b-v+1}F\right)^{-[(n-b-\bar{\bar{\alpha}})/2]+\mu+\nu}$$
$$\times\frac{\bar{\alpha}}{n-b-v+1} \qquad (0 < F < \infty).$$

Let us denote by F^* the random variable whose probability density function is obtained by integrating the factors involving $\bar{\theta}$ and $\bar{\bar{\theta}}$ out of (6.5.47) with respect to the Dirichlet distribution whose probability density function is given by (6.5.43). The exact null distribution of the F-statistic should be obtained by integrating factors involving $\bar{\theta}$ and $\bar{\bar{\theta}}$ out of (6.5.47) with respect to the permutation distribution $(\bar{\theta},\bar{\bar{\theta}})$ due to randomization. Thus problem 2 mentioned in 6.5 concerns the type of the asymptotic equivalence of the two variates F and F^* as $b \to \infty$.

To answer this question, Corollary (6.4.2) is useful, and reference should be made to [1]. Let us put

(6.5.49)
$$q(F|\bar{\theta},\bar{\bar{\theta}}) = p\left(F\left|\frac{\bar{\theta}}{n-b}, \frac{\bar{\bar{\theta}}}{n-b}\right.\right)$$

and

(6.5.50)
$$\phi_1(F|\bar{\theta},\bar{\bar{\theta}}) = \frac{\partial}{\partial\bar{\theta}}q(F|\bar{\theta},\bar{\bar{\theta}}) \quad \text{and}$$
$$\phi_2(F|\bar{\theta},\bar{\bar{\theta}}) = \frac{\partial}{\partial\bar{\bar{\theta}}}q(F|\bar{\theta},\bar{\bar{\theta}}).$$

Then, for the two random variables F and F^* to be asymptotically equivalent in the sense of the type $(\boldsymbol{M})_d$ as $b \to \infty$, it is sufficient that there exist for any given bounded subset belonging to $S_{(2)}$, the class of all rectangular subsets, left-closed and right-open, of the two-dimensional Euclidean space, a positive number $\eta = \eta(E)$ and a positive integer $b_0 = b_0(E)$ such that

$$\sup_{(\bar{\theta},\bar{\bar{\theta}})\in E} \int_0^\infty |\phi_1(F|\bar{\theta},\bar{\bar{\theta}})| \; dF \leqq \eta \tag{6.5.51}$$

and

$$\sup_{(\bar{\theta},\bar{\bar{\theta}})\in E} \int_0^\infty |\phi_2(F|\bar{\theta},\bar{\bar{\theta}}) \quad dF \in \eta \tag{6.5.52}$$

for all $b \geqq b_0$.

In the present case these conditions are seen to be satisfied, as will be seen below.

In the first place, one can see that

$$\int_0^\infty h_{\mu,\nu}(F) \; dF = 1.$$

Hence, it follows from (6.5.47) and (6.5.49) that

$$\int_0^\infty |\phi_1(F|\bar{\theta},\bar{\bar{\theta}})| dF \leqq A_1 + A_2, \tag{6.5.53}$$

where A_1 and A_2 are given by

$$A_1 = exp\left(-\frac{\Delta}{2\sigma^2}\right) \sum_{l=0}^{\infty} \frac{(\Delta/2\sigma^2)^l}{l!} \sum_{\mu+\nu+\kappa=l} \frac{l!}{\mu!\nu!\kappa!} \frac{\mu}{n-b} \tag{6.5.54a}$$
$$\times \left(\frac{\bar{\theta}}{n-b}\right)^{\mu-1} \left(\frac{\bar{\bar{\theta}}}{n-b}\right)^{\kappa} \left(1 - \frac{\bar{\theta}+\bar{\bar{\theta}}}{n-b}\right)^{\kappa},$$

(6.5.54b)
$$A_2 = exp\left(-\frac{\Delta}{2\sigma^2}\right) \sum_{l=0}^{\infty} \frac{(\Delta/2\sigma^2)^l}{l!} \sum_{\mu+\nu+\kappa=l} \frac{\nu}{n-b} \times \left(\frac{\bar{\theta}}{n-b}\right)^{\mu} \left(\frac{\bar{\bar{\theta}}}{n-b}\right)^{\kappa} \left(1 - \frac{\bar{\theta}+\bar{\bar{\theta}}}{n-b}\right)^{\nu-1}.$$

After a little algebra, one sees that

$$A_1 = \frac{\Delta}{2(n-b)\sigma^2} \quad \text{and} \quad A_2 = \frac{\Delta}{2(n-b)\sigma^2}.$$

Thus one obtains

$$\int_0^{\infty} |\phi_1(F|\bar{\theta},\bar{\bar{\theta}})| \; dF \leqq \frac{\Delta}{(n-b)\sigma^2}$$

for all possible values of $\bar{\theta}$ and $\bar{\bar{\theta}}$. Since $\Delta/\{(n-b)\sigma^2\} \to \Delta_0/\{(k-1)\sigma^2\}$ as $b \to \infty$, for a certain positive constant η, one obtains

$$\sup_{0<\bar{\theta},\bar{\bar{\theta}}:\bar{\theta}+\bar{\bar{\theta}}<1} \int_0^{\infty} |\phi_1(F|\bar{\theta},\bar{\bar{\theta}})| \; dF \leqq \eta$$

for sufficiently large values of b. This confirms (6.5.51). In a similar way one can confirm condition (6.5.52). Whence it follows that $F \sim F^*, (\boldsymbol{M})_d$ under the uniformity condition (6.5.27).

The probability element of F^* is obtained by integrating the factors $\bar{\theta}^{\mu}\bar{\bar{\theta}}^{\kappa}(1 - \bar{\theta} - \bar{\bar{\theta}})^{\nu}$ with respect to the Dirichlet distribution (6.5.10) out of (6.5.8) as follows :

(6.5.55)
$$\frac{\Gamma\left(\frac{n-b-\bar{\bar{\alpha}}}{2}\right)}{\Gamma\left(\frac{\bar{\alpha}}{2}\right)\cdot\Gamma\left(\frac{n-b-v+1}{2}\right)} \left(\frac{\bar{\alpha}}{n-b-v+1}F^*\right)^{(\bar{\alpha}/2)-1} \times \left(1 + \frac{\bar{\alpha}}{n-b-v+1}F^*\right)^{-(n-b-\bar{\bar{\alpha}})/2} d\left(\frac{\bar{\alpha}}{n-b-v+1}F^*\right).$$

This is the familiar central F-distribution with $(\bar{\alpha}, n - b - v + 1)$ degrees of freedom.

Summarizing the results so far obtained, we can conclude this section by stating the following theorem :

Theorem 6.5.2. Under the uniformity conditions (6.5.27), the null distribution of the F-statistic (6.5.7) for testing a partial null hypothesis $H_{0(h)}$ is asymptotically equivalent in the sense of the type $(\boldsymbol{M})_d$ to the central F-distribution with $(\bar{\alpha}, n - b - v + 1)$ degrees of freedom under the limiting process explained by (6.5.12). In particular, the null distribution of the F-statistic for testing the total null hypothesis $H_0 : \underline{\tau} = 0$ is asymptotically equivalent in the sense of the type $(\boldsymbol{M})_d$ to the central F-distribution with $(v - 1, n - b - v + 1)$ degrees of freedom.

6.6. The Asymptotic Non Null Distribution of the F-Statistic for Testing a Partial Null Hypothesis in a Randomized PBIB Design with m Associate Classes under the Neyman Model.

We are concerned with the power function of the F-statistic occurring in the analysis of variance of a PBIB design with m associate classes, where there are v treatments with an association of m associate classes being defined among them, b blocks of size k each, and r replications of each treatment, the number of incidence of any pair of treatments being λ_u if they are uth associates. The randomization procedure is applied in allocating k treatments to the k units in each block, independently from block to block.

Let us take the special labeling of all the $n = vr = bk$ experimental units in such a way that the ith unit in the pth block bears the number $f = (p - 1)k + i$. We shall fix this labeling throughout this section.

Let Φ and Ψ be the incidence matrices of the treatments and blocks, respectively, and put $B = \Psi\Psi'$ and $N = \Phi'\Psi$, where N is the incidence matrix of the design under consideration. Furthermore, let $\underline{\tau} = (\tau_1,\ldots,\tau_v)'$ and $\underline{\beta} = (\beta_1,\ldots,\beta_b)'$ be the treatment effects subject to the restrictions

$$\sum_{\alpha=1}^{v} \tau_\alpha = 0 \quad \text{and} \quad \sum_{p=1}^{b} \beta_p = 0,$$

respectively, and let $\underline{\pi} = (\pi_1,\ldots,\pi_n)'$ be the unit-error vector being subject to the restriction

$$\sum_{i=1}^{k} \pi_i^{(p)} = 0 \qquad (p = 1,\ldots,b),$$

where $\pi_f = \pi_i^{(p)}$ if $f = (p - 1)k + i$. In matrix notation one can write this as $\Psi'\underline{\pi} = 0$.

The Neyman model assuming no interaction between treatments and experimental units is given by

$$(6.6.1) \qquad \underline{x} = \gamma\underline{1} + \Phi\underline{\tau} + \Psi\underline{\beta} + \underline{\pi} + \underline{e}.$$

where $\underline{x} = (x_1,\ldots,x_n)'$ is the observation vector, γ is the general mean, and $\underline{e} = (e_1,\ldots,e_n)'$ stands for the technical-error vector, which is assumed to be distributed as $N(0,\sigma^2 I_n)$ with unknown variances σ^2. This Neyman model includes the Fisher model and the normal regression model as its special cases :

Fisher model : $\underline{x} = \gamma\underline{1} + \Phi\underline{\tau} + \Psi\underline{\beta} + \underline{\pi}$.

Normal regression model : $\underline{x} = \gamma\underline{1} + \Phi\underline{\tau} + \Psi\underline{\beta} + \underline{e}$.

The null hypothesis to be tested is

$$(6.6.2) \qquad H_{0(h)} : A_u^{\#}\underline{\tau} = 0 \qquad (u = 1,\ldots,h),$$

where h is any given integer such that $1 \leq h \leq m$, and $A_0^{\#} = G_v/v$,

$A_1^{\#},\ldots,A_m^{\#}$ are $m + 1$ mutually orthogonal idempotent matrices of the association algebra. This null hypothesis is called a partial null hypothesis and reduces to the total null hypothesis $H_0 : \underline{\tau} = 0$ if $h = m$.

To test the null hypothesis $H_{0(h)}$, one uses the F-statistic given by

$$F = \frac{n - b - v + 1}{\bar{\alpha}} \frac{S^2_{t(h)}}{S^2_e}, \tag{6.6.3}$$

where $\bar{\alpha} = \alpha_1 + \cdots + \alpha_h$, α_u is the rank of the matrix $A_u^{\#}$, $u = 1,\ldots,m$, and

$$S^2_{t(h)} = \underline{x}'\left(\sum_{u=1}^{h} V_u^{\#}\right)\underline{x},$$

(6.6.4)

$$S^2_e = \underline{x}'\left(I_n - \frac{1}{k}B - \sum_{u=1}^{m} V_u^{\#}\right)\underline{x};$$

that is, $S^2_{t(h)}$ and S^2_e are the sums of squares due to treatments adjusted by blocks and due to errors, respectively. Here we have put

$$V_u^{\#} = \left(I_n - \frac{1}{k}B\right)\Phi\left(c_u A_u^{\#}\right)\Phi'\left(I_n - \frac{1}{k}B\right) \qquad (u = 1,\ldots,m) \tag{6.6.5}$$

and

$$c_u = \frac{k}{rk - \rho_u} \qquad (u = 1,\ldots,m), \tag{6.6.6}$$

where ρ_u, $u = 0,1,\ldots,m$, are the characteristic roots of NN', with the respective multiplicities $\alpha_0 = 1$, α_u, $u = 1,\ldots,m$. It is known as

$$\sum_{u=1}^{m} \alpha_u = v - 1, \qquad \sum_{u=0}^{m} A_u^{\#} = I_v,$$

and

$$NN' = rkA_0^{\#} + \rho_1 A_1^{\#} + \cdots + \rho_m A_m^{\#}.$$

If $h = m$, the F-statistic (6.6.3) reduces to the usual

$$F = \frac{n - b - v + 1}{v - 1} \frac{S_t^2}{S_e^2}. \tag{6.6.7}$$

In the preceding section the asymptotic null distribution of the F-statistic given by (6.6.3) was discussed rigorously from the point of the theory of asymptotic equivalence. It was shown that the null distribution of the F-statistic after randomization is asymptotically equivalent in the sense of the type $(\boldsymbol{M})_d$ to the usual F-distribution, which is to be obtained under the normal regression model without the unit errors. One may say that one can get rid of the unit errors (nuisance parameters) asymptotically by means of the randomization procedure.

Based on the same standpoint as in the null case, we show that a non central F-distribution that is to be obtained under the normal regression model is asymptotically equivalent in the sense of the type $(\boldsymbol{M})_d$ to the non-null distribution of the F-statistic (6.6.3) under the Neyman model.

In this section we shall first derive the non-null distribution of the F-statistic before randomization, and that involves the conditioning random variables $(\xi,\bar{\eta},\bar{\bar{\eta}})$ as parameters. Then we shall discuss the asymptotic behaviour of the permutation distribution of $(\xi,\bar{\eta},\bar{\bar{\eta}})$ due to the randomization procedure and shown that $\xi/(\Delta + T)$, $\bar{\eta}/[\bar{T}/(\Delta + T)]$, $\bar{\bar{\eta}}/[\bar{\bar{T}}/(\Delta + T)]$ converges in probability to $(1,1,1)$ in a certain limiting process under consideration. The last two subsections are devoted to the derivation of the asymptotic power function of the F-statistic based on a theorem of the theory of asymptotic equivalence. In the penultimate subsection we show that the conditions of the theorem are satisfied in our present case. In the last subsection a non-central F-distribution where the non centrality parameter involves $\Delta = \underline{\pi}'\underline{\pi}$ and $\bar{T}$ is derived and is shown to be asymptotically equivalent in the sense of the type

$(\boldsymbol{M})_d$ to the power function of the F-statistic (6.6.3). We then show that a non central F-distribution whose non-centrality parameter is dependent only on $\bar{T}$ and which is to be obtained under the normal regression model is asymptotically equivalent in the sense of the type $(\boldsymbol{M})_d$ to the non-central F-distribution involving Δ and $\bar{T}$.

The Non Null Distribution of the F-Statistic before Randomization. Since $S^2_{t(h)}$ becomes

$$S^2_{t(h)} = (\Phi\underline{\tau} + \underline{\pi})'\left(\sum_{u=1}^{h} V^{\#}_u\right)(\Phi\underline{\tau} + \underline{\pi}) + 2(\Phi\underline{\tau} + \underline{\pi})'$$

$$\times \left(\sum_{u=1}^{h} V^{\#}_u\right)\underline{e} + \underline{e}'\left(\sum_{u=1}^{h} V^{\#}_u\right)\underline{e}$$

under the Neyman model (6.6.1), and the matrix $\sum_{u=1}^{h} V^{\#}_u$ is of rank $\bar{\alpha}$, the non null distribution of the variate

(6.6.8) $$\bar{\chi}^2_1 = \frac{S^2_{t(h)}}{\sigma^2}$$

before randomzation is the non-central χ^2 distribution with $\bar{\alpha}$ degrees of freedom, and with non-centrality parameter $\bar{\delta}_1/\sigma^2$, where

(6.6.9) $$\bar{\delta}_1 = (\Phi\underline{\tau} + \underline{\pi})'\left(\sum_{u=1}^{h} V^{\#}_u\right)(\Phi\underline{\tau} + \underline{\pi}).$$

Hence its probability element is given by

(6.6.10) $$exp\left(-\frac{\bar{\delta}^2_1}{2\sigma^2}\right)\sum_{\mu=0}^{\infty}\frac{(\bar{\delta}_1/2\sigma^2)^{\mu}}{\mu!}\frac{\left(\frac{\bar{\chi}^2_1}{2}\right)^{(\bar{\alpha}/2)+\mu-1}}{\Gamma\left(\frac{\bar{\alpha}}{2}+\mu\right)}exp\left(-\frac{\bar{\chi}^2_1}{2}\right)d\left(\frac{\bar{\chi}^2_1}{2}\right).$$

Similarly the non-null distribution of the variate

(6.6.11) $$\chi^2_2 = \frac{S^2_e}{\sigma^2}$$

before randomization is seen to be the non-central χ^2 distribution with $n - b - v + 1$ degrees of freedom and with the non centrality parameter δ_2/σ^2, where

$$(6.6.12) \qquad \delta_2 = (\Phi\underline{\tau} + \underline{\pi})'\left(I_n - \frac{1}{k}B - \sum_{u=1}^{m} V_u^{\#}\right)(\Phi\underline{\tau} + \underline{\pi}).$$

Hence its probability element is given by

$$(6.6.13) \qquad exp\left(-\frac{\delta^2}{2\sigma^2}\right)\sum_{\nu=0}^{\infty}\frac{(\delta_2/2\sigma^2)^{\nu}}{\nu!}\frac{\left(\frac{\chi_2^2}{2}\right)^{[(n-b-v+1)/2]+\nu-1}}{\Gamma\left(\frac{n-b-v+1}{2}+\nu\right)}$$
$$\times\ exp\left(-\frac{\chi_2^2}{2}\right)d\left(\frac{\chi_2^2}{2}\right).$$

Since the two variates $\bar{\chi}_1^2$ and χ_2^2 are stochastically independent before randomization, the non-null distribution of the F-statistic (6.6.3) is seen to be a non-central F-distribution with $(\bar{\alpha}, n - b - v + 1)$ degrees of freedom. The probability element of the F-statistic before randomization is given by

(6.6.14)

$$exp\left(-\frac{\bar{\delta}_1 + \delta_2}{2\sigma^2}\right)\sum_{\mu=0}^{\infty}\sum_{\mu=0}^{\infty}\frac{(\delta_1/2\sigma^2)^{\mu}}{\mu!}\frac{(\delta_2/2\sigma^2)^{\nu}}{\nu!}$$

$$\times\ \frac{\Gamma\{[(n-b-\bar{\bar{\alpha}})/2]+\mu+\nu\}}{\Gamma[(\bar{\alpha}/2)+\mu]\cdot\Gamma\{[(n-b-v+1)/2]+\nu\}}$$

$$\times\left(\frac{\bar{\alpha}}{n-b-v+1}F\right)^{(\bar{\alpha}/2)+\mu-1}\left(1+\frac{\bar{\alpha}}{n-b-v+1}F\right)^{-\{[(n-b-\bar{\bar{\alpha}})/2]+\mu+\nu\}}$$

$$\times \frac{\Gamma\left(\frac{n - b - \bar{\bar{\alpha}}}{2} + \mu + \nu\right)}{\Gamma\left(\frac{\bar{\alpha}}{2} + \mu\right) \cdot \Gamma\left(\frac{n - b - v + 1}{2} + \nu\right)} \left(\frac{\bar{\alpha}}{n - b - v + 1} F\right)^{(\bar{\alpha}/2)+\mu-1}$$

$$\times \left(1 + \frac{\bar{\alpha}}{n - b - v + 1} F\right)^{-\{[(n-b-\bar{\alpha})/2]+\mu+\nu\}} d\left(\frac{\bar{\alpha}}{n - b - v + 1} F\right),$$

where we have put

$$\bar{\bar{\alpha}} = v - 1 - \bar{\alpha},$$

$$\bar{\bar{\delta}}_1 = \delta_1 - \delta_1 \quad \text{with} \quad \delta_1 = (\Phi\underline{\tau} + \underline{\pi})' \left(\sum_{u=1}^{m} V_u^{\#}\right) (\Phi\underline{\tau} + \underline{\pi}). \tag{6.6.15}$$

$$\xi = \delta_1 + \delta_2, \quad \bar{\eta} = \frac{\bar{\delta}_1}{(\delta_1 + \delta_2)}, \quad \bar{\bar{\eta}} = \frac{\bar{\bar{\delta}}_1}{(\delta_1 + \delta_2)}.$$

It should be remarked that in the case of $h = m$, the probability element of the F-statistic (6.6.7) before randomization is given by

$$exp\left(- \frac{\xi}{2\sigma^2}\right) \sum_{l=0}^{\infty} \frac{(\xi/2\sigma^2)^l}{l!} \sum_{\mu+\nu=l} \frac{l!}{\mu!\nu!} \eta^{\mu}(1 - \eta)^{\nu} \tag{6.6.16}$$

$$\times \frac{\Gamma\left(\frac{n - b}{2} + \mu + \nu\right)}{\Gamma\left(\frac{v - 1}{2} + \mu\right) \Gamma\left(\frac{n - b - v + 1}{2} + \nu\right)}$$

$$\times \left(\frac{v - 1}{n - b - v + 1} F\right)^{[(v-1)/2]+\mu-1}$$

$$\times \left(1 + \frac{v - 1}{n - b - v + 1} F\right)^{-\{[(n-b)/2]+\mu+\nu\}} d\left(\frac{v - 1}{n - b - v + 1} F\right),$$

where

$$\eta = \frac{\delta_1}{(\delta_1 + \delta_2)} = \bar{\eta} + \bar{\bar{\eta}}. \tag{6.6.17}$$

The probability element of the power function of the F-statistic (6.6.3) after randomization should be obtained by taking the mathematical expectation of (6.6.14) with respect to the permutation distribution of $(\xi,\bar{\eta},\bar{\bar{\eta}})$ due to randomization.

Asymptotic Behaviour of the Permutation Distribution of $(\xi,\bar{\eta},\bar{\bar{\eta}})$ Due to the Randomization Procedure. Let us denote the permutation associated with randomization within the pth block by

$$\sigma_p = \begin{bmatrix} 1 & 2 & \cdots & k \\ \sigma_p(1) & \sigma_p(2) & \cdots & \sigma_p(k) \end{bmatrix}$$

and let us put

$$U_\sigma = \begin{Vmatrix} S_{\sigma_1} & & & 0 \\ & S_{\sigma_2} & & \\ & & \ddots & \\ 0 & & & S_{\sigma_b} \end{Vmatrix}, \tag{6.6.18}$$

where S_{σ_p} is a $k \times k$ permutation matrix corresponding to the permutation σ_p, $p = 1,\ldots,b$.

The incidence matrix of the treatments becomes a random variable through randomization, and that takes one of the $(k!)^b$ values $\{U_\sigma\Phi\}$ with equal probability $1/(k!)^b$, where Φ is any one fixed incidence matrix of treatments.

Now, since

$$\bar{\delta}_1 = (\Phi\underline{\tau})'\left(\sum_{u=1}^{h} V_u^{\#}\right)(\Phi\underline{\tau}) + 2(\Phi\underline{\tau})'\left(\sum_{u=1}^{h} V_u^{\#}\right)\underline{\pi} + \underline{\pi}'\left(\sum_{u=1}^{h} V_u^{\#}\right)\underline{\pi},$$

and

$$(\Phi\underline{\tau})'\left(\sum_{u=1}^{h} V_u^{\#}\right)(\Phi\underline{\tau}) = \underline{\tau}'\Phi'\left(I_n - \frac{1}{k}B\right)\Phi\left(\sum_{u=1}^{h} c_u A_u^{\#}\right)\Phi'\left(I_n - \frac{1}{k}B\right)\Phi\underline{\tau}$$

$$= \underline{\tau}'\left(rI_v - \frac{1}{k}NN'\right)\left(\sum_{u=1}^{h} c_u A_u^{\#}\right)\left(rI_v - \frac{1}{k}NN'\right)\underline{\tau}$$

$$= \underline{\tau}'\left(\sum_{u=1}^{m} \frac{1}{c_u} A_u^{\#}\right)\left(\sum_{u=1}^{h} c_u A_u^{\#}\right)\left(\sum_{u=1}^{m} \frac{1}{c_u} A_u^{\#}\right)\underline{\tau}$$

$$= \underline{\tau}'\left(\sum_{u=1}^{m} \frac{1}{c_u} A_u^{\#}\right)\underline{\tau} = \left(\sum_{u=1}^{h} A_u^{\#}\underline{\tau}\right)'\left(\sum_{u=1}^{m} \frac{1}{c_u} A_u^{\#}\right)\left(\sum_{u=1}^{h} A^{\#}\underline{\tau}\right),$$

and

$$(\Phi\underline{\tau})'\left(\sum_{u=1}^{h} V_u^{\#}\right)\underline{\tau} = \underline{\tau}'\Phi'\left(I_n - \frac{1}{k}B\right)\Phi\left(\sum_{u=1}^{h} c_u A_u^{\#}\right)\Phi'\left(I_n - \frac{1}{k}B\right)\underline{\pi}$$

$$= \underline{\tau}'\left(rI_n - \frac{1}{k}NN'\right)\left(\sum_{u=1}^{h} c_u A_u^{\#}\right)\Phi'\underline{\pi}$$

$$= \underline{\tau}'\left(\sum_{u=1}^{m} \frac{1}{c_u} A_u^{\#}\right)\left(\sum_{u=1}^{h} c_u A_u^{\#}\right)\underline{\Phi}'\underline{\pi}$$

$$= \underline{\tau}'\left(\sum_{u=1}^{h} A_u^{\#}\right)\Phi'\underline{\pi},$$

we get

$$\bar{\delta}_1 = \bar{T} + 2\left(\sum_{u=1}^{h} A_u^{\#}\underline{\tau}\right)'\Phi'\underline{\pi} + \underline{\pi}'\left(\sum_{u=1}^{h} V_u^{\#}\right)\underline{\pi}, \tag{6.6.19}$$

where

$$\bar{T} = \left(\sum_{u=1}^{h} A_u^{\#}\underline{\tau}\right)'\left(\sum_{u=1}^{m} \frac{1}{c_u} A_u^{\#}\right)\left(\sum_{u=1}^{h} A_u^{\#}\underline{\tau}\right), \tag{6.6.20}$$

In a similar manner we also get

$$(6.6.21)\qquad \bar{\bar{\delta}}_2 = \bar{\bar{T}} + 2\left(\sum_{u=h+1}^{m} A_u^{\#}\underline{\tau}\right)'\Phi'\underline{\pi} + \underline{\pi}'\left(\sum_{u=h+1}^{m} V_u^{\#}\right)\underline{\pi},$$

where

$$(6.6.22)\qquad \bar{\bar{T}} = \left(\sum_{u=h+1}^{m} A_u^{\#}\underline{\tau}\right)'\left(\sum_{u=1}^{m}\frac{1}{c_u}A_u^{\#}\right)\left(\sum_{u=h+1}^{m} A_u^{\#}\underline{\tau}\right).$$

Here it should be noted that

$$T = \bar{T} + \bar{\bar{T}},$$

where

$$(6.6.23)\qquad T = \underline{\tau}'\left(\sum_{u=1}^{m}\frac{1}{c_u}A_u^{\#}\right)\underline{\tau}.$$

From (6.6.19) and (6.6.21) it follows that

$$(6.6.24)\qquad \delta_1 = \bar{\delta}_1 + \bar{\bar{\delta}}_1 = T + 2\underline{\tau}'\Phi'\underline{\pi} + \underline{\pi}'\left(\sum_{u=1}^{m} V_u^{\#}\right)\underline{\pi}.$$

Since

$$\begin{aligned}\delta_1 + \delta_2 &= (\Phi\underline{\tau} + \underline{\pi})'\left(I_n - \frac{1}{k}B\right)(\Phi\underline{\tau} + \underline{\pi})\\ &= \underline{\tau}'\left(rI_n - \frac{1}{k}NN'\right)\underline{\tau} + 2\underline{\tau}'\Phi'\underline{\pi} + \underline{\pi}'\underline{\pi}\\ &= \underline{\tau}'\left(\sum_{u=1}^{m}\frac{1}{c_u}A_u^{\#}\right)\underline{\tau}' + 2\underline{\tau}'\Phi'\underline{\pi} + \underline{\pi}'\underline{\pi},\end{aligned}$$

it is also seen that

$$(6.6.25)\qquad \delta_2 = \underline{\pi}'\underline{\pi} - \underline{\pi}'\left(\sum_{u=1}^{m} V_u^{\#}\right)\underline{\pi} = \Delta - \underline{\pi}'\left(\sum_{u=1}^{m} V_u^{\#}\right)\underline{\pi}$$

From (6.6.15), (6.6.17), (6.6.19), (6.6.21), (6.6.24) and (6.6.25) it follows that

$$(6.6.26a)\qquad \xi = \Delta + T + 2\underline{\tau}'\Phi'\underline{\pi},$$

$$(6.2.26b) \quad \eta = \frac{\bar{T} + 2\underline{\tau}'\Phi'\underline{\pi} + \underline{\pi}'\left(\sum_{u=1}^{m} V_u^{\#}\right)\underline{\pi}}{\xi},$$

$$(6.2.26c) \quad \bar{\eta} = \frac{\bar{T} + 2\underline{\tau}'\left(\sum_{u=1}^{h} A_u^{\#}\right)\Phi'\underline{\pi} + \underline{\pi}'\left(\sum_{u=1}^{h} V_u^{\#}\right)\underline{\pi}}{\xi},$$

$$(6.2.26d) \quad \bar{\bar{\eta}} = \frac{\bar{\bar{T}} + 2\underline{\tau}'\left(\sum_{u=h+1}^{m} A_u^{\#}\right)\Phi'\underline{\pi} + \underline{\pi}'\left(\sum_{u=h+1}^{m} V_u^{\#}\right)\underline{\pi}}{\xi},$$

These quantities become random variables through the permutation distribution of the incidence matrix Φ of the treatments due to randomization. However, it should be noted that $T, \bar{T}, \bar{\bar{T}}$ and Δ are constant parameters.

For the sake of notational simplicity let us put

$$(6.6.27) \quad X = \underline{\tau}'\Phi_{\sigma}'\underline{\pi}, \qquad \bar{X} = \underline{\tau}'\left(\sum_{u=1}^{h} A_u^{\#}\right)\Phi_{\sigma}'\underline{\pi}, \qquad \bar{\bar{X}} = \underline{\tau}'\left(\sum_{u=h+1}^{m} A_u^{\#}\right)\Phi_{\sigma}'\underline{\pi},$$

and

$$(6.6.28) \quad Y = \underline{\pi}'\left(\sum_{u=1}^{m} V_u^{\#}\right)_{\sigma}\underline{\pi}, \qquad \bar{Y} = \underline{\pi}'\left(\sum_{u=1}^{h} V_u^{\#}\right)_{\sigma}\underline{\pi}, \qquad \bar{\bar{Y}} = \underline{\pi}'\left(\sum_{u=h+1}^{m} V_u^{\#}\right)_{\sigma}\underline{\pi},$$

$$\Phi_{\sigma} = U_{\sigma}\Phi, \qquad \left(\sum_{u=1}^{m} V_u^{\#}\right)_{\sigma} = \left(I_n - \frac{1}{k}B\right)\Phi_{\sigma}\left(\sum_{u=1}^{m} c_u A_u^{\#}\right)\Phi_{\sigma}'\left(I_n - \frac{1}{k}B\right).$$

Then the variates in (6.6.26) can be written as

$$(6.6.29a) \quad \xi = \Delta + T + 2X,$$

$$(6.6.29b) \quad \eta = \frac{(T + 2X + T)}{(\Delta + T + 2X)},$$

$$(6.6.29c) \quad \bar{\eta} = \frac{(\bar{T} + 2\bar{X} + \bar{Y})}{(\Delta + T + 2X)},$$

(6.6.29d) $$\bar{\bar{\eta}} = \frac{(\bar{\bar{T}} + 2\bar{\bar{X}} + \bar{\bar{Y}})}{(\Delta + T + 2X)},$$

Now we consider the limiting process

(6.6.30) $$b \to \infty \quad \text{but} \quad v,\ k,\ n_i,\ p^i_{jk} \quad (i,j,k = 0,1,\ldots,m)$$

are kept fixed, and denote this limiting process simply by $b \to \infty$. Under this limiting process, r and at least one λ_u must tend to infinity with the same order of magnitude as b. Suppose that one can find non-negative numbers ω_u such that

(6.6.31) $$\frac{rk - \rho_u}{b} \to \omega_u \quad \text{as} \quad b \to \infty, \qquad (u = 0,1,\ldots,m),$$

where $\omega_0 = r/b = k/v$. Furthermore, we assume that the following uniformity conditions of unit errors are satisfied :

(6.6.32) $$\bar{\Delta} \equiv \frac{1}{b}\sum_{p=1}^{b} \Delta_p \to \Delta_0 \quad \text{and} \quad \frac{1}{b}\sum_{p=1}^{b} |\Delta_p - \bar{\Delta}|^{1+\delta} \to 0 \quad \text{as} \quad b \to \infty,$$

where $\Delta_p = \sum_{i=1}^{k} \pi_i^{(p)2}$ and Δ_0 and δ are some positive constants.

We have shown in the preceding section that in such a situation the permutation distribution of the variates $\left[(k - 1)\bar{Y}/\Delta_0,\ (k - 1)\bar{\bar{Y}}/\Delta_0\right]$ converges in law to that of the mutually independent χ^2 variates $\left(\chi^2_{\underline{\alpha}}, \chi^2_{\underline{\underline{\alpha}}}\right)$ under the limiting process (6.6.30) provided that the conditions (6.6.31) and (6.6.32) are satisfied.

We shall now consider the asymptotic behavior of the variates X, $\bar{X}$, and $\bar{\bar{X}}$.

From (6.6.18) we can see that

$$\Phi'_\sigma \underline{\pi} = \Phi' U'_\sigma \underline{\pi} = \Phi' \begin{pmatrix} S'_{\sigma_1} \underline{\pi}^{(1)} \\ S'_{\sigma_2} \underline{\pi}^{(2)} \\ \vdots \\ S'_{\sigma_b} \underline{\pi}^{(b)} \end{pmatrix},$$

where $\underline{\pi}^{(p)} = \left(\pi_1^{(p)}, \ldots, \pi_k^{(p)}\right)'$. Let us put

$$\Pi^\sigma_{\alpha p} \equiv \sum_{i=1}^{k} \zeta_{\alpha(p-1)k+i} \pi^{(p)}_{\sigma_p(i)} \qquad (\alpha = 1,\ldots,v) \tag{6.6.33}$$

and

$$\underline{\Pi}^\sigma_p = \begin{pmatrix} \Pi^\sigma_{1p} \\ \vdots \\ \Pi^\sigma_{vp} \end{pmatrix} \qquad (p = 1,\ldots,b). \tag{6.6.34}$$

Furthermore let

$$\underline{\bar{\tau}} = \begin{pmatrix} \bar{\tau}_1 \\ \vdots \\ \bar{\tau}_v \end{pmatrix} = \sum_{u=1}^{h} A^{\#}_u \underline{\tau}, \quad \underline{\bar{\bar{\tau}}} = \begin{pmatrix} \bar{\bar{\tau}}_1 \\ \vdots \\ \bar{\bar{\tau}}_v \end{pmatrix} = \sum_{u=h+1}^{m} A^{\#}_u \underline{\tau}. \tag{6.6.35}$$

Then, from (6.6.27), (6.6.33), (6.6.34) and (6.6.35) it follows that

$$\bar{X} = \sum_{p=1}^{b} \underline{\bar{\tau}}' \underline{\Pi}^\sigma_p,$$

$$\bar{\bar{X}} = \sum_{p=1}^{b} \underline{\bar{\bar{\tau}}}' \underline{\Pi}^\sigma_p, \tag{6.6.36}$$

$$X = \sum_{p=1}^{b} \underline{\tau}' \underline{\Pi}^\sigma_p.$$

It is noted that the variates $\underline{\Pi}_p^{\sigma}$, $p = 1,\ldots,b$, given by (6.6.34) form a stochastically independent set of v-dimensional random vectors under the permutation distribution due to randomization. Furthermore, one can notice that

$$\underline{\bar{\tau}}'\underline{\bar{\bar{\tau}}} = 0, \quad \underline{\bar{\tau}}'A_{\underline{u}}^{\#}\bar{\bar{\tau}} = 0 \qquad (u = 1,\ldots,m),$$

and consequently $\underline{\bar{\tau}}'NN'\underline{\bar{\bar{\tau}}} = 0$.

Since

$$E\left[\pi^{(p)}_{\sigma_p(1)}\right] = 0,\ E\left[\pi^{(p)}_{\sigma_p(i)}\right] = \frac{1}{k}\Delta_p,\ E\left[\pi^{(p)}_{\sigma_p(i)}\pi^{(p)}_{\sigma_p(i)}\right] = \frac{-1}{k(k-1)}\Delta_p \qquad (i \neq j),$$

where E denotes the mathematical expectation with respect to the permutation.

(6.6.37)
$$E(\underline{\Pi}_p^{\sigma}) = 0 \qquad (p = 1,\ldots,b),$$
$$E(\underline{\Pi}_p^{\sigma}\,\underline{\Pi}_p^{\sigma}{}') = \frac{\Delta_p}{k(k-1)}\Delta_p \qquad (p = 1,\ldots,b),$$

where

(6.6.38)
$$\Lambda_p = \left\| \begin{matrix} (k-1)n_{1p} & -n_{1p}n_{2p} & \cdots & -n_{1p}n_{vp} \\ -n_{1p}n_{2p} & (k-1)n_{2p} & \cdots & -n_{2p}n_{vp} \\ \cdots & \cdots & \cdots & \cdots \\ -n_{1p}n_{vp} & -n_{2p}n_{vp} & \cdots & (k-1)n_{vp} \end{matrix} \right\|.$$

Notice that

(6.6.39)
$$\sum_{p=1}^{b} \Lambda_p = rkI_v - NN'.$$

From (6.6.30) and (6.6.37) it follows that

(6.6.40) $E(\bar{X}) = E(\bar{\bar{X}}) = E(X) = 0.$

The variance-covariance matrix of $(\bar{X},\bar{\bar{X}})$ is seen by (6.6.37) to be

(6.6.41)
$$D(\bar{X},\bar{\bar{X}}) = \frac{1}{k(k-1)} \sum_{p=1}^{b} \Delta_p \begin{Vmatrix} \bar{\underline{\tau}}'\Lambda_p\bar{\underline{\tau}} & \bar{\underline{\tau}}'\Lambda_p\bar{\bar{\underline{\tau}}} \\ \bar{\underline{\tau}}'\Lambda_p\bar{\bar{\underline{\tau}}} & \bar{\bar{\underline{\tau}}}'\Lambda_p\bar{\bar{\underline{\tau}}} \end{Vmatrix}$$

$$= \frac{\bar{\Delta}}{k-1} \begin{Vmatrix} \bar{T} & 0 \\ 0 & \bar{\bar{T}} \end{Vmatrix} + \frac{1}{k(k-1)} \sum_{p=1}^{b} (\Delta_p - \bar{\Delta}) \times \begin{Vmatrix} \bar{\underline{\tau}}'\Lambda_p\bar{\underline{\tau}} & \bar{\underline{\tau}}'\Lambda_p\bar{\bar{\underline{\tau}}} \\ \bar{\underline{\tau}}'\Lambda_p\bar{\bar{\underline{\tau}}} & \bar{\bar{\underline{\tau}}}'\Lambda_p\bar{\bar{\underline{\tau}}} \end{Vmatrix},$$

where $\bar{T}$ and $\bar{\bar{T}}$ are defined by (6.6.20) and (6.6.22) respectively. If we put

$$T_0 = \underline{\tau}' \sum_{u=1}^{m} \left(\frac{\omega_u}{k}\right) A_u^{\#} \underline{\tau},$$

(6.6.42)
$$\bar{T}_0 = \underline{\tau}' \sum_{u=1}^{h} \left(\frac{\omega_u}{k}\right) A_u^{\#} \underline{\tau},$$

$$\bar{\bar{T}}_0 = \underline{\tau}' \sum_{u=h+1}^{m} \left(\frac{\omega_u}{k}\right) A_u^{\#} \underline{\tau},$$

then, under conditions (6.6.31) and (6.6.32) we obtain

$$\frac{\bar{\Delta}}{b(k-1)} \begin{Vmatrix} \bar{T} & 0 \\ 0 & \bar{\bar{T}} \end{Vmatrix} \to \frac{\Delta_0}{k-1} \begin{Vmatrix} \bar{T}_0 & 0 \\ 0 & \bar{\bar{T}}_0 \end{Vmatrix} \quad \text{as} \quad b \to \infty,$$

From (6.6.38), by using Hölder's inequality, it follows that

$$\left|\frac{1}{b}\sum_{p=1}^{b}(\Delta_p - \bar{\Delta})\bar{\underline{\tau}}'\Lambda_p\bar{\underline{\tau}}\right| \leqq \left(\frac{1}{b}\sum_{p=1}^{b}|\Delta_p - \bar{\Delta}|^{\mu}\right)^{1/\mu}\left(\frac{1}{b}\sum_{p=1}^{b}|\bar{\underline{\tau}}'\Lambda_p\bar{\underline{\tau}}|^{\eta}\right)^{1/\eta}$$

for any given μ, $\eta > 1$ such that $1/\mu + 1/\eta = 1$. But, since

$$|\bar{\underline{\tau}}'\Lambda_p\bar{\underline{\tau}}| = \left|(k - 1)\sum_{\alpha=1}^{v} n_{\alpha p}\bar{\tau}_{\alpha}^2 - \sum_{\alpha\neq\beta} n_{\alpha p}n_{\beta p}\bar{\tau}_{\alpha}\bar{\tau}_{\beta}\right| \leqq (k - 1)$$

$$\times\left(\sum_{\alpha=1}^{v} n_{\alpha p}|\bar{\tau}_{\alpha}|\right)^2 \leqq k^2(k - 1)\bar{\tau}_*^2 \qquad (p = 1,\ldots,b),$$

where $\bar{\tau}_* = \max_{1\leq\alpha\leq v}\left|\bar{\tau}_{\alpha}\right|$, one has

$$\left(\frac{1}{b}\sum_{p=1}^{b}|\bar{\underline{\tau}}'\Lambda_p\bar{\underline{\tau}}|^{\eta}\right)^{1/\eta} \leqq k^2(k - 1)\bar{\tau}_*^2,$$

Since $\bar{\tau}_*$ depends only on the parameters of the association under consideration, it is bounded independently of b. Thus, if one chooses μ so close to unity that $1 < \mu \leqq 1 + \delta$, it follows that

$$\frac{1}{b}\sum_{p=1}^{b}(\Delta_p - \bar{\Delta})\underline{\tau}'\Lambda_p\underline{\tau} \to 0 \quad \text{as} \quad b \to \infty,$$

provided conditions (6.6.32) are satisfied. In a similar manner one can show that the remaining elements of the second matrix of the right hand side of (6.6.41) tend to zero as $b \to \infty$. Hence

(6.6.43) $$\frac{1}{b}D(\bar{X},\bar{\bar{X}}) \to \frac{\Delta_0}{k - 1}\left\|\begin{matrix}\bar{T}_0 & 0\\ 0 & \bar{\bar{T}}_0\end{matrix}\right\| \quad \text{as} \quad b \to \infty$$

under conditions (6.6.31) and (6.6.32). In particular

$$\frac{1}{b}\,Var(X) \to \frac{\Delta_0}{k-1}T_0 \quad \text{as} \quad b \to \infty. \tag{6.6.44}$$

Now we are going to show that

$$\left[\xi^* = \frac{\xi}{\Delta + T},\ \bar{\eta}^* = \frac{\bar{\eta}}{\bar{T}/(\Delta + T)},\ \bar{\bar{\eta}}^* = \frac{\bar{\bar{\eta}}}{\bar{\bar{T}}/(\Delta + T)}\right] \to (1,1,1) \tag{6.6.45}$$

in probability, and hence

$$\left[\frac{\xi}{\Delta + T},\ \frac{\eta}{T(\Delta + T)}\right] \to (1,1) \quad \text{in probability.} \tag{6.6.46}$$

To prove (6.6.45) and hence (6.6.46) it will be sufficient to show that

$$\left.\begin{aligned} \eta &= \frac{T + 2X + Y}{\xi} \to \frac{T_0}{\Delta_0 + T_0} \\ \bar{\eta} &= \frac{\bar{T} + 2\bar{X} + \bar{Y}}{\xi} \to \frac{\bar{T}_0}{\Delta_0 + T_0} \\ \bar{\bar{\eta}} &= \frac{\bar{\bar{T}} + 2\bar{\bar{X}} + \bar{\bar{Y}}}{\xi} \to \frac{\bar{\bar{T}}_0}{\Delta_0 + T_0} \end{aligned}\right\} \text{in probability} \tag{6.6.47}$$

as $b \to \infty$. Indeed, since $\xi = \Delta + T + 2X$ and we have already shown that $X/b \to 0$ in probability it is clear that $\xi/b \to \Delta_0 + T_0$ in probability as $b \to \infty$. Since

$$\eta = \frac{T + 2X + Y}{\Delta + T + 2X} = 1 - \frac{\Delta - Y}{\Delta + T + 2X}$$

and we have seen in the preceding section that $(\boldsymbol{n} - b)Y$ is asymptotically equivalent in the sense of the type $(\boldsymbol{M})_d$ to χ^2_{v-1}. It is clear that $Y/b \to 0$. Thus we can see that

$$\eta \to 1 - \frac{\Delta_0}{\Delta_0 + T_0} = \frac{T_0}{\Delta_0 + T_0} \text{ in probability}$$

Validity of Certain Conditions in a Theorem from the Theory of Asymptotic Equivalence. In this section we shall show that certain conditions in a theorem on the asymptotic equivalence of two probability distributions are satisfied in our present situation. The proof given in this section will be useful in the next section too.

From (6.6.14), the conditional probability density function of the F-statistic (6.6.3), given ξ, $\bar{\eta}$ and $\bar{\bar{\eta}}$, is expressed as

$$p_b(F|\xi,\bar{\eta},\bar{\bar{\eta}}) = exp\left(\frac{-\xi}{2\sigma^2}\right) \sum_{l=0}^{\infty} \frac{(\xi/2\sigma^2)^l}{l!} \sum_{\mu+\nu+\gamma=l} \frac{l!}{\mu!\nu!\gamma!} \bar{\eta}^{\mu}\bar{\bar{\eta}}^{\gamma}(1 - \bar{\eta} - \bar{\bar{\eta}})^{\nu}$$

$$\times \frac{\bar{\alpha}}{n - b - v + 1} \frac{\Gamma\left(\frac{n - b - \bar{\bar{\alpha}}}{2} + \mu + \nu\right)}{\Gamma\left(\frac{\bar{\alpha}}{2} + \mu\right)\Gamma\left(\frac{n - b - v + 1}{2} + \nu\right)}$$

(6.6.48)

$$\times \left(\frac{\bar{\alpha}}{n - b - v + 1}F\right)^{(\bar{\alpha}/2)+\mu-1}$$

$$\times \left(1 + \frac{\bar{\alpha}}{n - b - v + 1}F\right)^{-\{[(n-b-\bar{\bar{\alpha}})/2]+\mu+\nu\}}$$

and hence, the conditional cumulative distribution function of the F-statistic given ξ, $\bar{\eta}$ and $\bar{\bar{\eta}}$, is given by

$$P_b(F|\xi,\bar{\eta},\bar{\bar{\eta}}) = \int_0^F p_b(x|\xi,\bar{\eta},\bar{\bar{\eta}})\, dx. \tag{6.6.49}$$

Let us put

$$Q_b(F|\xi^*,\bar{\eta}^*,\bar{\bar{\eta}}^*) = P_b\left[F\Big|(\Delta+T)\xi^*, \frac{T}{\Delta+T}\bar{\eta}^*, \frac{T}{\Delta+T}\bar{\bar{\eta}}^*\right] \tag{6.6.50}$$

Then Theorem 6.4.6 states that the distribution of the F-statistic where cumulative distribution function is given by expected value of (6.6.49) with respect to $(\xi, \bar{\eta}, \bar{\bar{\eta}})$, that is, the distribution of the F-statistic after randomization, is asymptotically equivalent in the sense of the type $(\boldsymbol{M})_d$ to the distribution whose cumulative distribution function is given by

$$(6.6.51)\qquad P_b\left(F \middle| \Delta + T, \frac{\bar{T}}{\Delta + T}, \frac{\bar{\bar{T}}}{\Delta + T}\right) \quad \text{as} \quad b \to \infty$$

if, in addition to condition (6.6.45), the following condition is also satisfied.

For any given $\varepsilon > 0$, there exist a positive number $\delta = \delta(\varepsilon)$ and a positive integer $b_0 = b_0(\varepsilon)$ such that

$$|(\xi^*, \bar{\eta}^*, \bar{\bar{\eta}}^*) - (1,1,1)| < \delta$$

implies

$$(6.6.52)\qquad \sup_F |Q_b(F|\xi^*, \bar{\eta}^*, \bar{\bar{\eta}}^*) - Q_b(F|1,1,1)| < \varepsilon$$

for all $b \geqq b_0$.

We are now going to show the above condition (6.6.52) is satisfied in the present case.

For simplicity let us put

$$(6.6.53a)\qquad u_{\mu,\nu,\gamma}(\bar{\eta}, \bar{\bar{\eta}}) = \frac{l!}{\mu!\nu!\gamma!} \bar{\eta}^{\mu} \bar{\bar{\eta}}^{\gamma} (1 - \bar{\eta} - \bar{\bar{\eta}})^{\nu}$$

and

$$(6.6.53b)\qquad H^b_{\mu,\nu}(F) = \int_{0 \leqq x \leqq F} \frac{\Gamma\left(\frac{n - b - \bar{\bar{\alpha}}}{2} + \mu + \nu\right)}{\Gamma\left(\frac{\bar{\alpha}}{2} + \mu\right)\Gamma\left(\frac{n - b - v + 1}{2} + \nu\right)}$$

$$\times \left(\frac{\bar{\alpha}}{n - b - v + 1}x\right)^{(\bar{\alpha}/2)+\mu-1}$$

$$\times \left(1 + \frac{\bar{\alpha}}{n - b - v + 1}x\right)^{-\{[(n-b-\bar{\bar{\alpha}})/2]+\mu+\nu\}} d\left(\frac{\bar{\alpha}}{n - b - v + 1}x\right).$$

Then it holds that

$$P_b(F|\xi,\bar{\eta},\bar{\bar{\eta}}) = exp\left(-\frac{\xi}{2\sigma^2}\right)\sum_{l=0}^{\infty}\frac{(\xi/2\sigma^2)^l}{l!} \times \sum_{\mu+\nu+\gamma=l} u_{\mu,\nu,\gamma}(\bar{\eta},\bar{\bar{\eta}})\, H^b_{\mu,\nu}(F). \tag{6.6.54}$$

It is not difficult to see that $H^b_{\mu,\nu}(F)$ is a monotone increasing function of b for any fixed values of μ, ν, and F, and that

$$H^b_{\mu,\nu}(F) \to H^{\infty}_{\mu}(F) \quad \text{as} \quad b \to \infty \tag{6.6.55}$$

where

$$H^{\infty}_{\mu}(F) = \int_0^F \frac{1}{\Gamma\left(\frac{\alpha}{2} + \mu\right)} \left(\frac{\alpha}{2}x\right)^{(\bar{\alpha}/2)+\mu-1} exp\left(-\frac{\bar{\alpha}x}{2}\right) d\left(\frac{\alpha x}{2}\right). \tag{6.6.56}$$

Notice that convergence (6.6.55) is of the type $(\mathcal{B})_d$. Furthermore, for Γ distribution (6.6.56), it holds that $H^{\infty}_{\mu}(F)$ is a monotone decreasing function with increasing μ for any fixed F and that

$$H^{\infty}_{\mu}(F) \to 0 \quad \text{as} \quad \mu \to \infty \tag{6.6.57}$$

Replacing $H^b_{\mu,\nu}(F)$ in the right-hand side of (6.6.53) by $H^{\infty}_{\mu}(F)$, one gets

$$P_{\infty}(F|\xi,\bar{\eta}) = exp\left(-\frac{\xi}{2\sigma^2}\right)\sum_{l=0}^{\infty}\frac{(\xi/2\sigma^2)^l}{l!} \sum_{\mu+\nu=l} u_{\mu,\nu}(\bar{\eta})H^{\infty}_{\mu}(F), \tag{6.6.58}$$

where

(6.6.59) $$u_{\mu,\nu}(\bar{\eta}) = \frac{l!}{\mu!\nu!}\bar{\eta}^{\mu}(1-\bar{\eta})^{\nu}.$$

Now, in the first step, we show that

(6.6.60) $$\sup_{F,\xi,\bar{\eta},\bar{\bar{\eta}}}|P_b(F|\xi,\bar{\eta},\bar{\bar{\eta}}) - P_{\infty}(F|\xi,\bar{\eta})| \to 0 \quad \text{as} \quad b \to \infty.$$

First, we note that

(6.6.61) $$0 \leqq P_b(F|\xi,\bar{\eta},\bar{\bar{\eta}}) \leqq P_{\infty}(F|\xi,\bar{\eta}) \leqq 1$$

for any given values of b, F, ξ, $\bar{\eta}$, and $\bar{\bar{\eta}}$, and these functions are all continuous with respect to F, ξ, $\bar{\eta}$ and $\bar{\bar{\eta}}$. We calculate the difference

(6.6.62) $$1 - P_b(F|\xi,\bar{\eta},\bar{\bar{\eta}}) = \exp\left(-\frac{\xi}{2\sigma^2}\right)\sum_{l=0}^{\infty}\frac{(\xi/2\sigma^2)^l}{l!}\sum_{\mu+\nu+\gamma=l} \times u_{\mu,\nu,\gamma}(\bar{\eta},\bar{\bar{\eta}})\left[1 - H^b_{\mu,\nu}(F)\right].$$

Since the function $1 - H^b_{\mu,\nu}(F)$ is monotone decreasing with increasing b for any fixed F, μ, and ν, it can be seen that

(6.6.63) $$1 - H^{b_0}_{\mu,\nu}(F) \geqq 1 - H^b_{\mu,\nu}(F) \geqq 0$$

for all $b \geqq b_0$. Since

(6.6.64) $$\int_0^{\infty} F\,dH^{b_0}_{\mu,\nu}(F) = \frac{n_0 - b_0 - v + 1}{\bar{\alpha}} \times \frac{(\bar{\alpha}/2) + \mu}{[(n_0 - b_0 - \bar{\bar{\alpha}})/2] + \mu + \nu} \quad (n_0 = b_0 k)$$

is bounded uniformly for all μ and ν, the Markov inequality assures us that, for any given $\varepsilon > 0$, there exists a positive number

$F_0 = F_0(\varepsilon)$ such that

$$1 - H_{\mu,\nu}^{b_0}(F_0) < \varepsilon,$$

and hence by (6.6.63)

$$0 \leqq 1 - H_{\mu,\nu}^{b}(F) < \varepsilon \tag{6.6.65}$$

uniformly for all $F \geqq F_0$, $b \geqq b_0$, μ and ν.

From (6.6.61), (6.6.62), and (6.6.65) it follows that

$$\sup_{\substack{\xi,\bar{\eta},\bar{\bar{\eta}} \\ F \geqq F_0}} \left| P_b(F|\xi,\bar{\eta},\bar{\bar{\eta}}) - P_\infty(F|\xi,\bar{\eta}) \right| < \varepsilon \tag{6.6.66}$$

for all $b \geqq b_0$.

Second, we have to examine the case where $F \leqq F_0$. We are assured by (6.6.55) and (6.6.57) that there exists a positive integer $H_{\mu_0}^{\infty}(F_0) < \varepsilon$ and that

$$\left| H_{\mu,\nu}^{b}(F) - H_{\mu}^{\infty}(F) \right| \leqq H_{\mu_0}^{\infty}(F_0) < \varepsilon \tag{6.6.67}$$

uniformly for all $F \leqq F_0$, $\mu \geqq \mu_0$, b, and ν.

For $\mu < \mu_0$, we have by (6.6.54)

$$\left| H_{\mu,\nu}^{b}(F) - H_{\mu}^{\infty}(F) \right| < \varepsilon \tag{6.6.68}$$

uniformly for all ν and $F \leqq F_0$, provided that $b \geqq b_0'$, where $b_0' = b'(\varepsilon)$ is some positive integer.

From (6.6.67) and (6.6.68) it now follows that

$$\sup_{\substack{\xi,\bar{\eta},\bar{\bar{\eta}} \\ F \leqq F_0}} \left| P_b(F|\xi,\bar{\eta},\bar{\bar{\eta}}) - P_\infty(F|\xi,\bar{\eta}) \right| < \varepsilon \tag{6.6.69}$$

for all $b \geqq b_0'$.

By combining (6.6.66) and (6.6.69) one gets (6.6.60) as was to be proved in the first step.

In the second step we shall show that for any given $\varepsilon > 0$, there exists a positive number $\xi_0 = \xi_0(\varepsilon)$ such that $\xi \geqq \xi_0$ and $\xi_1 \geqq \xi_0$ imply that

$$\sup_{\bar{\eta},F} \left| P_\infty(F|\xi,\bar{\eta}) - P_\infty(F|\xi_1,\bar{\eta}) \right| < \varepsilon \tag{6.6.70}$$

From (6.6.61), (6.6.62) and (6.6.66) it follows that

$$\sup_{\substack{\xi,\bar{\eta} \\ F > F_0}} \left| 1 - P_\infty(F|\xi,\bar{\eta}) \right| < \varepsilon, \tag{6.6.71}$$

and hence

$$\sup_{\substack{\bar{\eta} \\ F > F_0}} \left| P_\infty(F|\xi,\bar{\eta}) - P_\infty(F|\xi_1,\bar{\eta}) \right| < \varepsilon \tag{6.6.72}$$

for any ξ and ξ_1.

Suppose $F \leqq F_0$. First, we have

$$\frac{\partial}{\partial \xi} P_\infty(F|\xi,\bar{\eta}) = L_+(F|\xi,\bar{\eta}) - L_-(F|\xi,\bar{\eta}), \tag{6.6.73}$$

where

$$L_+(F|\xi,\bar{\eta})$$

$$= \frac{1}{2\sigma^2} \, exp\left(\frac{-\xi}{2\sigma^2}\right) \left\{ \sum_{\frac{\xi}{2\sigma^2} \leq l} \left[\frac{(\xi/2\sigma^2)^{l-1}}{(l-1)!} - \frac{(\xi/2\sigma^2)^{l}}{l!} \right] \right.$$

$$\left. \times \sum_{\mu+\nu=l} u_{\mu,\nu}(\bar{\eta}) H_\mu^\infty(F) + H_0^\infty(F) \right\},$$

(6.6.74)

$$L_-(F|\xi,\bar{\eta})$$

$$= \frac{1}{2\sigma^2} \, exp\left(\frac{-\xi}{2\sigma^2}\right) \left\{ \sum_{1 \leq l < \xi/2\sigma^2} \left[\frac{(\xi/2\sigma^2)^{l}}{l!} - \frac{(\xi/2\sigma^2)^{l-1}}{(l-1)!} \right] \right.$$

$$\left. \times \sum_{\mu+\nu=l} u_{\mu,\nu}(\bar{\eta}) H_\mu^\infty(F) \right\},$$

which are both positive for all values of F, ξ, and $\bar{\eta}$. Note that L_+ and $-L_-$ are both non-increasing functions of ξ for any given values of F and $\bar{\eta}$, and hence, for any given F and $\bar{\eta}$, $(\partial/\partial\xi)P_\infty(F|\xi,\bar{\eta})$ is a non-increasing function of ξ. Furthermore, $(\partial/\partial\xi)P_\infty(F|\xi,\bar{\eta}) \geqq 0$ at $\xi = 0$, with equality holding if and only if $F = 0$.

From (6.6.73) and (6.6.74) it is seen that

$$\frac{\partial}{\partial\xi} P_\infty(F|\xi,\bar{\eta}) \leqq \left| L_+(F|\xi,\bar{\eta}) \right| \leqq \frac{1}{2\sigma^2} \, exp(-\zeta) \left[\frac{\zeta^{\zeta-1}}{\Gamma(\zeta)} + 1 \right]$$

where we have put $\zeta = \xi/2\sigma^2$. Then, since

$$exp(-\zeta)\frac{\zeta^{\zeta-1}}{\Gamma(\zeta)} \sim c/\sqrt{\zeta} \quad \text{as} \quad \zeta \to \infty$$

by the Stirling formula, we get

$$\sup_{F,\bar{\eta}} \left| \frac{\partial}{\partial\xi} P_\infty(F|\xi,\bar{\eta}) \right| \to 0 \quad \text{as} \quad \xi \to \infty.$$

Thus, it can be argued that $(\partial/\partial\xi)P_\infty(F|\xi,\bar{\eta}) \geqq 0$ for all values of F and ξ, and hence $P_\infty(F|\xi,\bar{\eta})$ is a non-decreasing function of ξ. Since $P_\infty(F|\xi,\bar{\eta})$ is bounded uniformly for all F, ξ and $\bar{\eta}$, there exists a limit function, say $P_\infty(F|\bar{\eta})$, such that

$$\lim_{\xi\to\infty} P_\infty(F|\xi,\bar{\eta}) = \bar{P}_\infty(F|\bar{\eta}).$$

By the continuity of $P_\infty(F|\xi,\bar{\eta})$ as a function of $(F,\bar{\eta})$ for every fixed ξ and the monotonicity of the sequence $\{P_\infty(F|\xi,\bar{\eta})\}(\xi\to\infty)$ then limit function $\bar{P}_\infty(F|\bar{\eta})$ is continuous with respect to $(F,\bar{\eta})$ and hence uniformly continuous for all F and $\bar{\eta}$ such that $0 \leqq F \leqq F_0$ and $0 \leqq \bar{\eta} \leqq 1$. Again, from the monotonicity of the sequence, the continuity of the functions $P_\infty(F|\xi,\bar{\eta})$ and $\bar{P}_\infty(F|\bar{\eta})$, and the compactness of the domain of $(F,\bar{\eta})$ under consideration, it follows that the convergence

$$P_\infty(F|\xi,\bar{\eta}) \to \bar{P}_\infty(F|\bar{\eta}) \quad \text{as} \quad \xi \to \infty$$

is uniform in $(F,\bar{\eta})$, that is,

$$\sup_{\substack{\bar{\eta} \\ F\leqq F_0}} \left|P_\infty(F|\xi,\bar{\eta}) - P_\infty(F|\bar{\eta})\right| \to 0 \quad \text{as} \quad \xi \to \infty. \tag{6.6.75}$$

Hence there exists a positive number $\xi_0 = \xi_0(\varepsilon)$ such that

$$\sup_{\bar{\eta},F\leqq F_0} \left|P_\infty(F|\xi,\bar{\eta}) - P_\infty(F|\xi_1,\bar{\eta})\right| < \varepsilon \tag{6.6.76}$$

for any given values of ξ and ξ_1 such that ξ, $\xi_1 \geqq \xi_0$. From (6.6.72) and (6.6.76) one can conclude (6.6.70), as was to be proved in the second step.

Let us put, for the sake of simplicity,

$$c_b = \Delta + T,\ \bar{d}_b = \frac{\bar{T}}{\Delta + T},\ \text{and}\ \bar{\bar{d}}_b = \frac{\bar{\bar{T}}}{\Delta + T}. \tag{6.6.77}$$

Then it is clear that

$$(6.6.78) \qquad \frac{c_b}{b} \to \Delta_0 + T_0, \quad \bar{d}_b \to \frac{\bar{T}_0}{\Delta_0 + T_0}, \quad \bar{\bar{d}}_b \to \frac{\bar{\bar{T}}_0}{\Delta_0 + T_0} \quad \text{as} \quad b \to \infty$$

Furthermore, let us define

$$(6.6.79) \qquad Q_\infty^b(F|\xi^*,\bar{\eta}^*) = P_\infty(F|c_b\xi^*,\bar{d}_b\bar{\eta}^*).$$

Now, in the third step, we shall show that, for any given $\varepsilon > 0$, there exist a positive number $\delta = \delta(\varepsilon)$ and a positive integer $b_0 = b_0(\varepsilon)$ such that

$$|(\xi^*, \bar{\eta}^*) - (1,1)| < \delta$$

implies that

$$(6.6.80) \qquad \sup_F \left| Q_\infty^b(F|\xi^*, \bar{\eta}) - Q_\infty^b(F|1,1) \right| < 2\varepsilon$$

for all $b \geqq b_0$.

To show (6.6.80), we first consider the case $F > F_0$ where F_0 is the same as in (6.6.65). From (6.6.71) we see that

$$\sup_{\xi^*,\bar{\eta}^*,F>F_0} \left|1 - Q_\infty^b(F|\xi^*,\bar{\eta}^*)\right| \to 0 \quad \text{and} \quad \sup_{F>F_0} \left|1 - Q_\infty^b(F|1,1)\right| \to 0$$

$$\text{as} \quad b \to \infty.$$

Therefore

$$(6.6.81) \qquad \sup_{\xi^*,\bar{\eta}^*,F>F_0} \left|Q_\infty^b(F|\xi^*,\bar{\eta}^*) - Q_\infty^b(F|1,1)\right| \to 0 \quad \text{as} \quad b \to \infty.$$

Suppose that $F \leqq F_0$ in the second place. Since $P_\infty(F|\xi,\bar{\eta})$ is uniformly continuous with respect to $(F,\xi,\bar{\eta})$ in the domain $0 \leqq F \leqq F_0$, $0 \leqq \xi \leqq \xi_0$, $1 \leqq \bar{\eta} \leqq 1$, ξ_0 being the same as in (6.6.70), there can be found a positive number $\delta_0 = \delta_0(\varepsilon)$ such that

$$\left|(\xi,\bar{\eta}) - (\xi_1,\bar{\eta}_1)\right| < \delta_0$$

implies that

$$\sup_{F \leqq F_0} \left|P_\infty(F|\xi,\bar{\eta}) - P_\infty(F|\xi_1,\bar{\eta}_1)\right| < \varepsilon. \tag{6.6.82}$$

Let b_* be the minimum integer such that $2\xi_0 \leqq c_b$ and put

$$\delta = min\left\{\frac{1}{2}, \frac{\delta_0}{c_{b_*}}\right\}.$$

Then it is easy to see that $|(\xi^*_{?}\bar{\eta}^*) - (1,1)| < \delta$ implies $|(c_b\xi^*_{?}\bar{d}_b\bar{\eta}^*) - (c_b,\bar{d}_b)| < \delta_0$ for all $b \leqq b_*$, and if $b > b_*$, then $c_b\xi^* > \xi_0$ and $c_b > \xi_0$ so long as $|(\xi^*_{?}\bar{\eta}^*) - (1,1)| < \delta$. Hence from (6.6.82) and (6.6.76) it follows, for $(\xi^*_{?}\bar{\eta}^*)$ satisfying $|(\xi^*_{?}\bar{\eta}^*) - (1,1)| < \delta$, that

$$\sup_{F \leqq F_0} \left|Q^b_\infty(F|\xi^*_{?}\bar{\eta}^*) - Q^b_\infty(F|1,1)\right| < \varepsilon \quad \text{if} \quad b \leqq b_*,$$

$$\sup_{F \leqq F_0} \left|Q^b_\infty(F|\xi^*_{?}\bar{\eta}^*) - Q^b_\infty(F|1,\bar{\eta}^*)\right| < \varepsilon \quad \text{if} \quad b > b_*,$$

and from the uniform continuity one can see that

$$\sup_{F \leqq F_0} \left|Q^b_\infty(F|1,\bar{\eta}^*) - Q^b_\infty(F|1,1)\right| < \varepsilon$$

Thus one can show that (6.6.79) is true.

We are now in a position to show that condition (6.6.52) is satisfied in our present case.

$$\sup_F \left| Q_b(F|\xi,\bar{\eta}^*,\bar{\bar{\eta}}^*) - Q_b(F|1,1,1) \right| \leqq \sup_F \left| Q_b(F|\xi^*,\bar{\eta}^*,\bar{\bar{\eta}}^*) - \mathcal{Q}_\infty^b(F|\xi^*,\bar{\eta}^*) \right|$$

$$+ \sup_F \left| Q_b(F|1,1,1) - \mathcal{Q}_\infty^b(F|1,1) \right| + \sup_F \left| \mathcal{Q}_\infty^b(F|\xi^*,\bar{\eta}^*) - \mathcal{Q}_\infty^b(F|1,1) \right|.$$

For a given $\varepsilon > 0$, one can choose $b_0 = b_0(\varepsilon)$ such that

$$\sup_F \left| Q_b(F|\xi^*,\bar{\eta}^*,\bar{\bar{\eta}}^*) - \mathcal{Q}_\infty^b(F|\xi^*,\bar{\eta}^*) \right| \leqq \sup_{F^*,\xi^*,\bar{\eta}^*,\bar{\bar{\eta}}^*} \left| Q_b(F|\xi^*,\bar{\eta}^*,\bar{\bar{\eta}}^*) - \mathcal{Q}_\infty^b(F|\xi^*,\bar{\eta}^*) \right| < \frac{\varepsilon}{3}$$

for all $b \geqq b_0$, and hence

$$\sup_F \left| Q_b(F|1,1,1) - \mathcal{Q}_\infty^b(F|1,1) \right| < \frac{\varepsilon}{3}.$$

One can also choose δ such that

$$\sup_F \left| \mathcal{Q}_\infty^b(F|\xi^*,\bar{\eta}^*) - \mathcal{Q}^b(F|1,1) \right| < \frac{\varepsilon}{3}$$

so long as $|(\xi,\bar{\eta},\bar{\bar{\eta}}^*) - (1,1,1)| < \delta(\varepsilon)$ for all b. Hence there exists a positive number $\delta = \delta(\varepsilon)$ such that

$$|(\xi^*,\bar{\eta}^*,\bar{\bar{\eta}}^*) - (1,1,1)| < \delta(\varepsilon)$$

implies that

$$\sup_F \left| Q_b(F|\xi^*,\bar{\eta}^*,\bar{\bar{\eta}}^*) - Q_b(F|1,1,1) \right| < \varepsilon$$

for all $b \geqq b_0$, which is the condition (6.6.52).

Asymptotic Power Function of the F-Statistic after Randomization. As was mentioned at the beginning of the preceding subsection, the distribution of the F-statistic whose cumulative distribution function should be obtained by taking the mathematical expectation of (6.6.49) with respect to the permutation distribution of $(\xi,\bar{\eta},\bar{\bar{\eta}})$ due to randomization is asymptotically equivalent in the sense of the type $(\mathbf{M})_d$

as $b \to \infty$ to a distribution whose cumulative distribution function is given by (6.6.51) or equivalently a distribution with the following probability element :

$$\exp\left(-\frac{\Delta + T}{2\sigma^2}\right) \sum_{l=0}^{\infty} \frac{[(\Delta + T)/2\sigma^2]}{l!} \sum_{\mu+\nu+\gamma=l} \frac{l!}{\mu!\nu!\gamma!} \left(\frac{\overline{T}}{\Delta + T}\right)^{\mu} \times \left(\frac{\overline{\overline{T}}}{\Delta + T}\right)^{\gamma} \left(1 - \frac{T}{\Delta + T}\right)^{\nu} \frac{\Gamma\left(\frac{n - b - \overline{\overline{\alpha}}}{2} + \mu + \nu\right)}{\Gamma\left(\frac{\overline{\alpha}}{2} + \mu\right)\Gamma\left(\frac{n - b - v + 1}{2} + \nu\right)} \times \left(\frac{\overline{\alpha}}{n - b - v + 1}F\right)^{(\overline{\alpha}/2)+\mu-1} \times \left(1 + \frac{\overline{\alpha}}{n - b - v + 1}F\right)^{-\{[(n-b-\overline{\overline{\alpha}})/2]+\mu+\nu\}} d\left(\frac{\overline{\alpha}}{n - b - v + 1}F\right). \tag{6.6.83}$$

This can be rewritten as

$$\exp\left(-\frac{\Delta + \overline{T}}{2\sigma^2}\right) \sum_{\mu=0}^{\infty} \sum_{\nu=0}^{\infty} \frac{(\overline{T}/2\sigma^2)^{\mu}}{\mu!} \frac{(\Delta/2\sigma^2)^{\nu}}{\nu!} h_{\mu\nu}^{b}(F)\ dF, \tag{6.6.84}$$

where

$$h_{\mu,\nu}^{b}(F) = \frac{\overline{\alpha}}{n - b - v + 1} \frac{\Gamma\left(\frac{n - b - v + 1}{2} + \mu + \nu\right)}{\Gamma\left(\frac{\overline{\alpha}}{2} + \mu\right)\Gamma\left(\frac{n - b - v + 1}{2} + \nu\right)} \times \left(\frac{\overline{\alpha}}{n - b - v + 1}F\right)^{(\overline{\alpha}/2)+\mu-1} \times \left(1 + \frac{\overline{\alpha}}{n - b - v + 1}F\right)^{-\{[(n-b-\overline{\alpha})/2]+\mu+\nu\}} \tag{6.6.85}$$

Let us put

$$G_b(F;\Delta,\overline{T}) = \exp\left(-\frac{\Delta + \overline{T}}{2\sigma^2}\right) \sum_{\mu=0}^{\infty} \sum_{\nu=0}^{\infty} \frac{(\overline{T}/2\sigma^2)^{\mu}}{\mu!} \frac{(\Delta/2\sigma^2)^{\nu}}{\nu!} H_{\mu,\nu}^{b}(F), \tag{6.6.86}$$

where $H^{b}_{\mu,\nu}(F)$ is the same as in (6.6.53b). Furthermore, using the function $H^{\infty}_{\mu}(F)$ given by (6.6.56), we define

$$(6.6.87) \qquad G^{\infty}_{b}(F;\Delta,\overline{T}) = exp\left(-\frac{\Delta+\overline{T}}{2\sigma^2}\right)\sum_{\mu=0}^{\infty}\sum_{\nu=0}^{\infty}\frac{(\overline{T}/2\sigma^2)^{\mu}}{\mu!}\frac{(\Delta/2\sigma^2)^{\nu}}{\nu!}H^{\infty}_{\mu}(F).$$

It should be noticed that the first step of the proof of condition (6.6.52) implies that

$$(6.6.88) \qquad \sup_{F}\left|G_{b}(F;\Delta,T) - G^{\infty}_{b}(F;\Delta,T)\right| \to 0 \quad \text{as} \quad b \to \infty$$

for any given values of Δ and $\overline{T}$. Hence, in the special case $\Delta = 0$, one gets

$$(6.6.89) \qquad \sup_{F}\left|G_{b}(F;0,\overline{T}) - G^{\infty}_{b}(F;0,\overline{T})\right| \to 0 \quad \text{as} \quad b \to \infty.$$

Since $H^{\infty}_{\mu}(F)$ is independent of ν, we have by (6.6.87)

$$(6.6.90) \qquad G^{\infty}_{b}(F;\Delta,\overline{T}) = exp\left(-\frac{\overline{T}}{2\sigma^2}\right)\sum_{\mu=0}^{\infty}\frac{(\overline{T}/2\sigma^2)^{\mu}}{\mu!}H^{\infty}_{\mu}(F)$$

$$= G^{\infty}_{b}(F;0,\overline{T}).$$

Hence from (6.6.88) and (6.6.89), it follows that

$$(6.6.91) \qquad \sup_{F}\left|G_{b}(F;\Delta,\overline{T}) - G^{\infty}_{b}(F;0,\overline{T})\right| \to 0 \quad \text{as} \quad b \to \infty.$$

This means that the distribution whose cumulative distribution function being given by $G^{\infty}_{b}(F;0,\overline{T})$ is asymptotically equivalent in the sense of the type $(M)_d$ to the power function of the F-statistic after the randomization. Thus we conclude this section by stating the following theorem:

Theorem 6.6.1. The power function of the F-statistic given by

(6.6.3) after randomization is asympotically equivalent in the sense of the type $(\boldsymbol{M})_d$ to the non-central F-distribution whose probability element is given by

(6.6.92)
$$\exp\left(\frac{-\bar{T}}{2\sigma^2}\right)\sum_{\mu=0}^{\infty}\frac{(\bar{T}/2\sigma^2)^{\mu}}{\mu!}\frac{\Gamma\left(\frac{n-b-\bar{\bar{\alpha}}}{2}+\mu\right)}{\Gamma\left(\frac{\bar{\alpha}}{2}+\mu\right)\Gamma\left(\frac{n-b-v+1}{2}\right)}$$
$$\times\left(\frac{\bar{\alpha}}{n-b-v+1}F\right)^{(\bar{\alpha}/2)+\mu-1}$$
$$\times\left(1+\frac{\bar{\alpha}}{n-b-v+1}F\right)^{-\{[(n-b-\bar{\bar{\alpha}})/2]+\mu\}}d\left(\frac{\bar{\alpha}}{n-b-v+1}F\right),$$

under the limiting process (6.6.30), provided that conditions (6.6.31) and (6.6.32) are satisfied.

In the special case $h = m$ this becomes

(6.6.93)
$$\exp\left(\frac{-T}{2\sigma^2}\right)\sum_{\mu=0}^{\infty}\frac{(T/2\sigma^2)^{\mu}}{\mu!}\frac{\Gamma\left(\frac{n-b}{2}+\mu\right)}{\Gamma\left(\frac{v-1}{2}+\mu\right)\Gamma\left(\frac{n-b-v+1}{2}\right)}$$
$$\times\left(\frac{v-1}{n-b-v+1}F\right)^{[(v-1)/2]+\mu-1}$$
$$\times\left(1+\frac{v-1}{n-b-v+1}F\right)^{-\{[(n-b)/2]+\mu\}}d\left(\frac{v-1}{n-b-v+1}F\right).$$

It should be remarked here that the distribution having the probability element

(6.6.94)
$$\exp\left(\frac{-\bar{T}}{2\sigma^2}\right)\sum_{\mu=0}^{\infty}\frac{(\bar{T}/2\sigma^2)^{\mu}}{\mu!}\frac{1}{\Gamma[(\bar{\alpha}/2)+\mu]}\left(\frac{\bar{\alpha}}{2}F\right)^{(\bar{\alpha}/2)+\mu-1}\times\exp\left(-\frac{\bar{\alpha}}{2}F\right)d\left(\frac{\bar{\alpha}}{2}F\right),$$

that is, the non-central χ^2 distribution is also asymptotically equivalent in the sense of the type $(\boldsymbol{M})_d$ as $b \to \infty$ to the power function of the F-statistic.

However, it is not surprising if one notices that

$$\frac{\Gamma\left(\dfrac{n - b - \bar{\bar{\alpha}}}{2} + \mu\right)}{\Gamma\left(\dfrac{n - b - v + 1}{2}\right)} \left(\frac{\bar{\alpha}}{n - b - v + 1}F\right)^{(\bar{\alpha}/2)+\mu-1}$$

$$\times \left(1 + \frac{\bar{\alpha}}{n - b - v + 1}F\right)^{-\{[(n-b-\bar{\bar{\alpha}})/2]+\mu\}} d\left(\frac{\bar{\alpha}}{n - b - v + 1}F\right)$$

$$= \frac{\Gamma\left(\dfrac{n - b - \bar{\bar{\alpha}}}{2} + \mu\right)}{\Gamma\left(\dfrac{n - b - v + 1}{2}\right)} \left(\frac{2}{n - b - v + 1}\right)^{(\bar{\alpha}/2)+\mu} \left(\frac{\bar{\alpha}}{2}F\right)^{(\bar{\alpha}/2)+\mu-1}$$

$$\times \left(1 + \frac{2}{n - b - v + 1}\,\frac{\bar{\alpha}}{2}F\right)^{-\{[(n-b-\bar{\bar{\alpha}})/2]+\mu\}} d\left(\frac{\bar{\alpha}}{2}F\right)$$

$$\sim \frac{\left(\dfrac{n - b - \bar{\bar{\alpha}}}{2} + \mu - 1\right)^{[(n-b-\bar{\bar{\alpha}})/2]+\mu-(1/2)}}{\left(\dfrac{n - b - v + 1}{2} - 1\right)^{[(n-b-v+1)/2]-(1/2)}} \, exp\left(\frac{\bar{\alpha}}{2} + \mu\right)$$

$$\times \left(\frac{2}{n - b - v + 1}\right)^{(\bar{\alpha}/2)+\mu} \left(\frac{\bar{\alpha}}{2}F\right)^{(\bar{\alpha}/2)+\mu-1}$$

$$\times \left\{1 + \frac{2}{(n - b - v + 1)} \frac{\bar{\alpha}}{2}F\right\}^{-\{[(n-b-v+1)/2]+(\bar{\alpha}/2)+\mu\}} d\left(\frac{\bar{\alpha}}{2}F\right)$$

$$\sim \left(\frac{\bar{\alpha}}{2}F\right)^{(\bar{\alpha}/2)+\mu-1} exp\left(-\frac{\bar{\alpha}}{2}F\right) d\left(\frac{\bar{\alpha}}{2}F\right) \quad \text{as} \quad b \to \infty,$$

in the sense of the type $(\boldsymbol{B})_d$.

6.7. Appendix: Properties of Incidence Numbers

From the definition, it is evident that

$$\lambda_{pp}^{(1)00} = 0. \tag{6.7.1}$$

Consider

$$\lambda_{pp}^{(4)} = \sum_i t_{ii}^{(\alpha)pp} + \sum_{i \neq j} t_{ij}^{(\alpha)pp}.$$

If $\alpha = 0$, the first term of the right-hand side is equal to k and the second term vanished, and if $\alpha \neq 0$, the first term vanishes and the second term is equal to $\lambda_{pp}^{(1)\alpha\alpha}$. Thus, taking (6.7.1) into account, one gets

$$\lambda_{pp}^{(4)\alpha\alpha} = \delta_{\alpha 0}\, k + \lambda_{pp}^{(1)\alpha\alpha}, \qquad (\alpha \geqq 0), \tag{6.7.2}$$

and in particular

$$\lambda_{pp}^{(4)\alpha\alpha} = \lambda_{pp}^{(1)\alpha\alpha}, \qquad (\alpha \neq 0). \tag{6.7.2$'$}$$

Let $p = q$ in $\lambda_{pq}^{(5)\alpha\alpha}$. Then, in a similar way to the above, one can easily see that

$$\lambda_{pp}^{(5)\alpha\alpha} = \lambda_{pp}^{(1)\alpha\alpha} + 2\lambda_{pp}^{(2)\alpha\alpha} + \lambda_{pp}^{(3)\alpha\alpha}$$

in the case $\alpha = \beta \neq 0$, and

$$\lambda_{pp}^{(5)\alpha\beta} = 2\lambda_{pp}^{(2)\alpha\beta} + \lambda_{pp}^{(3)\alpha\beta}$$

in the case $\alpha \neq \beta \neq 0$. Finally, in the case $\alpha = 0$, it becomes

$$\lambda_{pp}^{(5)0\beta} = \begin{cases} k(k - 1), \text{ if } \beta = 0, \\ (k - 2)\lambda_{pp}^{(1)\beta\beta}, \text{ if } \beta \neq 0, \end{cases}$$

and in the case $\beta = 0$

$$\lambda_{pp}^{(5)\alpha 0} = \begin{cases} k(k - 1), \text{ if } \alpha = 0, \\ (k - 2)\lambda_{pp}^{(1)\alpha\alpha}, \text{ if } \alpha \neq 0. \end{cases}$$

These results can be put together in the following form:

$$\begin{aligned} \lambda_{pp}^{(5)\alpha\beta} &= \delta_{\alpha 0}\,\delta_{\beta 0}k(k - 1) \\ &\quad + (1 - \delta_{\alpha 0})\delta_{\beta 0}(k - 2)\lambda_{pp}^{(1)\alpha\alpha} + (1 - \delta_{\beta 0})\delta_{\alpha 0} \\ &\quad \times (k - 2)\lambda_{pp}^{(1)\beta\beta} + (1 - \delta_{\alpha 0})(1 - \alpha_{\beta 0}) \\ &\quad \times \left[\delta_{\alpha\beta}\lambda_{pp}^{(1)\alpha\beta} + 2\lambda_{pp}^{(2)\alpha\beta} + \lambda_{pp}^{(3)\alpha\beta}\right], \end{aligned} \tag{6.7.3}$$

$(\alpha,\beta \geqq 0)$,

and in particular

$$\lambda_{pp}^{(5)\alpha\beta} = \delta_{\alpha\beta}\lambda_{pp}^{(1)\alpha\beta} + 2\lambda_{pp}^{(2)\alpha\beta} + \lambda_{pp}^{(3)\alpha\beta}, \qquad (\alpha,\beta \neq 0). \tag{6.7.3)'$$

Let us put $p = q$ in $2\lambda_{pp}^{(6)\alpha\beta}$. In the same manner as in the preceding cases, one can easily see that

$$\lambda_{pp}^{(6)\alpha\beta} = \begin{cases} 2\lambda_{pp}^{(2)\alpha\beta}, & \text{if } \alpha,\beta \neq 0, \\ 2\lambda_{pp}^{(1)\beta\beta}, & \text{if } \alpha = 0, \\ 2\lambda_{pp}^{(1)\alpha\alpha}, & \text{if } \beta = 0, \\ 0, & \text{if } \alpha = \beta = 0, \end{cases}$$

or, putting these together, one has

$$\lambda_{pp}^{(6)\alpha\beta} = \delta_{\alpha 0}\lambda_{pp}^{(1)\beta\beta} + \delta_{\beta 0}\lambda_{pp}^{(1)\alpha\alpha} + (1 - \delta_{\alpha 0})(1 - \delta_{\beta 0}) \times \lambda_{pp}^{(2)\alpha\beta}, \qquad (\alpha,\beta \geqq 0), \tag{6.7.4}$$

and in particular

$$\lambda_{pp}^{(6)\alpha\beta} = \lambda_{pp}^{(2)\alpha\beta}, \qquad (\alpha,\beta \neq 0). \tag{6.7.4}$$

In the next place, let us consider the product

$$\begin{aligned} \lambda_{pp}^{(1)\alpha\alpha}\lambda_{pp}^{(1)\beta\beta} &= \sum_{i\neq j} t_{ij}^{(\alpha)pp} \sum_{l\neq s} t_{ls}^{(\beta)pp} \\ &= \sum_{i\neq j}\left(t_{ij}^{(\alpha)pp}t_{ij}^{(\beta)pp} + t_{ij}^{(\alpha)pp}t_{ji}^{(\beta)pp}\right. \\ &\quad + \sum_{i\neq j\neq l}\left(t_{ij}^{(\alpha)pp}t_{il}^{(\beta)pp} + t_{ij}^{(\alpha)pp}t_{li}^{(\beta)pp}\right. \\ &\quad\quad \left. + t_{ij}^{(\alpha)pp}t_{jl}^{(\beta)pp} + t_{ij}^{(\alpha)pp}t_{lj}^{(\beta)pp}\right) \\ &\quad + \sum_{i\neq j\neq l\neq s} t_{ij}^{(\alpha)pp}t_{ls}^{(\beta)pp} . \end{aligned}$$

Investigating the right-hand members in cases $\alpha = \beta \neq 0$, $\alpha \neq \beta \neq 0$ and $\alpha = 0$ or $\beta = 0$ separately, and putting them together, one can find the following

$$\lambda_{pp}^{(1)\alpha\alpha}\lambda_{pp}^{(1)\beta\beta} = (1 - \delta_{\alpha 0})(1 - \delta_{\beta 0})\left(2\delta_{\alpha\beta}\lambda_{pp}^{(1)\alpha\beta}\right. \tag{6.7.5}$$

$$+ 4\lambda_{pp}^{(2)\alpha\beta} + \lambda_{pp}^{(3)\alpha\beta}\Big] , \qquad (\alpha,\beta \geqq 0),$$

and in particular

$$(6.7.5)' \qquad \lambda_{pp}^{(1)\alpha\alpha}\lambda_{pp}^{(1)\beta\beta} = 2\delta_{\alpha\beta}\lambda_{pp}^{(1)\alpha\beta} + 4\lambda_{pp}^{(2)\alpha\beta} + \lambda_{pp}^{(3)\alpha\beta} , \qquad (\alpha,\beta \neq 0).$$

In the final place, let us consider the product

$$\lambda_{pq}^{(4)\alpha\alpha}\lambda_{pq}^{(4)\beta\beta} = \left(\sum_i t_{ii}^{(\alpha)pq} + \sum_{i\neq j} t_{ij}^{(\alpha)pq}\right)\left(\sum_l t_{ll}^{(\beta)pq} + \sum_{l\neq s} t_{ls}^{(\beta)pq}\right) .$$

This can be expanded and arranged in the following form:

$$\lambda_{pq}^{(4)\alpha\alpha}\lambda_{pq}^{(4)\beta\beta} = \delta_{\alpha\beta}\left(\sum_i t_{ii}^{(\alpha)pq} + \sum_{i\neq j} t_{ij}^{(\alpha)pq}\right)$$

$$+ \sum_{i\neq j}\Big(t_{ii}^{(\alpha)pq}t_{jj}^{(\beta)pq} + t_{ij}^{(\alpha)pq}t_{ji}^{(\beta)pq} +$$

$$+ t_{ij}^{(\alpha)pq}t_{ii}^{(\beta)pq} + t_{ij}^{(\alpha)pq}t_{jj}^{(\beta)pq}$$

$$+ t_{ii}^{(\alpha)pq}t_{ij}^{(\beta)pq} + t_{ii}^{(\alpha)pq}t_{ji}^{(\beta)pq}\Big)$$

$$+ \sum_{i\neq j\neq l}\Big(t_{ij}^{(\alpha)pq}t_{ll}^{(\beta)pq} + t_{ij}^{(\alpha)pq}t_{jl}^{(\beta)pq}$$

$$+ t_{ii}^{(\alpha)pq}t_{jl}^{(\beta)pq} + t_{ij}^{(\alpha)pq}t_{li}^{(\beta)pq}$$

$$+ t_{ij}^{(\alpha)pq}t_{il}^{(\beta)pq} + t_{ij}^{(\alpha)pq}t_{lj}^{(\beta)pq}\Big)$$

$$+ \sum_{i\neq j\neq l\neq s} t_{ij}^{(\alpha)pq}t_{ls}^{(\beta)pq} ,$$

from which one gets

$$\lambda_{pq}^{(4)\alpha\alpha}\lambda_{pq}^{(4)\beta\beta} = \delta_{\alpha\beta}\lambda_{pq}^{(4)\alpha\alpha} + \lambda_{pq}^{(5)\alpha\beta} + 2\lambda_{pq}^{(6)\alpha\beta}, \quad (\alpha,\beta \geqq 0). \tag{6.7.6}$$

In the case when $p = q$, it follows from (3.1.1), (3.1.2) and (3.1.5) that

$$\begin{aligned}\lambda_{pp}^{(4)\alpha\alpha}\lambda_{pp}^{(4)\beta\beta} = {} & \delta_{\alpha 0}\delta_{\beta 0}k + \delta_{\alpha 0}k\lambda_{pp}^{(1)\beta\beta} + \delta_{\beta 0}k\lambda_{pp}^{(1)\alpha\alpha} \\ & + (1 - \delta_{\alpha 0})(1 - \delta_{\beta 0})\left[2\delta_{\alpha\beta}\lambda_{pp}^{(1)\alpha\alpha} + 4\lambda_{pp}^{(2)\alpha\beta} + \lambda_{pp}^{(3)\alpha\beta}\right], \\ & (\alpha,\beta \geqq 0),\end{aligned} \tag{6.7.7}$$

and in particular,

$$\lambda_{pp}^{(4)\alpha\alpha}\lambda_{pp}^{(4)\beta\beta} = 2\delta_{\alpha\beta}\lambda_{pp}^{(1)\alpha\alpha} + 4\lambda_{pp}^{(2)\alpha\beta} + \lambda_{pp}^{(3)\alpha\beta}, \quad (\alpha,\beta \neq 0). \tag{6.7.7$'$}$$

Partitioning the incidence matrices of treatments and blocks, let us put

$$\Phi' = (\Phi_1', \Phi_2', \ldots, \Phi_b') = (\underline{\phi}_{11}, \ldots, \underline{\phi}_{1k}, \underline{\phi}_{21}, \ldots, \underline{\phi}_{2k}, \ldots\ldots, \underline{\phi}_{b1}, \ldots, \underline{\phi}_{bk})$$

and

$$\Psi' = (\Psi_1', \Psi_2', \ldots, \Psi_b') = (\underline{\psi}_{11}, \ldots, \underline{\psi}_{1k}, \underline{\psi}_{21}, \ldots, \underline{\psi}_{2k}, \ldots\ldots, \underline{\psi}_{b1}, \ldots, \underline{\psi}_{bk}).$$

Then, $\underline{\psi}_{pi}' = (0, \ldots, 0, \underset{p}{\hat{1}}, 0, \ldots, 0)$, $i = 1, \ldots, k$, and

$$N = \Phi'\Psi = \sum_{p=1}^{b} \Phi_p'\Psi_p \quad (= (\underline{n}_1, \underline{n}_2, \ldots, \underline{n}_b)),$$

where

$$\Phi_p'\Psi_p = \sum_{i=1}^{k} \underline{\phi}_{pi}\underline{\psi}_{pi} \ (= (\underline{0},\ldots,\underline{0}\ ,\underline{n}_p,\underline{0}\ ,\ldots,\underline{0}) \equiv N_p)\ .$$

It is easy to check that

$$t_{ij}^{(\alpha)pq} = \underline{1}_b'\underline{\psi}_{pi}\underline{\phi}_{pi}'A_\alpha\underline{\phi}_{qj}\underline{\psi}_{qj}'\underline{1}_b\ , \qquad (\alpha \geqq 0),$$

where $\underline{1}_b$ stands for the vector of order b whose components are all unity. Hence, it follows that

$$\lambda_{pq}^{(4)\alpha\alpha} = \sum_{i,j} t_{ij}^{(\alpha)pq} = \underline{1}_b'N_p'A_\alpha N_q\underline{1}_b = (N'A_\alpha N)_{p,q}\ ,$$

which is the (p,q)-element of the matrix $N'A_\alpha N$. Whence one obtains

$$\sum_{p,q}\lambda_{pq}^{(4)\alpha\alpha} = \underline{1}_b'N'A_\alpha N\underline{1}_b = r^2\underline{1}_v'A_\alpha\underline{1}_v = vr^2n_\alpha\ , \quad (\alpha \geqq 0).$$

It is also readily seen that

$$\begin{aligned}\sum_p \lambda_{pp}^{(4)\alpha\alpha} &= tr(N'A_\alpha N) = tr(NN'A_\alpha)\\ &= tr\left(\sum_{\beta=0}^{m}\lambda_\beta A_\beta A_\alpha\right)\\ &= \sum_{\beta=0}^{m}\lambda_\beta\sum_{\gamma=0}^{m}p_{\alpha\beta}^{\gamma}tr(A_\gamma)\\ &= \sum_{\beta=0}^{m}\lambda_\beta\, p_{\alpha\beta}^{0}v = v\lambda_\alpha n_\alpha\ , \qquad (\alpha \geqq 0).\end{aligned}$$

Hence, one gets

$$\text{(6.7.8)} \qquad \sum_{p\neq q}\lambda_{pq}^{(4)\alpha\alpha} = v(r^2 - \lambda_\alpha)n\ , \qquad (\alpha \geqq 0),$$

and it is also easy, by (3.1.2), to show that

$$\sum_p \lambda_{pp}^{(1)\alpha\alpha} = (1 - \delta_{\alpha 0})v\lambda_\alpha n_\alpha , \qquad (\alpha \geqq 0), \tag{6.7.9}$$

and in particular

$$\sum_p \lambda_{pp}^{(1)\alpha\alpha} = v\lambda_\alpha n_\alpha , \qquad (\alpha \neq 0). \tag{6.7.9'}$$

From (6.7.5) and (6.7.9), it follows that

$$\sum_p \lambda_{pp}^{(1)\alpha\alpha}\lambda_{pp}^{(1)\beta\beta} = (1 - \delta_{\alpha 0})(1 - \delta_{\beta 0})\left(2\delta_{\alpha\beta}v\lambda_\alpha n_\alpha + 4\Lambda_{\alpha\beta}^{(1)} + \Lambda_{\alpha\beta}^{(2)}\right), \qquad (\alpha,\beta \geqq 0), \tag{6.7.10}$$

and in particular

$$\sum_p \lambda_{pp}^{(1)\alpha\alpha}\lambda_{pp}^{(1)\beta\beta} = 2\delta_{\alpha\beta}v\lambda_\alpha n_\alpha + 4\Lambda_{\alpha\beta}^{(1)} + \Lambda_{\alpha\beta}^{(1)} ,(\alpha,\beta \neq 0), \tag{6.7.10'}$$

where we have put

$$\Lambda_{\alpha\beta}^{(1)} = \sum_p \lambda_{pp}^{(2)\alpha\beta} \text{ and } \Lambda_{\alpha\beta}^{(2)} = \sum_p \lambda_{pp}^{(3)\alpha\beta} . \tag{6.7.11}$$

Now, let us consider the sum of product $\lambda_{pq}^{(4)\alpha\alpha}\lambda_{pq}^{(4)\beta\beta}$ with respect to p and q. First, one can see by (6.7.7) and (6.7.9) that

$$\begin{aligned}\sum_p \lambda_{pp}^{(4)\alpha\alpha}\lambda_{pp}^{(4)\beta\beta} &= \delta_{\alpha 0}\delta_{\beta 0}vrk \\ &+ (1 - \delta_{\alpha 0})\{\delta_{\beta 0}k + (1 - \delta_{\beta 0})\delta_{\alpha\beta}\}v\lambda_\alpha n_\alpha \\ &+ (1 - \delta_{\beta 0})\{\delta_{\alpha 0}k + (1 - \delta_{\alpha 0})\delta_{\alpha\beta}\}v\lambda_\beta n_\beta \\ &+ (1 - \delta_{\alpha 0})(1 - \delta_{\beta 0})\left(4\Lambda_{\alpha\beta}^{(1)} + \Lambda_{\alpha\beta}^{(2)}\right), \\ &\qquad (\alpha,\beta \geqq 0),\end{aligned} \tag{6.7.12}$$

and, in the special case $\alpha,\beta \neq 0$,

$$(6.7.12)' \qquad \sum_p \lambda_{pp}^{(4)\alpha\alpha} \lambda_{pp}^{(4)\beta\beta} = 2\delta_{\alpha\beta} v\lambda_\alpha n_\alpha + 4\Lambda_{\alpha\beta}^{(1)} + \Lambda_{\alpha\beta}^{(2)} \;, (\alpha,\beta \quad 0).$$

In the second place, let us calculate the sum

$$\sum_{p,q} \lambda_{pq}^{(4)\alpha\alpha} \lambda_{pq}^{(4)\beta\beta} = \sum_{p,q} (N'A_\alpha N)_{p,q} (N'A_\beta N)_{p,q}$$

$$= tr(N'A_\alpha NN'A_\beta N) = tr(NN'A_\alpha NN'A_\beta) \; .$$

Since

$$NN'A_\alpha NN'A_\beta = \sum_{\sigma=0}^{m} \lambda_\sigma A_\sigma A_\alpha \sum_{\tau=0}^{m} \lambda_\tau A_\tau A_\beta$$

$$= \sum_{\sigma=0}^{m} \lambda_\sigma \sum_{\gamma=0}^{m} p_{\sigma\alpha}^{\gamma} A_\gamma \sum_{\tau=0}^{m} \lambda_\tau \sum_{\eta=0}^{m} p_{\tau\beta}^{\eta} A_\eta$$

$$= \sum_{\sigma,\gamma,\tau,\eta\geqq 0} \lambda_\sigma \lambda_\tau \, p_{\sigma\alpha}^{\gamma} \, p_{\tau\beta}^{\eta} \sum_{\xi=0}^{m} p_{\gamma\eta}^{\xi} \, A_\xi$$

and $tr(A_\xi) = \delta_{\xi 0} v$, it holds that

$$tr(NN'A_\alpha NN'A_\beta) = v \sum_{\sigma,\tau,\gamma\geqq 0} \lambda_\sigma \lambda_\tau \, p_{\sigma\alpha}^{\gamma} p_{\tau\beta}^{\gamma} \, n_\gamma \; .$$

But, since

$$\sum_{\gamma=0}^{m} p_{\sigma\alpha}^{\gamma} p_{\tau\beta}^{\gamma} n_\gamma = \sum_{\gamma=0}^{m} p_{\sigma\alpha}^{\gamma} p_{\gamma\beta}^{\tau} n_\tau$$

$$= n_\tau (P_\alpha P_\beta)_{\sigma,\tau}$$

$$= n_\tau \sum_{\gamma=0}^{m} p_{\alpha\beta}^{\gamma} (P_\gamma)_{\sigma,\tau}$$

$$= \sum_{\gamma=0}^{m} p_{\alpha\beta}^{\gamma} \, p_{\sigma\gamma}^{\tau} n_\tau = \sum_{\gamma=0}^{m} p_{\sigma\tau}^{\gamma} \, p_{\alpha\beta}^{\gamma} n_\gamma \; ,$$

one has the alternative expression

$$tr(NN'A_\alpha NN'A_\beta) = v \sum_{\sigma,\tau,\gamma\ 0} \lambda_\sigma \lambda_\tau \, p^\gamma_{\sigma\tau} p^\gamma_{\alpha\beta} \, n_\gamma ,$$

which is more convenient for later use. Note that this expression can directly be derived by calculating $tr(NN'NN'A_\alpha A_\beta)$, which is equal to $tr(NN'A_\alpha NN'A_\beta)$ becuase of the commutativity of the association algebra.

Thus, one gets

$$\sum_{p,q} \lambda^{(4)\alpha\alpha}_{pq} \lambda^{(4)\beta\beta}_{pq} = v \sum_{\sigma,\tau,\gamma\ 0} \lambda_\sigma \lambda_\tau \, p^\gamma_{\sigma\tau} p^\gamma_{\alpha\beta} \, n_\gamma \, , (\alpha,\beta \geqq 0). \tag{6.7.13}$$

It follows from (6.7.12) and (6.7.13) that

$$\begin{aligned} \sum_{p \neq q} \lambda^{(4)\alpha\alpha}_{pq} \lambda^{(4)\beta\beta}_{pq} = v \sum_{\sigma,\tau,\gamma \geqq 0} & \lambda_\sigma \lambda_\tau p^\gamma_{\sigma\tau} p^\gamma_{\alpha\beta} n_\gamma - \delta_{\alpha 0} \delta_{\beta 0} vrk \\ & - (1 - \delta_{\alpha 0})\{\delta_{\beta 0} k + (1 - \delta_{\beta 0}) \delta_{\alpha\beta}\} v \lambda_\alpha n_\alpha \\ & - (1 - \delta_{\beta 0})\{\delta_{\alpha 0} k + (1 - \delta_{\alpha 0}) \delta_{\alpha\beta}\} v \lambda_\beta n_\beta \\ & - (1 - \delta_{\alpha 0})(1 - \delta_{\beta 0}) \left(4\Lambda^{(1)}_{\alpha\beta} + \Lambda^{(2)}_{\alpha\beta}\right), \\ & \qquad (\alpha,\beta \geqq 0). \end{aligned} \tag{6.7.14}$$

In the next place, we shall consider the sum of $\lambda^{(6)\alpha\beta}_{pq}$ with respect to p and q. By the definition, we have

$$2 \sum_{p,q} \lambda^{(6)\alpha\beta}_{pq} = 2 \sum_{p \neq q} \lambda^{(6)\alpha\beta}_{pq} + 2 \sum_{p} \lambda^{(6)\alpha\beta}_{pq}$$

$$= \sum_{p \neq q} \left[\sum_{i \neq j} \left(t^{(\alpha)pq}_{ii} t^{(\beta)pq}_{ij} + t^{(\alpha)pq}_{ij} t^{(\beta)pq}_{ii} + t^{(\alpha)pq}_{ii} t^{(\beta)pq}_{ji} \right.\right.$$

$$+ t_{ji}^{(\alpha)pq} t_{ii}^{(\beta)pq}\Big)$$

$$+ \sum_{i \neq j \neq l} \left(t_{ij}^{(\alpha)pq} t_{il}^{(\beta)pq} + t_{ij}^{(\alpha)pq} t_{lj}^{(\beta)pq} \right) \Bigg]$$

$$+ \sum_{p} \Bigg[\sum_{i \neq j} \Big(t_{ii}^{(\alpha)pp} t_{ij}^{(\beta)pp} + t_{ij}^{(\alpha)pp} t_{ii}^{(\beta)pp} + t_{ii}^{(\alpha)pp} t_{ji}^{(\beta)pp} +$$

$$+ t_{ji}^{(\alpha)pp} t_{ii}^{(\beta)pp}\Big)$$

$$+ \sum_{i \neq j \neq l} \left(t_{ij}^{(\alpha)pp} t_{il}^{(\beta)pp} + t_{ij}^{(\alpha)pp} t_{lj}^{(\beta)pp} \right) \Bigg] .$$

Suppose first that $\alpha, \beta \neq 0$. In this case, it is easily noticed that the above quantity is twice the total number of such triplets $\{u, u', u''\}$ of treatments in the design that u and u' are α-th associates and u and u'' are β-th associates, which must be equal to $2vr \sum_{\gamma=1}^{m} \lambda_\gamma p_{\alpha\beta} n_\gamma$. (See Fig. 6.7.1).

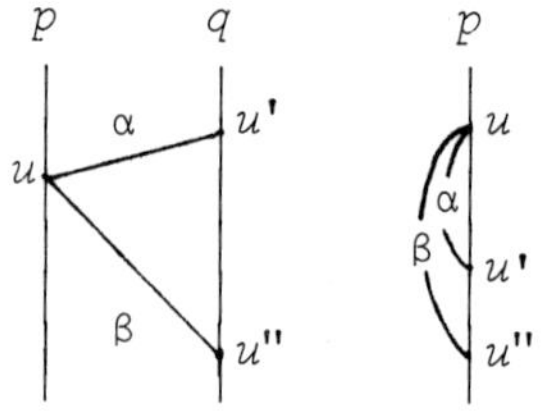

Fig 6.7.1

Secondly, consider the case $\alpha = 0$ and $\beta \neq 0$. In this case, the above expression of $2 \sum_{p,q} \lambda_{pq}^{(6)\alpha\beta}$ can slightly be simplified in the form:

$$2 \sum_{p,q} \lambda_{pq}^{(6)0\beta} = \sum_{p \neq q} \Bigg[\sum_{i \neq j} \Big(t_{ii}^{(0)pq} t_{ij}^{(\beta)pq} + t_{ij}^{(0)pq} t_{ii}^{(\beta)pq} + t_{ii}^{(0)pq} t_{ji}^{(\beta)pq}$$

$$+ t_{ji}^{(0)pq} t_{ii}^{(\beta)pq}\Big)$$

$$+ \sum_{i \neq j \neq l} \left(t_{ij}^{(0)pq} t_{il}^{(\beta)pq} + t_{ij}^{(0)pq} t_{lj}^{(\beta)pq} \right) \Bigg]$$

$$+ \sum_{p} \sum_{i \neq j} \left(t_{ij}^{(\beta)pp} + t_{ji}^{(\beta)pp} \right) ,$$

the first member of the right-hand side of which is equal to twice the number of such pairs $\{u,u'\}$ of β-th associates as shown in Fig. 6.7.2, which should be equal to $2v(r - 1)\lambda_\beta n_\beta$. The second member is twice the number of such pairs $\{u,u'\}$ of β-th associates as shown in Fig. 6.7.3, which must be equal to $2v\lambda_\beta n_\beta$.

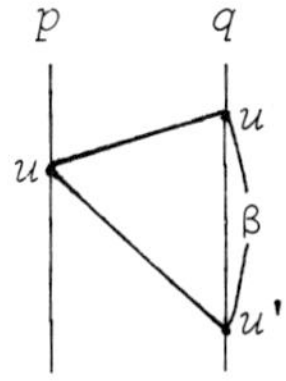

Fig. 6.7.2

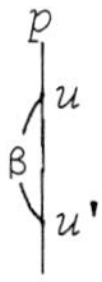

Fig. 6.7.3

Similar result is obtained in the case $\alpha \neq 0$ and $\beta = 0$.

Finally, in the case $\alpha = \beta = 0$, it is evident that $2 \sum_{p,q} \lambda_{pq}^{(6)00} = 0.$

Summarizing the results thus obtained, one gets

(6.7.15) $$\sum_{p,q} \lambda_{pq}^{(6)\alpha\beta} = vr \sum_{\gamma \geq 1} \lambda p_{\alpha\beta}^{\gamma} n_\gamma \quad , \qquad (\alpha,\beta \geqq 0),$$

where, we have used the relation

$$\sum_{\gamma \geqq 1} \lambda_\gamma p_{\alpha\beta}^{\gamma} n_\gamma = \begin{cases} \lambda_\beta n_\beta, & \text{if } \alpha = 0 \text{ and } \beta \neq 0, \\ \lambda_\alpha n_\alpha, & \text{if } \alpha \neq 0 \text{ and } \beta = 0, \\ 0, & \text{if } \alpha = \beta = 0. \end{cases}$$

Now, taking the sum of (6.7.4) with respect to p and making use of (6.7.9), one can easily get

(6.7.16) $$\sum_p \lambda_{pp}^{(6)\alpha\beta} = \delta_{\alpha 0}(1 - \delta_{\beta 0})v\lambda_\beta n_\beta + \delta_{\beta 0}(1 - \delta_{\alpha 0})v\lambda_\alpha n_\alpha + (1 - \delta_{\alpha 0})(1 - \delta_{\beta 0})\Lambda_{\alpha\beta}^{(1)}, \qquad (\alpha,\beta \geqq 0).$$

Hence, by (6.7.15) it holds that

(6.7.17) $$\sum_{p\neq q} \lambda_{pq}^{(6)\alpha\beta} = vr \sum_{\gamma\geqq 1} \lambda_\gamma p_{\alpha\beta}^\gamma n_\gamma - \delta_{\alpha 0}(1 - \delta_{\beta 0})v\lambda_\beta n_\beta - \delta_{\beta 0}(1 - \delta_{\alpha 0})v\lambda_\alpha n_\alpha - (1 - \delta_{\alpha 0})(1 - \delta_{\beta 0})\Lambda_{\alpha\beta}^{(1)}, \qquad (\alpha,\beta \geqq 0).$$

In the final place, let us calculate the sum of $\lambda_{pq}^{(5)\alpha\beta}$ with respect to different p and q. From (6.7.6) it follows that

$$\sum_{p\neq q} \lambda_{pq}^{(5)\alpha\beta} = \sum_{p\neq q} \lambda_{pq}^{(4)\alpha\alpha}\lambda_{pq}^{(4)\beta\beta} - \delta_{\alpha\beta}\sum_{p\neq q} \lambda_{pq}^{(4)\alpha\alpha} - 2\sum_p\sum_q \lambda_{pq}^{(6)\alpha\beta}$$

for all $\alpha,\beta \geqq 0$. Hence, by (6.7.8), (6.7.14) and (6.7.17), it is easy to see that

(3.2.11) $$\sum_{p\neq q} \lambda_{pq}^{(5)\alpha\beta} = v \sum_{\sigma,\tau,\gamma\geqq 0} \lambda_\sigma \lambda_\tau p_{\sigma\tau}^\gamma p_{\alpha\beta}^\gamma n_\gamma - 2vr \sum_{\gamma\geqq 1} \lambda_\gamma p_{\alpha\beta}^\gamma n_\gamma - \delta_{\alpha\beta} v(r^2 - \lambda_\alpha)n_\alpha - \delta_{\alpha 0}\delta_{\beta 0} vrk - (1 - \delta_{\alpha 0})\{\delta_{\beta 0}(k - 2) + (1 - \delta_{\beta 0})\delta_{\alpha\beta}\}v\lambda_\alpha n_\alpha - (1 - \delta_{\beta 0})\{\delta_{\alpha 0}(k - 2) + (1 - \delta_{\alpha 0})\delta_{\alpha\beta}\}v\lambda_\beta n_\beta - (1 - \delta_{\alpha 0})(1 - \delta_{\beta 0})\left(2\Lambda_{\alpha\beta}^{(1)} + \Lambda_{\alpha\beta}^{(2)}\right), \qquad (\alpha,\beta \geqq 0).$$

REFERENCE

1. S. Ikeda, *Ann. Inst. Stat. Math., Tokyo 20*, 339 (1968)

BIBLIOGRAPHY

ADERLMAN, S. and O. KEMPTHORNE, Some main effect plans and orthogonal arrays of strength two, *Ann. Math. Statist.*, *32*, 1167-1176 (1961).

BANERJEE, K. S., A note on fractional replication of factorial arrangements, *Sankhyā*, *10*, 87-94 (1950).

BANNERJEE, K. S., On estimation and construction in fractional replication, *Ann. Math. Statist.*, *37*; 1033-1039 (1966).

BOSE, R. C., Least square aspects of the analysis of variance. *Institute of Statistics, University of North Carolina, Mimeograph Series 9*, (1949).

BOSE, R. C., and K. R. NAIR, Partially balanced incomplete block designs, *Sankhyā*, *4*, 337-372 (1939).

BOSE, R. C., A note on the resolvability of balanced incomplete block designs, *Sankhyā*, *6*, 105-110 (1942).

BOSE, R. C., Mathematical theory of the symmetrical factorial design, *Sankhyā*, *8*, 107-166 (1947).

BOSE, R. C., The design of experiments: Presidential address to the Section of Statistics, 34th Indian Science Congress, 1947.

BOSE, R. C., A note on Fisher's inequality for balanced incomplete block designs, *Ann. Math. Statist.*, *20*, 619-620 (1949).

BOSE, R. C., Partially balanced incomplete block designs with two associate classes involving only two replications, *Calcutta Stat. Assoc. Bull.*, *3*, 120-125 (1950-1951).

BOSE, R. C., Group divisible incomplete block designs, *Ann. Math. Statist.*, *22*, 311 (1951) (abstract).

BOSE, R. C. and K. KISHEN, On the problem of confounding in general symmetrical factorial design, *Sankhyā*, *5*, 21-36 (1940).

BOSE, R. C. and D. M. MESNER, On linear associative algebras corresponding to association schemes of partially balanced designs, *Ann. Math. Statist.*, *30*, 21-38 (1959).

BOSE, R. C. and J. N. SRIVASTAVA, Mathematical theory of factorial designs, I. Analysis, II. Construction, *Bull. Int. Stat. Inst.*, *40*, 780-794 (1964).

BOSE, R. C., W. H. CLATWORTHY and S. S. SHRIKHANDE, *Tables of partially balanced designs with two associate classes*, North Carolina Agricultural Experimental Station, Technical Bulletin No. 107 (1954).

BOSE, R. C. and W. S. CONNOR, Combinatorial properties of group-divisible incomplete block designs, *Ann. Math. Statist.*, *23*, 367-383 (1952).

BOSE, R. C. and W. S. CONNOR, Analysis of fractionally replicated $2^m \times 3^n$ designs, *Bull. Int. Stat. Inst.*, *37*; 142-160 (1960).

BOSE, R. C., Combinatorial properties of partially balanced designs and association schemes, *Contribution to Statistics*, volume presented to Prof. P. C. Mahalanobis on his 70th birthday, (1963).

BOX, G. E. P., Some theorems on quadratic forms applied in the study of analysis of variance problems, I. Effect of inequality of variance in the one-way classification, II. Effects of inequality of variance and of correlation between errors in the two-way classification, *Ann. Math. Statist.*, *25*, 290-484 (1954).

BOX, G. E. P. and J. S. HUNTER, The 2^{k-p} fractional factorial designs, I. *Technometrics*, *3*, 311-352 (1961).

BOX, G. E. P. and J. S. HUNTER, The 2^{k-p} fractional factorial designs, II. *Technometrics*, *3*, 449-458 (1961 b).

BRAUD, D., Main effect analysis of the general non-orthogonal layout with any number of factors, *Ann. Math. Statist.*, *36*, 88 (1965).

BROWNLEE, K. A., B. K. KELLY and P. K. LORAINE, Fractional replication arrangements for factorial experiments with factors at two levels, *Biometrika*, *25*, 268-282 (1948).

BURTON, R. C. and W. S. CONNOR, On the identity relationship for fractional replicates in the 2^n series, *Ann. Math. Statist.*, *28*, 762-767 (1957).

CHAKRAVARTI, I. M., Fractional replication in asymmetrical factorial designs and partially balanced arrays, *Sankhyā*, *17*, 143-164 (1956).

CHAKRAVARTI, I. M., Orthogonal and partially balanced arrays and their applications in design of experiments, *Metrika*, *7*, 231-243 (1961)

CHAKRAVARTI, M. C., *Mathematics of Design and Analysis of Experiments*, Asia Publishing House, Bombay (1962).

CLATWORTHY, W. H., Partially balanced incomplete block designs with two associate classes and two treatments per block, *J. Res. Natl. Bur. Std.*, *54*, 177-190 (1955).

CLATWORTHY, W. H., *Contributions on partially balanced incomplete block designs with two associate classes*, National Bureau of Standards, Applied Math. Series, No. 47, Washington, D.C., 1956.

COCHRAN, W. G., The distribution of quadratic forms in a normal system, with applications to the analysis of covariance, *Proc. Camb. Phil. Soc.*, *30*, 178-191 (1934).

COCHRAN, W. G. and G. M. COX, *Experimental Designs*, 2nd Ed., Wiley, New York, 1957.

CONNOR, W. S., On the structure of balanced incomplete block designs, *Ann. Math. Statist.*, *23*, 57-71 (1952).

CONNOR, W. W., Some relations among the blocks of symmetrical group-divisible designs, *Ann. Math. Statist., 23,* 602-609 (1952).

CONNOR, W. S., The uniqueness of the triangular association scheme, *Ann. Math. Statist., 29,* 262-266 (1958).

CONNOR, W. S. and M. ZELEN, *Fractional Factorial Experiment Design for Factors at Three Levels,* National Bureau of Standard Applied Applied Mathematics Series No. 54, Washington, D.C., 1959.

CONNOR, W. S., Construction of fractional factorial designs of the mixed $2^m 3^n$ series in *Contributions to Probability and Statistics,* Stanford University Press, Stanford, Calif., 1960.

CONNOR, W. S. and S. YOUNG, *Fractional Factorial Designs for Experiments with Factors at Two and Three Levels,* National Bureau of Standards Applied Mathematics Series No. 58, Washington, D.C., 1961.

COX, D. R., *Planning of Experiments,* Wiley, New York, 1958.

COX, D. R., The interpretation of the effects of non-additivity in the latin square, *Biometrika, 45,* 69 (1958).

COX, G. M., Enumeration and construction of balanced incomplete block configurations, *Ann. Math. Statist., 11,* 72-85 (1940).

CRAIG, A. T., On the independence of certain estimates of variance, *Ann. Math. Statist.,* Ay. pp. 9, 45-55 (1938).

CRAIG, A. T., Note on the independence of certain quadratic forms, *Ann. Math. Statist., 14,* 195-197 (1943).

DAS, M. N., Fractional replication as asymmetrical factorial designs, *J. Indian Soc. Agric. Stat., 12,* 159 (1960).

DAVIES, O. L., *The Designs and Analysis of Experiments,* Hafner, New York, 1945.

DAVIES, O. L., G. E. P. BOX, L. R. CONNOR, W. R. COUSINS, F. R. HIMSWORTH and G. P. SILLITO, In *The Design and Analysis of Industrial Experiments,* O. L. Davis, ed., Oliver and Boyd, Edinburgh, 1954.

DURBIN, J., Incomplete blocks in ranking experiments, *Brit. J. Psychology (Stat. Sec.) 4,* 85-90(1951).

FEDERER, W. T., *Experimental Design: Theory and Application,* Macmillan, New York, 1955.

FEDERER, W. T. and L. N. BALAAM, *Bibliography on Experiment and Treatment Design, PRE - 1968,* Oliver & Boyd, Edinburgh, (1972).

FINNEY, D. J., The fractional replication of factorial arrangements, *Ann. Eugenics, 12,* 291-301 (1945).

FINNEY, D. J., Recent developments in the design of field experiments, I. split-plot confounding, *J. Agric. Sci. 36,* 56-62 (1946).

FINNEY, D. J., *Experimental Design and Its Statistical Baisis,* Chicago University Press, Chicago, 1955.

FINNEY, D. J., *An Introduction to The Theory of Experimental Design,* Chicago University Press, Chicago, 1955.

FISHER, R. A., *Statistical Methods of Research Workers,* Oliver and Boyd, Edinburgh, 1925.

FISHER, R. A., The arrangement of field experiment, *J. Ministry Agric., 23,* 503-513 (1926).

FISHER, R. A., *The Design of Experiments,* Oliver and Boyd, Edinburgh, 1935.

FISHER R. A., An examination of the different possible solutions of a problem in incomplete blocks, *Ann. Eugenics, 10,* 52-75 (1940).

FISHER, R. A., The theory of confounding in factorial experiments in relation to the theory of groups, *Ann. Eugenics, 11,* 341-353 (1942).

FISHER, R. A., A system of confounding for factors with more than two alternatives giving completely orthogonal cubes and higher powers, *Ann. Eugenics, 12,* 283-290 (1945).

FISHER, R. A., *The Design of Experiments,* 4th ed., Oliver and Boyd, Edinburgh, 1947.

FISHER, R. A., *Collected Papers*, Vol. I, 1912-24, The University of Adelaide (1971).

FISHER, R. A., *Collected Papers*, Vol. II, 1925-31, The University of Adelaide (1972).

FISHER, R. A., *Collected Papers*, Vol. III, 1932-36, The University of Adelaide (1973).

FISHER, R. A. and F. YATES, *Statistical Tables*, 3rd ed., Hafner, New York, 1948.

FRY, R. E., Finding new fractions of factorial experimental designs, *Technometrics, 3*, 359-370 (1961).

GRAYBILL, F. A., *An Introduction to Linear Statistical Models*, Vol. I, McGraw-Hill, New York, 1961.

GRAYBILL, F. A. and G. MARSAGLIA, Idempotent matrices and quadratic forms in the general linear hypothesis, *Ann. Math. Statist., 28*, 678-686 (1957).

GRAYBILL, F. A. and D. L. WEEKS, Combining interblock and intra-block information in balanced incomplete blocks, *Ann. Math. Statist. 30*, 799-805 (1959).

GRAYBILL, F. A. and R. B. DEAL, Combining unbiased estimators, *Biometrics, 15*, 543-550 (1959).

GRAYBILL, F. A. and D. L. SESHADRI, On the unbiasedness of Yate's method of estimation using interblock information, *Ann. Math. Statist. 31*, 786-787 (1960).

GREENBERG, V. L., Robust estimation in incomplete blocks designs. *Ann. Math. Statist., 37*, 1331, (1966).

GRUNDY, P. M. and M. J. R. HEALY, Restricted randomization and Quasi-Latin squares, *J. Roy. Stat. Soc.*, Ser B, 12, 286-291 (1950).

HARSHBARGER, B., Preliminary report on rectangular lattices, *Biometrics Bulletin*, 2, 115-119 (1946).

HARSHBARGER, B., Rectangular lattices, *Virginia Agricultural Experiment Station Memoir, 1,* (1947).

HOFFMAN, A. J., On the uniqueness of the triangular association scheme, *Ann. Math. Statist., 31,* 492-497 (1960).

HOTELLING, H., Some improvements in weighing and other experimental techniques, *Ann. Math. Statist., 15,* 297-306 (1944).

HOTELLING, H., On a matric theorem of A. T. Craig, *Ann. Math. Statist., 15,* 427-429 (1944).

IKEDA, S., Asymptotic equivalence of probability distributions with applications to some problems of asymptotic independence, *Ann. Inst. Stat. Math., Tokyo, 15,* 87-116 (1963).

IKEDA, S., Asymptotic equivalence of real probability distributions, *Ann. Inst. Stat. Math., Tokyo, 20,* 339-362 (1968).

JAMES, A. T., The relationship algebra of an experimental design, *Ann. Math. Statist., 28,* 993-1002 (1957).

JAMES, G. S., Notes on a theorem of Cochran, *Proc. Comb. Phil. Soc. 48,* 443-446 (1952).

JOHN, J. A., Cyclic incomplete block designs, *J. Roy. Stat. Soc. B28,* 345-360 (1966).

JOHN, P. W. M., An extension of the triangular association scheme to three associate classes., *J. Roy. Stat. Soc., B28,* 361-365 (1966).

JOHN, P. W. M., *Statistical Design and Analysis of Experiments,* The Macmillan Company, New York, 1971.

KAPADIA, C. H. and D. WEEKS, On the analysis of group-divisible designs, *J. Amer. Stat. Assoc. 59,* 1217-1219 (1964).

KEMPTHORNE, O., A simple approach to confounding the fractional replication in factorial experiments, *Biometrika, 34,* 255-272 (1947).

KEMPTHORNE, O., *The Design and Analysis of Experiments,* Wiley, New York, 1952.

KEMPTHORNE, O., A class of experimental designs using blocks of two plots, *Ann. Math. Statist.*, *24*, 76-84 (1953).

KEMPTHORNE, O., The randomization theory of experimental inference, *J. Amer. Stat. Assoc.*, *50*, 946-967 (1955).

KEMPTHORNE, O., The efficiency factor of an incomplete block design, *Ann. Math. Statist.*, *27*, 846-849 (1956).

KENDALL, M. G. and A. STUART, *The Advanced Theory of Statistics, Volume 3, Design and Analysis, and Time-Series, 2nd ed.*, Hafner, New York, 1968.

KIEFER, J., On the nonrandomized optimality and randomized non-optimality of symmetrical designs, *Ann. Math. Statist.*, *29*, 675 (1958).

KIEFER, J., Optimum experimental designs, *J. Roy. Stat. Soc.*, *B 21*, 272 (1959).

KIEFER, J. and J. WOLFOWITZ, Optimum designs in regression problems, *Ann. Math. Statist.*, *30*, 271 (1954).

KISHEN, K. and J. N. SRIVASTAVA, Mathematical theory of confounding in asymmetrical and symmetrical factorial designs, *Journal of the Indian Society of Agricultural Statistics*, *11*, 73-110(1959).

KRAMER, C. Y. and R. A. BRADLEY, Example of intrablock analysis of factorials in group-divisible, partially balanced, incomplete block designs, *Biometrics*, *13*,]97-224 (1957).

KSHIRSAGAR, A. M., Balanced factorial designs, *J. Roy. Stat. Soc.*, *B28*, 559-567 (1966).

KURKJIAN, B. and M. ZELEN, A calculus of factorial arrangements. *Ann. Math. Stat.* *33*, 600-619 (1962).

LANCASTER, H. O., Traces and cumulants of quadratic forms in normal variables, *J. Roy. Stat. Soc.*, *B16*, 247-254 (1954).

LANCASTER, H. O., *The Chi-squared Distribution*, John Wiley & Sons, Inc., New York, London, Sydney, Toronto (1969).

MADOW, W. G., The distribution of quadratic forms in non-central normal random variables, *Ann. Math. Statist. 11*, 100-103 (1940).

MANN, H. B., The algebra of a linear hypothesis, *Ann. Math. Statist., 31*, 1-15 (1960).

MATSUSHITA, K., On the independence of certain estimates of variances, *Ann. Inst. Stat. Math., Tokyo, 1*, 79-82 (1949).

McCARTHY, M. D., On the application of the Z-test to randomized blocks, *Ann. Math. Statist., 10*, 337-359 (1939).

MILLER, R. G., Jr., *Simultaneous Statistical Inference*, McGraw-Hill, New York, 1966.

MOTE, V. L., On a minimax property of balanced incomplete block designs, *Ann. Math. Statist., 29*, 910-914 (1958).

MURTY, V. N., An inequality for balanced incomplete block designs, *Ann. Math. Statist., 32*, 908-909 (1961).

NAIR, K. R., The recovery of interblock information in incomplete block designs, *Sankhyā, 6*, 383-390 (1944).

NAIR, K. R., Rectangular lattices and partially balanced incomplete block designs, *Biometrics, 7*, 145-154 (1952).

NAIR K. R., Analysis of partially balanced incomplete block designs illustrated on the simple square and rectangular lattices, *Biometrics, 8*, 122-155 (1952).

NAIR, K. R. and C. R. RAO, Confounding in asymmetrical factorial experiments, *J. Roy. Stat. Soc., B10*, 109-131 (1948).

NAIR, K. R. and C. R. RAO, Confounded designs for asymmetrical factorial experiments, *Science and Culture, 7*, 361-362 (1941).

NAIR, K. R. and C. R. RAO, A note on partially balanced incomplete block designs, *Science and Culture, 7*, 568-569 (1942).

NAIR, K. R. and C. R. RAO, Confounded designs for $k \times p^m \times q^n \times \cdots$ type of factorial experiment, *Science and Culture 7*, 361-362 (1942).

NATIONAL BUREAU OF STANDARDS, *Fractional Factorial Experiment Designs for Factors at 2 Levels*, National Bureau of Standards Applied Mathematics Series, 48, 1957.

NELDER, J. A., Indentification of contrasts in fractional replication of 2 experiments, *Applied Stat.*, *12*, 38-43 (1963).

OGAWA, J., On the sampling distribution of classical statistics in multivariate analysis, *Osaka Math. J.*, *5*, 1-52 (1953).

OGAWA, J., On the randomization of the Latin-square design, *Proc. Inst. Stat. Math., Tokyo*, *10*, 1-16 (1962).

OGAWA, J., On the null-distribution of the F-statistic in a randomized balanced incomplete block design under the Neyman model, *Ann. Math. Statist.*, *34*, 1558-1568 (1963).

OGAWA, J., S. IKEDA and M. OGASAWARA, On the null-distribution of the F-statistic in a randomized PBIB design with two associate classes under the Neyman model, *Essays in Probability and Statistics*, The University of North Carolina Monograph Series in Probability and Statistics, No. 3, 1964, pp.517-548.

OGAWA, J. and G. ISHII, The relationship algebra and the analysis of a PBIB design, *Ann. Math. Statist.*, *36*, 1815-1828 (1965).

OGAWA, J., S. IKEDA and M. OGASAWARA, On the null-distribution of the F-statistic for testing a partial null-hypothesis in a randomized PBIB design with m associate classes under the Neyman model, *Ann. Inst. Stat. Math., Tokyo*, *19*, 313-330 (1967).

OGAWA, J. and S. IKEDA, On the asymptotic null-distribution of the F-statistic for testing a partial null-hypothesis in a randomized PBIB design with m associate classes under the Neyman model, *Canadian J. Statist.*, *1*, 1-24 (1973).

OGAWA, J. and S. IKEDA, On the asymptotic non-null distribution of the F-statistic for testing a partial null-hypothesis in a randomized PBIB design with m associate classes under the Neyman model, *Ann. Inst. Stat. Math., Tokyo*, *25*, 239-259(1973).

OGAWA, J. and S. IKEDA, On the randomization of block designs, *A Survey of Combinatorial Theory*, J. N. SRIVASTAVA et al. (ed.), North Holland Pub. Comp. (1973).

PLAY, R. E. A.C., On orthogonal matrices, *J. Math. Phys. 12*, 311-320 (1933).

PITMAN, E. J. G., Significant tests which can be applied to samples from any population, IV, The analysis of variance test, *Biometrika, 29*, 70-90 (1937).

PLACKETT, R. L., Some generalization in multifactorial designs, *Biometrika, 33*, 328-332 (1946).

PLACKETT, R. L., Some theorems in least squares, *Biometrika 37*, 149-157 (1950).

PLACKETT, R. L., *Regression Analysis*, Oxford University Press, London (1960).

PLACKETT, R. L., Models in the analysis of variance, *J. Roy. Stat. Soc., B22*, 195 (1960).

QUENOUILLE, M. H., *The Design and Analysis of Experiment*, Griffin, London, 1953.

RAGHAVARAO, D., Some optimum weighing designs, *Ann. Math. Statist., 30*, 295-303 (1959).

RAGHAVARAO, D., A generalization of group-divisible designs, *Ann. Math. Statist., 31*, 756-765 (1960).

RAGHAVARAO, D., On the block structure of certain PBIB designs with two associate classes having triangular and L_2 association schemes, *Ann. Math. Statist., 31*, 787-791 (1960).

RAGHAVARAO, D., Some aspects of weighing designs, *Ann. Math. Statist., 31*, 878-884 (1960).

RAGHAVARAO, D., *Construction and Combinatorial Problems in Designs of Experiments*, John Wiley & Sons, Inc., New York 1971.

RAGHAVARAO, D. and K. CHANDRASEKHARARAO, Cubic designs, *Ann. Math. Statist., 35,* 389-398 (1964).

RAO, C. R., General methods of analysis for incomplete block designs, *J. Amer. Stat. Assoc., 42,* 511-561 (1947).

RAO, C. R., Factorial experiments derivable from combinatorial arrangements of arrays, *J. Roy. Stat. Soc., B9,* 128-139 (1947).

RAO, C. R., The theory of fractional replication in factorial experiments, *Sankhyā, 10,* 81-86 (1950).

RAO, C. R., A note on generalization inverse of a matrix with applications to problems in mathematical statistics, *J. Roy. Stat. Soc., B24,* 152-158 (1962).

RAO, C. R., *Linear Statistical Inference and Its Applications,* Wiley, New York, 1965, pp. 4, 7, 118, Multivariate central limit theorem.

RAO, C. R., Group-divisible family of PBIB designs, *J. Indian Stat. Assoc., 4,* 14-28 (1966).

ROY, J., On the efficiency factor of block designs, *Sankhyā, 19,* 181-188 (1958).

ROY, J. and R. G. LAHA, On partially balanced linked block designs, *Ann. Math. Statist., 28,* 488-493 (1957).

ROY, S. N., R., GNANADESIKAN and J. N. SRIVASTAVA, *Analysis and Design of Certain Quantitative Multiresponse Experiments,* Pergamon, Press, New York, 1971.

STTERTHWAITE, F. E., Random balance experimentation, *Technometrika, 1,* 111 (1959).

SCHEFFE, H., A mixed model for the analysis of variance, *Ann. Math. Statist., 27,* 23-36 (1956).

SCHEFFE, H., *The Analysis of Variance,* Wiley, New York, 1959.

SESHADRI, V., Combining unbiased estimators, *Bioretrics, 19,* 163-170 (1963).

SHAH, B. V., On balancing in factorial experiments, *Ann. Math. Statist.*, *29*, 766-779 (1958).

SHAH, B. V., A generalization of partially balanced incomplete block designs, *Ann. Math. Statist.*, *30*, 1040-1050 (1959).

SHAH, B. V., Balanced factorial experiments, *Ann. Math. Statist.*, *31*, 502-514 (1960).

SHAH, B. V., On a 5×2^2 factorial design, *Biometrics*, *16*, 115-118 (1960).

SHRIKHANDE, S. S., Designs for two-way elimination of heterogeneity, *Ann. Math. Statist.*, *11*, 235-247 (1951).

SHRIKHANDE, S. S., On the dual of some balanced incomplete block designs, *Biometrics*, *8*, 66-72 (1952).

SHRIKHANDE, S. S., On a characterization of the triangular association scheme, *Ann. Math. Statist.*, *30*, 39-47 (1959).

SHRIKHANDE, S. S., The uniqueness of the L_2 association scheme, *Ann. Math. Stat.* *30*, 781-798 (1959).

SHRIKHANDE, S. S., Relations between certain incomplete block designs, In *Contributions to Probability and Statistics*, Stanford University Press, Stanford, Calif. 1960, pp. 388-395.

SMITH, C. A. B., and H. O. HARTLEY, The construction of Youden squares, *J. Roy. Stat. Soc.*, *B10*, 262-263 (1948).

SPROTT, D. A., A note on combined interblock and intrablock estimation in incomplete block designs, *Ann. Math. Statist.*, *27*, 633-641 (1956). [Erratum in Vol. 28, 269 (1957)].

SRIVASTAVA, J. N. and R. C. BOSE, Some economic partially balanced 2^m factorial fractions, *Ann. Inst. Stat. Math., Tokyo*, *18*, 57-73 (1966).

STEIN, C., An approach to the recovery of interblock information in balanced incomplete block designs, In *Research Papers in Statistics*, F. N. David, Wiley, New York, 1966.

STEVENS, W. L., Statistical analysis of a non-orthogonal trifactorial experiment, *Biometrika, 35,* 346 (1948).

TANG, P. C., The power function of the analysis of variance tests with tables and illustrations of their use, *Stat. Res. Mem. 2,* 126-149 (1938).

THARTHARE, S. K., Generalized right angular designs, *Ann. Math. Statist., 36,* 1535-1553 (1965).

THARTHARE, S. K., Right angular designs, *Ann. Math. Statist., 34,* 1057-1067 (1963).

THOMPSON, H. R. and I. D. DICK, Factorial designs in small blocks derivable from orthogonal latin squares, *J. Roy. Soc., 2,* 181-247 (1951).

TOCHER, K. D., The design and analysis of block experiments, *J. Roy. Stat. Soc., B14,* 45 (1952).

VAJDA, S., *The Mathematics of Experimental Design: Incomplete Block Designs and Latin Squares,* Griffin's Statistical Monographs and Courses, No. 23, Griffin, London (1967).

VARTAK, M. N., On an application of Kronecker product of matrices to statistical designs, *Ann. Math. Statist., 26,* 420-438 (1955).

VARTAK, M. N., The non-existence of certain PBIB designs, *Ann. Math Statist., 30,* 1051-1062 (1959).

WILK, M. B., The randomized analysis of a generalized randomized block design, *Biometrika, 42,* 70-79 (1955).

WILK, M. B. and O. KEMPTHORNE, Fixed, mixed and random models, *J. Amer. Stat. Assoc., 50,* 1144-1167 (1955).

WILK, M. B. and O. KEMPTHORNE, Some aspects of the analysis of factorial experiments in a completely randomized design, *Ann. Math. Statist., 27,* 950-985 (1956).

YAMAMOTO, S. and Y. FUJII, Analysis of partially balanced incomplete block designs, *J. Sci. Hiroshima Univ. A.27,* 119-135 (1963).

YAMAMOTO, S. T., FUKUDA and N. HAMADA, Composition of some series of association algebras, *J. Sci. Hiroshima Univ. A29*, 181-215 (1966).

YATES, F., Complex experiments, *J. Roy. Stat. Soc. Suppl. 2*, 181-247 (1935).

YATES, F., Incomplete randomized blocks, *Ann. Eugenics*, *7*, 121-140 (1936).

YATES, F., *The Design and Analysis of Factorial Experiments*, Imperial Bureau of Soil Science, Harpende, England, Technical Communication No. 35, 1937.

YATES, F., The recovery of interblock information in varietal trials arranged in a three-dimensional lattice, *Ann. Eugenics*, *9*, 136-156 (1939).

YATES, F., The recovery of interblock information in balanced incomplete block designs, *Ann. Eugenics*, *10*, 317-325 (1940).

YATES, F., Lattice square, *Agric. Sci.*, *30*, 672-687 (1940).

YOUDEN, W. G., Use of incomplete block replication in estimating tobacco virus, *Contribution Boyce Thompson Inst.*, *9*, 317-326. (1937).

YOUDEN, W. G., Partial confounding in fractional replication, *Technometrics*, *3*, 353-358 (1961).

YOUDEN, W. G. and J. S. HUNTER, Partially replicated latin squares, *Biometrics*, *11*, 399-405 (1955).

ZELEN, M., A note on partially balanced designs, *Ann. Math. Statist.*, *25*, 599-602 (1954).

ZELEN, M., The use of group-divisible designs for confounded asymmetrical factorial arrangements, *Ann. Math. Statist.*, *29*, 22-40 (1958).

INDEX